Saint Peter's University Library
Withdrawn

Modules in Applied Mathematics: Volume 3

Edited by William F. Lucas

Modules in Applied Mathematics

Volume 1
Differential Equation Models
Martin Braun, Courtney S. Coleman, and Donald A. Drew, *Editors*

Volume 2
Political and Related Models
Steven J. Brams, William F. Lucas, and Philip D. Straffin, Jr., *Editors*

Volume 3
Discrete and System Models
William F. Lucas, Fred S. Roberts, and Robert M. Thrall, *Editors*

Volume 4
Life Science Models
Helen Marcus-Roberts and Maynard Thompson, *Editors*

Discrete and System Models

Edited by
William F. Lucas, Fred S. Roberts,
and Robert M. Thrall

With 133 Illustrations

Springer-Verlag
New York Heidelberg Berlin

William F. Lucas
School of Operations Research
Cornell University
Ithaca, NY 14853
USA

Fred S. Roberts
Department of Mathematics
Rutgers University
New Brunswick, NJ 08903
USA

Robert M. Thrall
Department of Mathematical Sciences
Rice University
Houston, TX 77001
USA

AMS Subject Classifications: 00A69; 05C01; 90B05, 10; 90C05, 50

Library of Congress Cataloging in Publication Data

Modules in applied mathematics.
Includes bibliographies.
Contents: — v. 2. Political and related models / edited by Steven J. Brams, William F. Lucas, and Philip D. Straffin, Jr. — v. 3. Discrete and system models / edited by William F. Lucas, Fred S. Roberts, and Robert M. Thrall.
1. Mathematics—1961. 2. Mathematical models. I. Lucas, William F., 1933– .
QA37.2.M6 1982 510 82-10439

This book was prepared with the support of NSF grants Nos. SED77-07482, SED75-00713, and SED72-07370. However, any opinions, findings, conclusions, and/or recommendations herein are those of the authors and do not necessarily reflect the views of NSF.

Typeset by Asco Trade Typesetting Ltd., Hong Kong.
Printed and bound by R. R. Donnelley & Sons, Harrisonburg, VA.
Printed in the United States of America.

9 8 7 6 5 4 3 2 1

ISBN 0-387-90724-6 Springer-Verlag New York Heidelberg Berlin
ISBN 3-540-90724-6 Springer-Verlag Berlin Heidelberg New York

Preface

The purpose of this four volume series is to make available for college teachers and students samples of important and realistic applications of mathematics which can be covered in undergraduate programs. The goal is to provide illustrations of how modern mathematics is actually employed to solve relevant contemporary problems. Although these independent chapters were prepared primarily for teachers in the general mathematical sciences, they should prove valuable to students, teachers, and research scientists in many of the fields of application as well. Prerequisites for each chapter and suggestions for the teacher are provided. Several of these chapters have been tested in a variety of classroom settings, and all have undergone extensive peer review and revision. Illustrations and exercises are included in most chapters. Some units can be covered in one class, whereas others provide sufficient material for a few weeks of class time.

Volume 1 contains 23 chapters and deals with differential equations and, in the last four chapters, problems leading to partial differential equations. Applications are taken from medicine, biology, traffic systems and several other fields. The 14 chapters in Volume 2 are devoted mostly to problems arising in political science, but they also address questions appearing in sociology and ecology. Topics covered include voting systems, weighted voting, proportional representation, coalitional values, and committees. The 14 chapters in Volume 3 emphasize discrete mathematical methods such as those which arise in graph theory, combinatorics, and networks. These techniques are used to study problems in economics, traffic theory, operations research, decision theory, and other fields. Volume 4 has 12 chapters concerned with mathematical models in the life sciences. These include aspects of population growth and behavior, biomedicine (epidemics, genetics and bio-engineering), and ecology.

These four volumes are the result of two educational projects sponsored by The Mathematical Association of America (MAA) and supported in part by the National Science Foundation (NSF). The objective was to produce needed material for the undergraduate curriculum. The first project was undertaken by the MAA's Committee on the Undergraduate Program in Mathematics (CUPM). It was entitled Case Studies and Resource Materials for the Teaching of Applied Mathematics at the Advanced Undergraduate Level, and it received financial support from NSF grant SED72-07370 between September 1, 1972 and May 31, 1977. This project was completed under the direction of Donald Bushaw. Bushaw and William Lucas served as chairmen of CUPM during this effort, and George Pedrick was involved as the executive director of CUPM. The resulting report, which appeared in late 1976, was entitled *Case Studies in Applied Mathematics*, and it was edited by Maynard Thompson. It contained nine chapters by eleven authors, plus an introductory chapter and a report on classroom trials of the material.

The second project was initiated by the MAA's Committee on Institutes and Workshops (CIW). It was a summer workshop of four weeks duration entitled Modules in Applied Mathematics which was held at Cornell University in 1976. It was funded in part by NSF grant SED75-00713 and a small supplemental grant SED77-07482 between May 1, 1975 and September 30, 1978. William F. Lucas served as chairman of CIW at the time of the workshop and as director of this project. This activity lead to the production of 60 educational modules by 37 authors.

These four volumes contain revised versions of 9 of the 11 chapters from the report *Case Studies in Applied Mathematics*, 52 of the 60 modules from the workshop Modules in Applied Mathematics, plus two contributions which were added later (Volume 2, Chapters 7 and 14), for a total of 63 chapters. A preliminary version of the chapter by Steven Brams (Volume 2, Chapter 3), entitled "One Man, *N* Votes," was written in connection with the 1976 MAA Workshop. The expanded version presented here was prepared in conjunction with the American Political Science Association's project Innovation in Instructional Materials which was supported by NSF grant SED77-18486 under the direction of Sheilah K. Mann. The unit was published originally as a monograph entitled *Comparison Voting*, and was distributed to teachers and students for classroom field tests. This chapter was copyrighted by the APSA in 1978 and has been reproduced here with its permission.

An ad hoc committee of the MAA consisting of Edwin Beckenbach, Leonard Gillman, William Lucas, David Roselle, and Alfred Willcox was responsible for supervising the arrangements for publication and some of the extensive efforts that were necessary to obtain NSF approval of publication in this format. The significant contribution of Dr. Willcox throughout should be noted. George Springer also intervened in a crucial way at one point. It should be stressed, however, that any opinions or recommendations are those of the particular authors, and do not necessarily reflect the views of NSF, MAA, the editors, or any others involved in these project activities.

There are many other individuals who contributed in some way to the realization of these four volumes, and it is impossible to acknowledge all of them here. However, there are two individuals in addition to the authors, editors and people named above who should receive substantial credit for the ultimate appearance of this publication. Katherine B. Magann, who had provided many years of dedicated service to CUPM prior to the closing of the CUPM office, accomplished the production of the report *Case Studies in Applied Mathematics.* Carolyn D. Lucas assisted in the running of the 1976 MAA Workshop, supervised the production of the resulting sixty modules, and served as managing editor for the publication of these four volumes. Without her efforts and perseverance the final product of this major project might not have been realized.

July 1982 W. F. LUCAS

Preface for Volume 3

This volume contains a rather broad variety of problems and approaches which illustrate the nature of mathematical modeling applied to current problems. The main emphasis is on discrete models, but not exclusively so. The problems presented arise in economics, traffic theory, operations research, decision theory, and other areas. Techniques from graph theory, combinatorics, and optimization are frequently introduced and employed. In several cases, detailed algorithms for obtaining optimal solutions are presented. Some chapters provide concise introductions to basic theories in discrete mathematics, whereas other chapters discuss quite novel new approaches to mathematical modeling which should prove to be fundamental methods in the future. The wealth of material in this volume should prove most useful for courses in discrete mathematics as well as mathematical modeling courses at various levels.

The initial chapter by Frauenthal and Saaty provides some thirty rather elementary problems which can be solved by either quick mental insights or else by more involved calculations or lengthy argument. This interesting collection illustrates the need for imagination and creativity in modeling and problem solving.

Chapter 2 through 5 present several elementary models or approaches which are not normally found in the standard text books. Baker and Marrero use techniques from analytic geometry and calculus to deal with five nautical models and an inventory problem, respectively. Simulation methods are employed by Packel to analyze the best form of traffic control at a street intersection. Greenspan provides a discrete approach to study the fall of a body due to gravity. Such arithmetical models can frequently provide alternatives to the more continuous methods usually employed in the physical sciences, and in light of modern digital computers they may well have major impact on future analytical methods and modeling.

Chapters 6 through 9 provide a selection of recent applications of modern finite mathematics. Topics from combinatorics, graph theory and network flows are applied to a broad variety of contemporary modeling problems. Tucker and Bodin discuss methods for obtaining efficient routes for sweeping a city's streets. They provide algorithms and computational procedures as well. Prather presents a survey of covering problems which arise frequently and in various fields, and he provides both branching and algebraic methods for solution. Perry makes use of pulse processes through directed graphs to analyze athletic financing at a major university. Peterson describes recently developed techniques for predicting the level of traffic congestion during urban and metropolitan rush hours.

Chapters 10 and 14 by Zahavi are concerned with efficiency in electric power generation systems. These units consider an optimal mix problem in capacity expansion, and cost and reliability calculations given supply and demand uncertainty. Some knowledge of nonlinear programming is presumed in Chapter 10, and some basic probability theory is used in Chapter 14.

Chapters 11 to 13 describe in some detail three fundamental approaches to recently developed theories which are important in a broad spectrum of applications. Saaty describes his new theory of analytical hierarchies which is designed to obtain a ratio scale for a decisionmaker's preferences. Techniques for obtaining such measures of intensity are necessary in modeling many problems arising throughout the social, decisional and managerial sciences. Weber discusses multiple-choice testing and considers different weighting and scoring schemes as well as optimal response strategies and systems for eliciting subjective probabilities. Todd gives an introduction to the recently developing field of path following algorithms for computing fixed points of mappings. He provides several applications of these algorithms, including the determination of equilibrium points in economic models.

July, 1982

William F. Lucas
Fred S. Roberts
Robert M. Thrall

Contents

Contents of the Companion Volumes

VOLUME 1. DIFFERENTIAL EQUATION MODELS

CHAPTER 1

Foresight—Insight—Hindsight[1]

James C. Frauenthal*
Thomas L. Saaty**

Introduction

Frequently, mathematical models describing problems stated in words appear trivial in hindsight. This is because the crucial insight needed to formulate the model is divulged by the structure of the model.

To be able to teach this critical inductive step in the same fashion as the deductive steps in mathematical analysis would be desirable but is probably impossible. However, to see this step detailed for a number of apparently difficult and confusing problems may be instructive. The problems here are chosen in most cases so that, once the insight is gained, the model and its solution are trivial and can be analyzed without pencil and paper. At best, the examples provided will illustrate an incisive way of thinking which can be carried over to more difficult problems; at worst, the student will be left with insight into several unimportant problems.

The format of the chapter is as follows. First, a simple problem is stated in words. Next, what appears to be a natural way of looking at the problem is given with whatever hidden constraining or unnecessary assumptions it may have. The next step is to give the critical insight. Finally, the model and its solution are described. The process of modeling may be explained in terms of a simple picture as depicted in Figure 1.1. The cycle can then be made concrete by using some of the problems in the module as illustrations. Finally, some of the unused problems can be assigned to students as home-

[1] An adaptation of this article has been published in the *Two-Year College Mathematics Journal*, vol. 10, pp. 245–254, 1979.

* Department of Applied Mathematics and Statistics, State University of New York, Stony Brook, NY 11794.

** 2025 C. L., University of Pittsburgh, Pittsburgh, PA 15260.

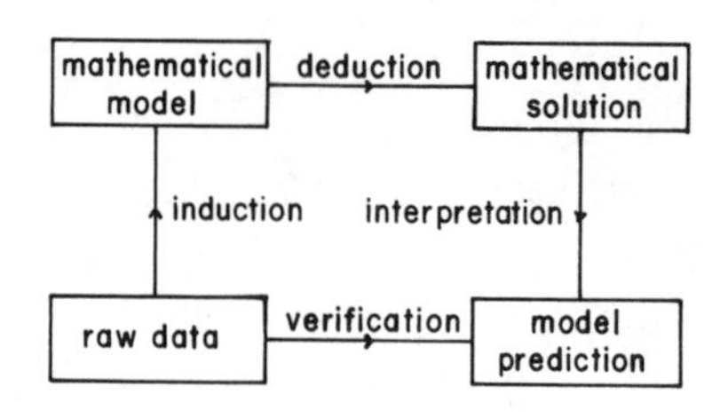

Figure 1.1

work to reinforce the idea that insight is much harder to acquire than hindsight.

It is our feeling that the creative act, as Koestler wrote, "is the defeat of habit by originality." Those engaged in the art of modeling may find it useful to attune the student's mind to subtleties of thought which may be large in number but are only accessible through examples. The hope is to improve intuition by alerting the student to principles operating in a domain apparently more refined than that encountered in daily discourse and in common thought.

We would like to acknowledge our sources. However, these examples have been with us for so long that their origins are now forgotten. Our only claim to originality regards the representation given here of solutions and observations.

The Problems

1. You Too Should Be Able To Do It: Columbus and the Egg

Christopher Columbus was heckled by the Spanish nobility about how easy it was to discover America. He had gone west to go to the east. After all, he had to arrive somewhere by going west since he did not go over the edge of the earth into the dark everlasting abyss. "Anyone could have done the same," they said.

Columbus gave them a hard boiled egg and asked them to stand it on its end. They all tried, but the egg was unstable and fell over on its side. They gave up saying, "Let us see you do it." So he turned it over on one end and crushed that end so the egg could stand up. "Oh," they said, "Anyone could do that." He said, "Yes, but only after you learned how it can be done could you say that."

The moral: If at first you don't succeed,

(1) copy someone else
(2) try and try again
(3) stop and think
(4)

2. Symmetry—Analytic versus Synthetic Thinking: The Cup of Coffee and the Cup of Milk

Imagine that you are given a cup of coffee and a cup of milk, with equal amounts of liquid in the two cups. A spoonful of milk is transferred from the milk cup to the coffee cup and stirred. Then a spoonful of the mixture is returned to the milk cup so, at the end, the amount of liquid in the two cups is again equal. Is there more milk in the coffee cup or more coffee in the milk cup or what?

Most people say there is more milk in the coffee cup, a few say the reverse and still fewer say they are equal. The feeling about this problem is that the first transfer of milk to the coffee cup so dilutes the spoonful in coffee that the next transfer of the mixture cannot take back much of it, hence leaving more milk in the coffee cup than coffee in the milk cup. Of course, not being able to take back much of it should make it possible to take a lot more coffee in the spoonful, but people don't think of it that way.

Insight. Notice that whatever amount of milk is missing from the milk cup is replaced by an equal amount of coffee (and vice versa) because at the end each cup has the same amount of liquid with which it started.

Solution. It is therefore obvious that there are equal amounts of coffee in the milk and milk in the coffee.

One can verify this with algebra but with more effort. However, the algebra assumes homogeneity of the mixture in the coffee cup after stirring. This assumption is artificial but, unfortunately, is needed to carry out the algebraic argument that the second spoonful has the same ratio of coffee to milk as is found in the entire coffee cup.

3. Symmetry in Action: Mountain Climbing

A man goes up a mountain path starting at his home base at 8:00 a.m. and arriving at the top at 5:00 p.m. The next day he comes down the same path starting at 8:00 a.m. and arriving at his home base at 5:00 p.m. Prove that there is a point on the path that he reaches at the same time on the second day as he did on the first day.

The conditions in the problem sound impossibly vague. No mention is made of rates or doubling back or rest stops. The only condition is that he is constrained to the path. Note that this is an existence example. We don't have to say where the point is. In fact more than one point may exist if he doubles back on the path.

Insight. The easy way to see the solution is to have two men on the same day starting at the same time, one going up and the other coming down the path.

Solution. The two men must then meet at some place, and this occurs at the same time of day for both of them. This is the required idea for the solution, though the problem statement with one man and two days tends to disguise it.

4. Continuity but Not Circular Reasoning: The Ski Area Problem

You are the manager of a ski area. Your customers complain that they are spending too much time waiting in line to ride the one chairlift. State law requires a lapse of 15 seconds between chairs. Since you don't have time to install a new chairlift this season, how should you modify your chairlift operating policy to satisfy the customers?

Most people suggest running the chairlift more quickly. This is, however, the wrong thing to do. The state mandated waiting time would then require that the chairs be moved further apart with the result that no greater number of skiers would be carried up the hill. In fact, the number of skiers transported up the hill is absolutely limited by the waiting time.

Insight. The proper way to view this problem is as a closed cycle for any given skier. Each individual does one of three things: ski, ride the lift, or wait on line to ride the lift.

Solution. Making the reasonable assumption that the time to ski from the top of the hill to the bottom is unaffected by the speed at which the lift is operated, the time not spent skiing is clearly divided between sitting on the lift and standing, waiting for the lift. If the lift runs more slowly, with the chairs more closely spaced, skiers will spend more time riding the lift and less time waiting in line. While this solution may appear unsatisfactory, it does respond to the customers' complaints.

5. Stability from Continuity: The Stable Table

Consider a square table with four legs whose lengths are equal. Suppose that the ground is not a smooth, flat surface but a wavy, humpy one (not too much relative to the length of the table's legs). Show that a position always exists where the table could stand so that each leg rests on the ground (i.e., the table has no wobble although it may be tilted).

Insight. Note that standing the table so that three legs touch the ground is always possible. To see this, hold two adjacent legs up and tilt the table so that the other two legs touch. Then try to put down the two legs; one of them will touch first. So there is always a pair of legs on one of the two diagonals which touches the ground. Let x be the sum of the heights of these legs above the ground. Since both of them touch the ground, we have $x = 0$. Let y be the sum of the heights of the two legs on the other diagonal (one of them contributes zero height because three legs always touch the ground). Let us rotate the table 90 degrees so that each diagonal goes into the other. Now x and y are functions of the angle of rotation θ, and we have

$$x(0) = 0,\ y(0) > 0$$

$$x\left(\frac{\pi}{2}\right) > 0,\ y\left(\frac{\pi}{2}\right) = 0$$

$$x(\theta)y(\theta) = 0,\ 0 \leq \theta \leq \frac{\pi}{2}.$$

Solution. From what we have just proved, three legs must always touch the ground for any θ; that is why we have the third relation. Consider the value of θ at which $x(\theta)$ just changes from zero to positive. At that value $y(\theta)$ must just turn zero. Thus, at that value of θ, both x and y vanish, and all four legs must rest on the ground.

6. The Strength of Parity: The Checkerboard Problem

Imagine that you are presented with an ordinary checkerboard from which two diagonally opposite corner squares have been cut off. The board therefore has a total of 62 squares. You also have a pile of 31 dominoes, each with dimensions one square by two squares. Construct a simple proof that to completely cover the deleted board with nonoverlapping dominoes is not possible.

Insight. The critical step is made by thinking about the color pattern of a checkerboard. Since the colors of adjacent squares are of opposite colors, diagonally opposite corners must be the same color.

Solution. The deleted board has 30 squares of one color and 32 of the other, but each domino must cover one square of each color. Therefore, the suggested tiling is impossible.

Extension. What if the undeleted board is 7×7 instead of 8×8 and, again, the opposite corners are removed?

Another Extension. Can you fill a $12 \times 12 \times 12$ cube with $2 \times 4 \times 8$ bricks? Hint: Color each brick half black and half white. Subdivide the cube into twenty-seven $4 \times 4 \times 4$ cubes and color them alternately black and white.

7. The Weakness of Imparity: The National Football League Problem

Until recently the NFL consisted of two conferences of 13 teams each. The rules of the league specified that during the 14-week season, each team would play 11 games against teams in its own conference and three games against teams in the opposite conference. Prove that this is impossible.

The initial method for solution which comes to mind is to represent each team by a point (vertex) and then to show games by lines between pairs of points. Such a model is called a finite graph. While essentially a correct beginning, trying to prove the impossibility of the above schedule by enumeration is very lengthy.

Insight. Think about the games within and across conferences separately. Clearly, the lines representing games within one conference must connect two different points (teams) within the conference. Thus an even number of ends-of-lines must exist, since each line has two ends.

Solution. The schedule demands that, within each conference, 13 teams each play a game against 11 other teams in the conference. This makes for $(13)(11) = 143$ ends-of-lines. Since this is an odd number, the schedule is impossible.

Note that, as one would expect, the NFL never succeeded in satisfying its own scheduling rule. What is the minimum number of teams that would have to violate the scheduling rule in order to make a feasible schedule?

8. The Pigeon-Hole Principle: Duplicate Acquaintances

If each of a group of N people knows at most $n < N$ members of the group, then at least two people know the same number of people.

Insight. The critical observation for solving this problem is that the N people in the group must be partitioned (pigeon-holed) into $n < N$ subgroups.

Solution. We put those who know one person in a group. If more than one such person exists, we are finished. If not, we consider all those who know two people and so on. Since $n < N$, there must be duplication.

9. Exhaustion: The Efficient Secretary

A secretary types letters and envelopes to four different individuals and then accidentally inserts the letters randomly into the envelopes. What is the probability that exactly one letter is in the wrong envelope?

One might at first reason as follows. There are 24 different ways to place the letters into the envelopes, since any of the four letters could go into the first envelope, any of the remaining three into the second envelope, etc., thus $(4)(3)(2)(1) = 24$. Then, three of the four ways of stuffing the first envelope are wrong, and one is correct, and of course there is only one correct way to stuff the remaining three envelopes. This implies that there are three ways in 24 or a probability of $3/24 = 1/8$ of satisfying the conditions of the problem. This answer is wrong!

Insight. If three envelopes are stuffed correctly, only one letter and one envelope remain, and they must go together.

Solution. Clearly, it is not possible to have just one letter in the wrong envelope.

10. Standing Back To See the Answer: The Tennis Tournament

A tennis tournament director wishes to organize an elimination-type tennis tournament. The names of all the people who enter are put into a hat and then drawn out in pairs. The two people in each pair play against each other, with the loser retiring from the tournament. The names of all the winners are put back in the hat, pairs are drawn out again, and the play-off takes place. If at any time an odd number of names appears in the hat, the name left over after the drawing of the pairs remains in the hat until the next round (i.e., that person gets a "bye" in that round). The director wishes to know how many matches will have to be played if N people enter his tournament.

Let us, for example, consider the following cases:

$$N = 31, 32, 33, 34.$$

The analytic solution for the problem may be obtained as follows. For

$$\left\{\begin{array}{l} N = 31 \\ N = 32 \\ N = 33 \\ N = 34 \end{array}\right. \text{ the number of matches is } \left\{\begin{array}{l} 15 + 8 + 4 + 2 + 1 = 30 \\ 16 + 8 + 4 + 2 + 1 = 31 \\ 16 + 8 + 4 + 2 + 1 + 1 = 32 \\ 17 + 8 + 4 + 2 + 1 + 1 = 33. \end{array}\right.$$

Insight. We may argue as follows. Everyone but the winner loses exactly one match and is then out of the tournament.

Solution. Therefore, $N - 1$ matches must be played.

11. Measure Counter Measure: The Coin Game

If you and an opponent are to alternate placing pennies on a rectangular table so that they do not overlap (and all lie flat), and if the object of the game is to be the last player to be able legally to place a penny, would you rather go first or second?

Insight. The feeling here is that one would like to be in a position to rebut any move made by the opponent and prevent him from doing the same on the last move.

Solution. By symmetry, any position of the table has a symmetric position with respect to a line through the center. The center is symmetric with respect to itself. So the scheme would be to go first in the center and then respond symmetrically to the opponent's moves with respect to the center.

12. Conditional Thinking: Prospecting for Gold

You are given a chest with three drawers, the first containing two gold coins, the second a gold and a silver coin, and the third two silver coins. A drawer is selected at random and a coin is taken out. The coin is a gold one. What is the probability that the second coin in that drawer is a gold one?

Many people argue as follows. Since a gold coin was found in the selected drawer, it must be either the drawer with two gold coins or the drawer with one gold and one silver coin. Thus a 50 : 50 chance exists that the remaining coin is gold. This is wrong!

Insight. List all of the possible ways of drawing two coins, denoted by G_i and S_i ($i = 1, 2, 3$), out of a drawer.

Drawer I	*Drawer II*	*Drawer III*
$\underline{G_1G_2}$	$\underline{G_3S_1}$	S_2S_3
$\underline{G_2G_1}$	S_1G_3	S_3S_2

Solution. Of the six possible realizations, only three have a gold coin drawn first (underlined). Of these three, two have a gold drawn second. Thus Prob $\{\text{second coin in drawer is gold} \mid \text{first coin in drawer is gold}\} = 2/3$.

13. What Is Random? The Subway Paradox

The Red Line Subway in Boston (Figure 1.2) runs from Harvard Square to Quincy Center via Park Street. Each day you arrive at the Park Street Station at a random time, go down to the platform between the sets of tracks,

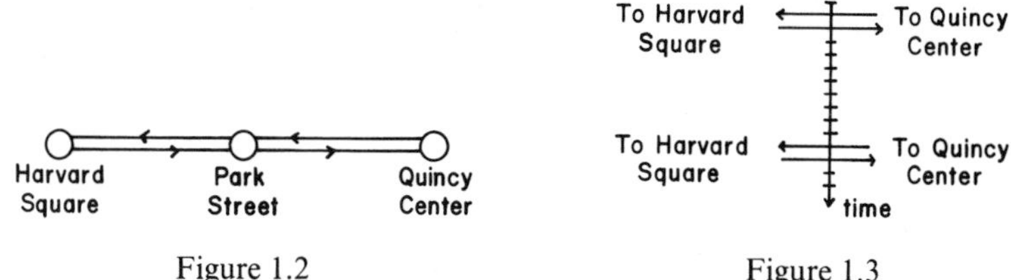

Figure 1.2

Figure 1.3

and get on the first Red Line train to arrive. Nine days out of ten you end up at Harvard Square. How can this be?

Most people's intuition tells them that one should go to Quincy Center as often as one goes to Harvard Square. This is not necessarily the case. The error results from considering the wrong property as being random.

Insight. Notice that it is you, and not the trains which arrive at random. Imagine, for example, that trains from Harvard Square to Quincy Center run every 10 minutes on a regular schedule. Also, to avoid having a buildup of trains at one end of the line or the other, trains also run from Quincy Center to Harvard Square every ten minutes.

Solution. To resolve the paradox, imagine that each train bound for Quincy Center arrives one minute after the train bound for Harvard Square. (Figure 1.3)

Clearly, if your random arrival time falls anywhere between the time of a Quincy Center bound train and a Harvard Square bound train (9 minutes) you will end up at Harvard Square, while if your arrival falls between the time of Harvard Square bound train and a Quincy Center Bound train (1 minute) you will end up at Quincy Center.

14. All Things Being Equal: The True–False Exam

Two students are given a two-question, true–false exam. One gets up to sharpen his pencil and on the way to his seat glances at the teacher's desk. He sees that the answer to one of the questions is "True" but he cannot tell if it is the first question or the second. He answers "True" for both questions. Reasoning as follows,

	Possible Outcomes
Saw answer to first question:	(T, T) or (T, F)
Saw answer to second question:	(T, T) or (F, T)

he concludes that since (T, T) occupies half of the cases, he has a probability of 1/2 of having 100% on the exam.

The second student does not see the teacher's notes, but knows from his friends who have had the teacher before that the teacher never gives an exam with all answers "False". He also answers "True" for both questions. Reasoning as follows,

Possible outcomes
(T, T) or (T, F) or (F, T)

he concludes that since (T, T) occupies one third of the cases, he has a probability of 1/3 of having 100% on the exam.

Both students seem to have the same information and both have reasoned correctly, yet they have come to different conclusions. How can this be?

Insight. The two students have come to different conclusions as a result of assuming different things to be "equally likely".

Solution. The first student has assumed that, for the exam question whose solution he did not see, it is "equally likely" that the answer will be "True" or "False". The second student has assumed the three possible outcomes (T, T), (T, F), and (F, T) are "equally likely". Note that for this to be the case, each question has probability 2/3 of being "True".

15. Symmetry and Randomness: The Tape Problem

What is the expected length of tape (in a tape recording machine) between two positions selected at random on the tape?

Solution. Let x_1 and x_2 denote the distance of the two positions from the starting position of the tape. Without loss of generality, assume that the tape has unit length. Then the average length is given by

$$\int_0^1 \int_0^1 |x_1 - x_2|\,dx_1\,dx_2 = \int_0^1 \int_0^{x_1} (x_1 - x_2)\,dx_2\,dx_1 + \int_0^1 \int_{x_2}^1 (x_2 - x_1)\,dx_2\,dx_1 = 1/3.$$

Comment. A large number of people guess the answer to be 1/2. They think that the two points are supposed to be at opposite sides of the tape each having 1/4 of the length between it and the nearby endpoint.

16. Intuitive Geometric Optimization: A Tale of Two Cities

Consider two cities on opposite sides of a river (parallel lines for shores). We wish to connect them with the shortest path across a bridge of finite length perpendicular to the river. Where should the bridge be located?

The first idea is to connect the cities with a straight line, mark its inter-

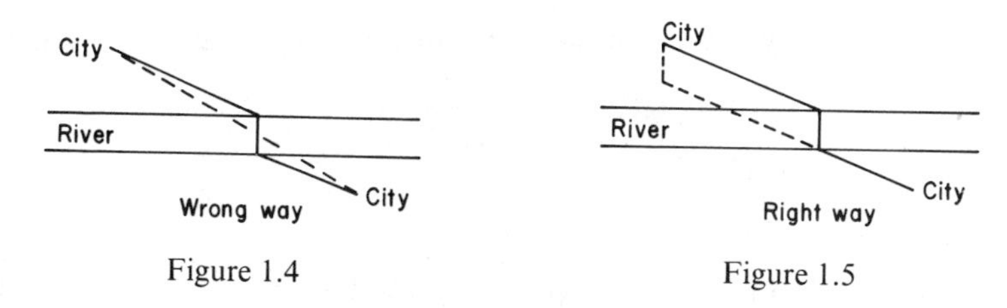

Figure 1.4

Figure 1.5

section with the middle of the river, build the bridge to pass through that point, then connect the cities to the bridge (Figure 1.4). The straight line was to assure a basis for the shortest distance, but in fact the total distance is not the shortest possible.

Insight. The straight line idea is a good one, but we note that when proceeding from either city, what is certain is that a bridge whose length is the width of the river is to be crossed and then a shortest distance is needed.

Solution. As we just said, we start by eliminating the width of the river, and drawing a straight line for the rest of the distance. This is most easily accomplished by shifting the upper city downward by the width of the river, connecting it with a straight line to the lower city as if the river were not there. We put the river back, keep the portion of the line from the lower city to the edge of the river, put the bridge at that position, and connect its upper point to the original position of the upper city. This line segment is parallel to the one resulting from the shifted position of the upper city and is equal to it in length (Figure 1.5).

17. Playing with Numbers

They Can Be Very Nice. The story is told of Gauss as a child of a few years who found the sum $1 + 2 + 3 + \cdots + 100$ in a matter of seconds. The numbers cooperated with his intuition as he lined them up as follows:

$$\begin{array}{lrrrcr} & 1\,+ & 2\,+ & 3\,+ & \cdots & +\;100 \\ & 100\,+ & 99\,+ & 98\,+ & \cdots & +\;\;\;1 \\ \hline \text{Sum} & 101\,+ & 101\,+ & 101\,+ & \cdots & +\;101 = 100 \times 101. \end{array}$$

The answer is obviously half of this.

They Can Do Unexpected Things. Given $0 < p < 1$, what is your estimate of $p^{\binom{k}{2}}$ as k gets larger and larger? Now what is your estimate of $(1 - p^{\binom{k}{2}})^{\binom{n}{k}}$ for n large? You may be inclined to say 1. Because $p^{\binom{k}{2}}$ was approximated by zero. But this is not always true. One can write

$$[1 - p^{\binom{k}{2}}]^{\binom{n}{2}} = e^{-\binom{n}{k}p^{\binom{k}{2}}}$$

When the exponent is near zero the above result is near one. For example,

$$\left[1 - \frac{1}{2}^{\binom{16}{2}}\right]^{\binom{1000}{16}} = e^{-0.0319} \approx 0.9686 \approx 1.$$

On the other hand, for example,

$$\left[1 - \frac{1}{2}^{\binom{15}{2}}\right]^{\binom{1000}{15}} \approx e^{-16.96} \approx 4.308 \times 10^{-8} \approx 0.$$

This is a somewhat surprising result. When we are dealing with large numbers, things cannot always be treated uniformly.

Looking for a helpful insight and trusting your intuition are not the same thing.

18. Using a Sledgehammer To Kill a Fly: Getting a Drink from the River

You wish to go from point I to point II, stopping along the way to get a drink from the river (Figure 1.6). You are in a hurry and wish to minimize your travel distance. Determine the point you should aim for on the river bank for the distances shown.

The sledgehammer approach is as follows. Let x locate the point on the river bank as shown, and let D be the total distance traveled, thus

$$D = \sqrt{a^2 + x^2} + \sqrt{b^2 + (d - x)^2}.$$

Now minimize D by setting $dD/dx = 0$ and solve for x:

$$\frac{dD}{dx} = \frac{x}{\sqrt{a^2 + x^2}} + \frac{-(d - x)}{\sqrt{b^2 + (d - x)^2}} = 0$$

$$x\sqrt{b^2 + (d - x)^2} = (d - x)\sqrt{a^2 + x^2}$$

$$x^2[b^2 + (d - x)^2] = (d - x)^2\sqrt{(a^2 + x^2)}$$

$$x^2b^2 = (d - x)^2a^2$$

$$xb = (d - x)a$$

$$x = \frac{ad}{a + b}.$$

By now the reader is probably dead of thirst.

Insight. Treat the river bank as a mirror and reflect point II in the mirror (Figure 1.7). Then connect I and II with a straight line and use similar triangles.

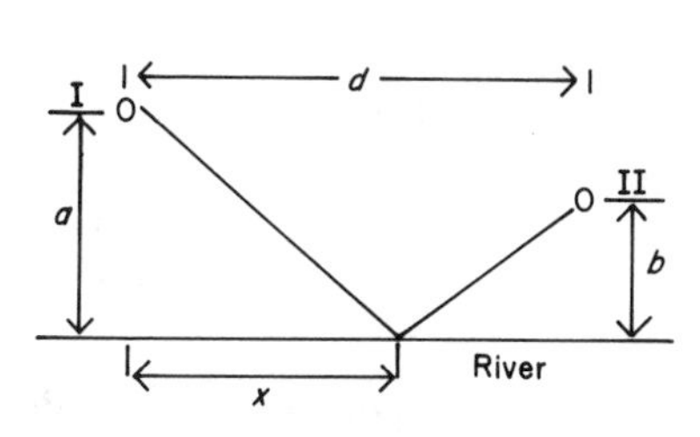

Figure 1.6

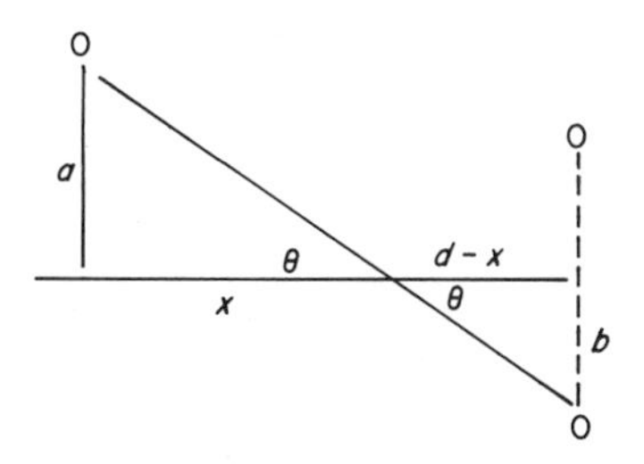

Figure 1.7

Solution.

$$\frac{a}{x} = \frac{b}{d - x}$$

$$x = \frac{ad}{a + b}.$$

19. Solution from the Core—Going Out from Within: Cutting a Cube

Imagine that you are given a solid cube, three units on a side. Your task is to cut the cube into 27 cubes, each one unit on a side (Figure 1.8).

After each cut, you may stack the resulting pieces in any way you desire before making the next cut. What is the minimum number of cuts required to divide the large cube into 27 small cubes?

The naive method for approaching this problem is very tedious, as it involves hunting for a "best" method without any guidelines. Problems of this type can usually be solved by this sort of enumeration, but the risk of error is great.

Insight. Think about the one unit on-a-side cube which is cut from the middle of the large cube. All six of its faces must be cut.

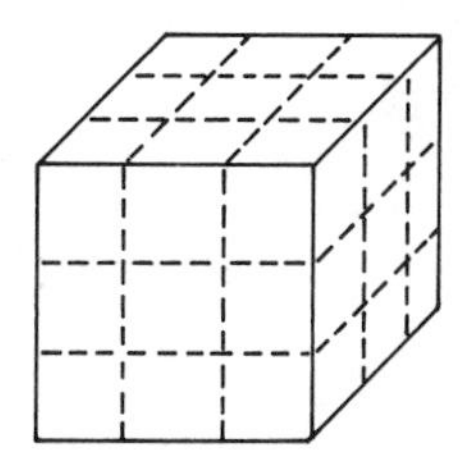

Figure 1.8

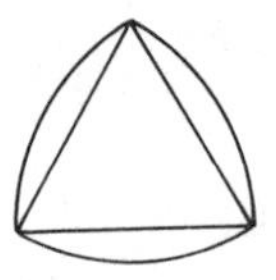

Figure 1.9

Solution. Clearly, no amount of juggling before cutting will permit the six faces of the interior small cube to be cut with less than six passes of the knife (Fig. 1.8). As is also clear from the illustration, six cuts are sufficient without any restacking of the pieces. Thus six cuts is both necessary and sufficient.

20. To Everything There Is a Reason: Why Not?

Why are manhole covers round?

Insight. Being accidentally dropped down the hole would be extremely dangerous.

Solution. The circle is a region of constant diameter. Somewhat surprisingly, other such regions exist. Figure 1.9 illustrates one (each arc is part of the circle centered at the opposite vertex); others can be similarly constructed on regular $(2k - 1)$-gons, and even less symmetric ones can be generated.

21. Inventing Restrictions Which Are Not There: The Loophole Principle

Can you draw four straight lines through the nine points in Figure 1.10(a) without lifting the pencil off the paper?

Solving the problem is not possible if all the lines start at one of the points and end at another. When given a problem, we are inclined to isolate the problem completely from its surroundings in our attempt to solve it.

Insight. The solution of this problem depends on the realization that the nine points are embedded in the plane of the paper so that we can, if we want, get out of the confines of the nine points. In that case the four lines

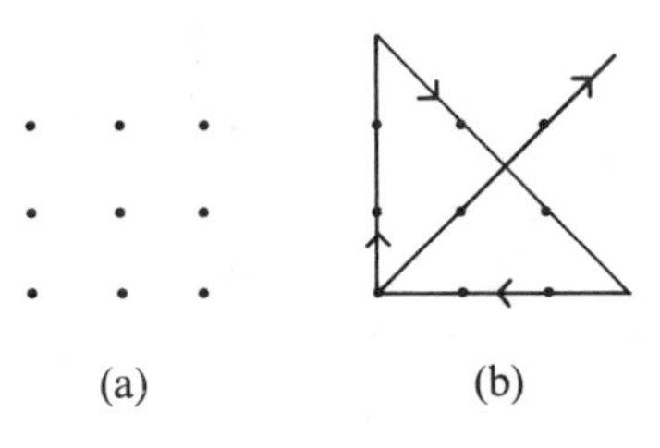

Figure 1.10

can start and end anywhere on the paper as long as together they pass through all the points.

Solution. The solution is shown in Figure 1.10(b). It is not unusual to find someone who regards this solution as dishonest because he has invented the constraint of staying within the grid.

A time was when people did not pay taxes. When it was decided to levy taxes, the government put down rules for paying them. Some people did not agree with the idea but obligingly complied with the rules. These rules were found to be lacking in thoroughness. Those who disagreed followed the rules precisely and carefully and found loopholes. Others closed the loopholes with their imagination and created a conviction that they had to be complied with in their own imagined way. Who is right and who is wrong became a volatile moral question.

Convictions about things that are more imagined than real can get one into difficulty with people who live in a fuller dimension than one is accustomed to. The abundance of their lives is sometimes judged as depravity.

22. The Wrong Feeling that with a Little More We Can Get What We Want: Very High Speed

A man goes from A to B at 30 mph. How fast would he have to return from B to A to average 60 mph over the time taken in the entire trip?

Insight. A great mathematician once guessed the answer to the problem to be 90 mph from B to A. He misinterpreted the (impossible) requirement of the problem. The average is not being taken over the two velocities but over the total time.

Solution. The total time required to go from A to B and back again at 60 mph is used up in going from A to B at 30 mph. Thus the return trip must be done at infinite speed.

This is reconfirmed by the following arithmetic model:

$$\frac{d}{t_1} = 30 \longrightarrow t_1 = \frac{d}{30}$$

$$\frac{d}{t_2} = v \longrightarrow t_2 = \frac{d}{v}$$

$$\frac{2d}{t_1 + t_2} = 60 \longrightarrow$$

$$2d = 60(t_1 + t_2) = 60\left(\frac{d}{30} + \frac{d}{v}\right)$$

$$2d = 2d + 60\frac{d}{v} \Rightarrow v = \infty.$$

Extension. In which of the following cases does our average speed equal 60 mph?

(1) We travel equal distances at 30 mph and 90 mph.
(2) We travel equal times at 30 mph and 90 mph.

23. Perspectives from the Outside: Spatial Visualization

The front and top views of a solid object are shown in Figure 1.11. (Note that there are no invisible edges.) Draw either a side view or an isometric (diagonal) projection to show that you understand what the object looks like. Note that several solutions are possible.

Most people have a very hard time with this problem, in part because they have difficulty visualizing in three dimensions, and in part because they attempt to construct the picture with all surfaces either horizontal or vertical.

Insight. The object must have at least one face which is inclined. In fact, the inclined face can be either planar or gently curved. The fact that two views do not determine the geometry helps explain why engineering drawings ordinarily include three views.

Solution. One solution is shown below in Figure 1.12; see if you can think of another.

24. Standing Back To See the Method of Solution: Boy and Dog Story

A boy walks home at 3 mph. He starts at a distance of 3 miles. His dog starting with him, runs to the house at 5 mph and immediately turns around and runs back to the boy and again back to the house, and so on and so forth until the boy reaches home. How far does the dog travel?

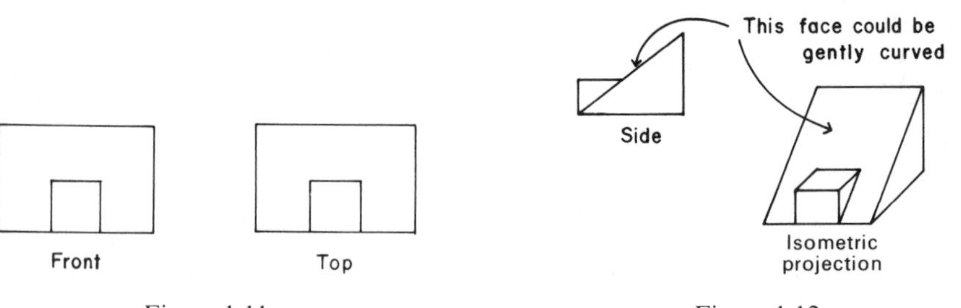

Figure 1.11

Figure 1.12

Insight. To the literal-minded who like to work things out in detail in search of elegant solutions, this problem looks a little tough, but if one insists on it, a series formulation and solution is possible. This is in fact what von Neumann (and one of the authors of this paper) did although von Neumann's brilliant mind got the answer in a short time. The easy way is to notice that it took the boy one hour to get home.

Solution. How far do you suppose the dog walked during the hour? 5 miles?

25. Too Little Information: The Early Commuter Problem

A man usually takes the 5:30 p.m. train, arriving at his station at 6:00, and his wife picks him up and drives him home. One day he took the 5:00 p.m. train, arriving at 5:30 p.m. at the station. He began walking home. His wife, starting out from home to meet him at the usual time met him on the way and brought him home 10 minutes before the usual time. How long did the man walk?

The first feeling about this problem is that not enough information is provided to solve it. Nothing is said about how fast the man walked or how fast the wife drove or when she left home every day to meet him at the station. However, we assume that no ambiguity exists in the information, so an answer is possible.

Insight. Think about the man's walking time. After arriving a half-hour earlier than usual, he meets his wife driving to the station before 6:00 p.m. Suppose she were to go on to the station and then back to the point where she first met him. She then picks him up and takes him home. Then they would arrive home at the usual time.

Solution. They arrived home 10 minutes earlier than usual, which represents twice the time it takes to drive the distance he has walked from the station. Thus the distance he walked would have taken five minutes to drive. So he walked 25 minutes, from 5:30 to 5:55 meeting his wife at a point where had she proceeded to the station she would have met the 6:00 p.m. train on time. Magic! Not really. See if you can solve this problem using either a graph or algebra.

26. Focus on the Precarious End: Stacking Bricks

If bricks are to be stacked upon a single brick as base, how much of an "offset" can be achieved between the top and the bottom of the stack?

Most people would start off by focusing on the brick resting on the table. They would next put a brick on top and shift it as far as possible. They would next add another brick, but this requires readjusting the second brick. Each successive brick added requires the readjustment of all previous bricks. Though obviously this is how one would have to build the stack, it is not how to build the model.

Insight. Any given part of the stack above a point is unaffected by what is below it. Therefore work from the top down.

Solution. Assume all bricks are one unit long and weigh one unit. Consider first just two bricks as shown in Figure 1.13. Clearly, the bottom brick is always stable, and the top brick can be situated so that its center of mass is just over the end of the bottom brick (at arrow). Thus $x = 1/2$.

Now consider adding another brick to the bottom of a pile of n bricks (Figure 1.14).

Since the pile of n bricks was stable, the center of mass of the first through the $(n - 1)$th bricks is above the right-hand end of the nth brick; with the $(n + 1)$th brick in place, the center of mass of the top n bricks must be above the arrow. Thus

$$\left(\frac{1}{2} - z\right) = (n - 1)z \to z = \frac{1}{2n}.$$

Thus, by induction, we have found that the distance d_{n+1} from the right-hand end of the $(n + 1)$th brick to the right-hand end of the top brick is

$$d_{n+1} = \frac{1}{2} + \frac{1}{4} + \cdots + \frac{1}{2n}$$

$$= \frac{1}{2} \sum_{j=1}^{n} \frac{1}{j}.$$

Thus in the limit as $n \to \infty$

$$d_\infty = \frac{1}{2} \sum_{j=1}^{\infty} \frac{1}{j} = \infty.$$

In other words, the pile of bricks extends infinitely far to the right as it becomes infinitely tall.

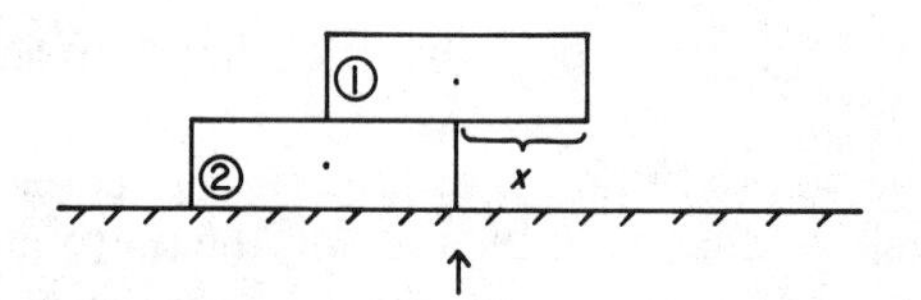

Figure 1.13

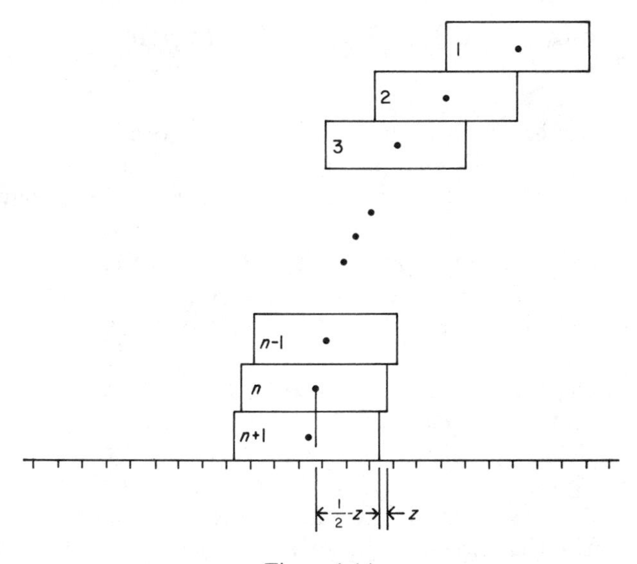

Figure 1.14

27. Obtaining Information When None Seems To Exist: Cheating at Cards

A man is in the habit of looking at the bottom card of the portion which goes on top of a cut deck of cards before putting the portion there. All the cards are dealt out, as in bridge. Does he gain information from this seemingly innocuous act?

Insight. He knows what the top card is *not*, and he knows that this top card, which he does not know, goes to his opponents.

Solution. The top card of the deck is selected from among 51 unknown cards of the deck instead of 52. After he looks at his hand, if he does not have that card, he considers his partner having it with probability greater than 1/3. Without additional information about where that card is in the deck he could take odds of 2 to 1 that his partner has it and in the long run he would have a slight advantage on this bet.

Wrong: Further Insight. Consider two cutters one of whom makes the cut randomly between the 23rd and the 29th card and the other between the 22nd and the 30th. The probability that the partner gets the card in question, given that it does not go to the dealer, is less than pure chance ($1/5 < 1/3$) in the first case and greater than pure chance ($3/7 > 1/3$) in the second. Therefore, although cheating may give useful information, one cannot be certain how to use it unless one knows the direction of the bias. Intuition triumphs as it does not see the usefulness of the information in this problem.

28. A Stitch in Time: The Water Lily Problem

You are in charge of maintaining a pond in which water lilies grow. Each day the area of the plants doubles. You decided to do nothing until the pond is half covered. Only then will you cut the water lilies back. Given that the entire pond is just covered on the twentieth day, on what day do you start to cut?

The usual response to this problem is that you start to cut on the tenth day, which is wrong. The error results from assuming that the lilies grow at a fixed rate which is independent of their present number.

Insight. Not enough information is given to solve this problem "forward" in time. It can, however, be solved "backward" in time.

Solution. The pond is just covered on the twentieth day. Each day the area covered doubles, thus going backward in time, on the nineteenth day, the pond was half covered. According to the problem statement, this is the day you start to cut.

Philosophical Comment. That old adage, "A stitch in time saves nine," may be applied here. You could save yourself a lot of work if you cut the lilies back on the 18th day when only a quarter of the pond is covered. Perhaps the 17th day is even better, when an eighth of the pond must be cleared. We note that the animal population (including human) grows like the water lilies in proportion to the present population size.

29. Difficulties of Asymmetry—Sufficient Information!? A Dizzy Dog Story, A Paradox

A boy, a girl, and a dog go for a walk on a straight road. All start together, but the girl walks at 2 mph, the boy at 3 mph, and the dog at 5 mph. The dog always remains between the people. When it reaches the boy it turns around and heads toward the girl, and when it reaches the girl, it turns again and heads back toward the boy. The dog keeps repeating this process each time it reaches the people. Determine where the dog is or devise a simple way to prove that it is impossible to determine where the dog is at the end of one hour.

The fact that this problem has no solution might at first appear odd, since the dog has obviously traveled 5 miles in the hour. However, any attempt at a straightforward solution fails.

Insight. Let us look at the situation one hour after the walk begins, and then let time run backwards.

Solution. Clearly, after one hour the boy has walked 3 miles, and the girl two. The dizzy dog going around in circles is somewhere between. Now let time run backwards. Regardless of where the dog begins the reverse time problem, he remains between the people, and all three cross the starting line when time has run backwards to zero. Therefore every point between the boy and the girl is a solution to the backward problem. What this suggests is that, although the laws of motion are determined, the information provided to describe the initial instant of the problem is inadequate for a unique solution.

30. Outsmarting Ourselves: The Apparent Irrationality of Increasing Complexity

At times, our ingenuity can go astray. For example, consider the following paradox. Given a triangle ABC, we find the midpoints of the sides A_1, B_1, C_1 (Figure 1.15).

Clearly, from simple geometry,

$$BA + AC = BC_1 + C_1A_1 + A_1B_1 + B_1C.$$

Now repeat the process as shown in Figure 1.16, this time on triangles A_1BC_1 and A_1B_1C.

Once again, the length of the zig-zag line segments equals $BA + AC$. This is true as we continually apply this procedure, but also, the zig-zag line segments approach the line BC. It would be easy, though obviously incorrect, to conclude that the length of the zig-zag line approaches the length of BC.

The mathematician Henri Lebesque recounts that while his high school friends took this as a joke he was disturbed by it. He could not see the difference between this type of argument and other proofs he got every day in his geometry classes for lengths and areas of curved figures.

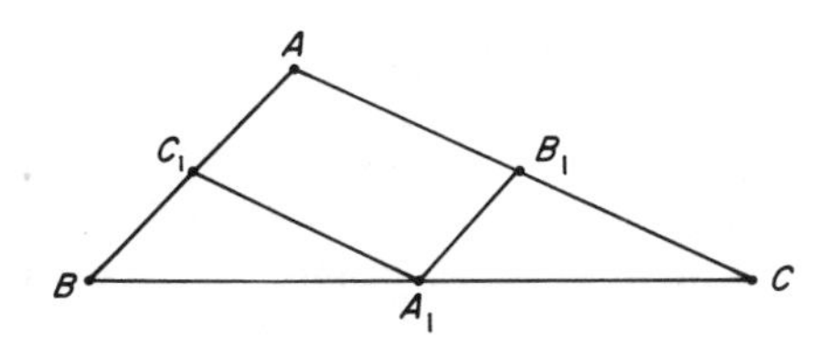

Figure 1.15

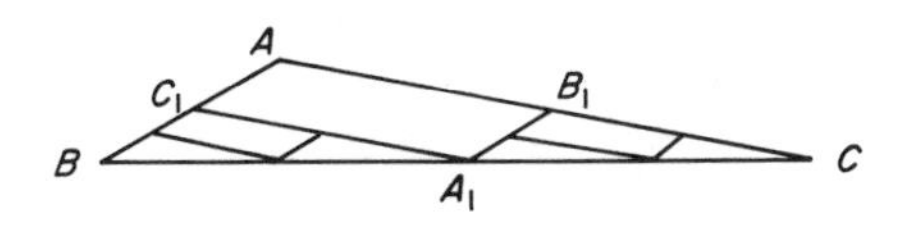

Figure 1.16

31. Quick and Dirty but Effective: The Gordian Knot

When traveling through Asia Minor during his conquests, Alexander was presented with the Gordian knot which no one was able to untie. They probably thought this might cut him down to size, but he was not to be outdone. He pulled out his sword and slashed the knot open. He solved the problem.

Insight. To Alexander, the Gordian knot was not a matter of finesse. It was a low priority problem for a conqueror to waste his time on, and he showed them how he would dismiss it from his mind.

Notes for the Instructor

Objectives. Frequently, mathematical models which attempt to describe problems stated in words appear trivial in hindsight. This is the case because the crucial insight needed to formulate the model is divulged by the structure of the model. Teaching this critical inductive step in the same fashion that the deductive steps in mathematical analysis are taught would be desirable. Regrettably, this is probably impossible. However, seeing this step detailed for a number of apparently difficult and confusing problems may be instructive.

Prerequisites. None. The problems in the chapter are chosen in most cases such that once the insight is gained, the model and its solution are trivial and can be analyzed without pencil and paper. As such, this chapter should be useful at almost any level of high school or college education in mathematics.

Time. The chapter consists of about thirty short problems, each of which should require perhaps 10 minutes to discuss in detail, so the amount of time devoted to the chapter is flexible.

Remarks. The chapter can probably be most effectively employed by distributing some problems for students to study in advance, discussing some problems in detail in class, and using some problems for homework. In this way, the critical nature of insight in model formulation should be made clear.

CHAPTER 2

Five Nautical Models

Robert L. Baker*

1. Introduction

This unit contains five models concerning the naval environment and using some of the more geometric ideas of the calculus. A panel of officers in the Mathematics Department at the U.S. Naval Academy suggested these problems to arouse the professional interest of midshipman studying calculus. The problems are not, however, strictly navy-oriented; the word naval should be interpreted in its broader context and appeal to all students who can assume, at least temporarily, an interest in nautical affairs.

This chapter makes no attempt to teach the basics of calculus. We assume that these notes are being used to supplement one of the many texts available. Having a calculator will be convenient for some of the exercises, and in a couple of instances projects are suggested which would require access to a computer.

The geometric flavor of calculus in recent times has been neglected by many instructors, who view calculus strictly as preparation for courses in higher level analysis. Perhaps, in a small way, this collection of problems is a protest against that trend.

2. Distance to the Horizon

One of the important facts that an underway sailor should keep in mind is the distance to the horizon from his position on his ship. This distance to the horizon is the distance d along the line of sight from an observer whose

* Sanders Associates, Inc., 95 Canal St., Nashua, NH 03061

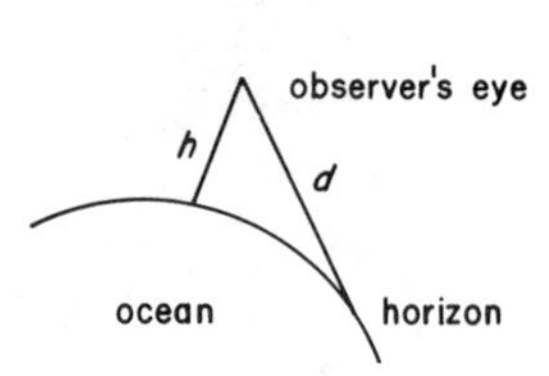

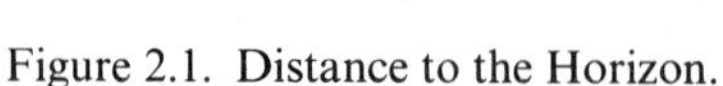

Figure 2.1. Distance to the Horizon.

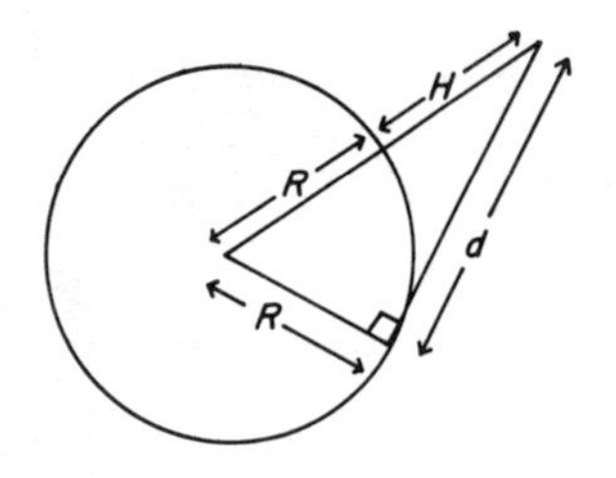

Figure 2.2. Model of the Earth.

eyes are a height h above the ocean to the horizon, i.e., from the observer's eyes to the point of tangency of this line of sight with the surface of the ocean (Figure 2.1).

Knowledge of the distance to the horizon lets the observer know the distance at which he could first see a floating object and gives him a standard to use in estimating distances to other ships and boats. It is surprising that among naval officers the distance to the horizon is often a subject of debate, even though it can be easily calculated using high-school level mathematics.

Problem 1

We want to find an approximate formula for the distance d in nautical miles (nmi) to the horizon as a function of the height h in feet of the observer's eyes above the ocean.

Symbols Used.

- R 3480, radius of earth in nautical miles; 1 nmi $\doteq$ 6076 ft,
- h height of eye above earth's surface in feet,
- H height of eye above the earth's surface in nautical miles,
- d distance of the observer to horizon in nautical miles.

Let us model the earth as spherical and look at a representative plane through the earth containing the observer's eyes and the center of the earth (Figure 2.2). By the Pythagorean theorem,

$$R^2 + d^2 = (R + H)^2 = R^2 + 2RH + H^2,$$

so that

$$d^2 = 2RH + H^2.$$

Since $R \gg H$, it follows that $2RH \gg H^2$ (do you see why?), and we drop the H^2 term in comparison with $2RH$ giving the approximations

$$d^2 \doteq 2RH, \qquad d \doteq \sqrt{2RH}.$$

(The symbol $\gg$ means "much greater than".)

One nautical mile is roughly 6000 ft, so that

$$H \text{ nmi} \doteq h \text{ ft}/6000 \text{ ft/nmi}$$

and thus

$$d \doteq \sqrt{2RH} \doteq \sqrt{2Rh/6000}$$

$$d \doteq \sqrt{3480h/3000}$$

$$d \doteq 1.08\sqrt{h}.$$

If we are willing to approximate a little more, we could say that

$$d \doteq \sqrt{h},$$

an easy rule to remember, where h is in *feet* and d in *nautical miles.*

As a numerical example, if $h = 20$ ft, then the approximation $d \doteq \sqrt{h}$ gives $d = 4.472$ nmi, whereas the exact formula (see Exercise 1) gives $d = 4.786$ nmi. The difference is 0.314 nmi which, when divided by the exact value, gives a relative error of 6.6%.

Exercises

1. Do the calculation above *without* the simplifying assumptions, i.e., a) do not neglect H^2, and b) use 1 nmi = 6076 ft. Make a table comparing values of $d = d_1$ from your more exact formula with $d = d_2 = \sqrt{h}$ for various realistic values of h. Also tabulate the absolute error $d_1 - d_2$ and the relative error $(d_1 - d_2)/d_1$.

2. The officier of the deck of a destroyer sights the radar mast of an aircraft carrier just as the radar antenna clears the horizon. What is the distance to the carrier, assuming the officer's height of eye is 36 ft and the carrier's antenna is 144 ft above sea level? Hint: Redraw Figure 2.2 to include two observers, both of whom see the same point on the horizon. (Answer: 18 nmi from $d = \sqrt{h}$; 19.44 nmi from $d = 1.08\sqrt{h}$.)

3. What area (square nautical miles) can an aviator see from 10,000 ft? Hint: Assume he sees a flat circular area. (Answer: $\pi(1.08\sqrt{h})^2 \doteq 36{,}400 \text{ nmi}^2$.)

3. Rendezvous (with Fixed Speeds)

Problem 2

An aircraft carrier has detached one of its protective destroyers to recover a downed aviator. The destroyer's navigator knows that, after the destroyer picks up the flyer and reports in, the carrier will radio its position at that time and its new course and speed and will order the destroyer to rejoin the carrier *as soon as possible.* The navigator wishes to determine quickly the course to rendezvous and the estimated time of arrival at the rendezvous point. The carrier and destroyer travel both in a fixed direction at fixed speeds until rendezvous.

Symbols Used.

v_1	aircraft carrier's speed in knots (1 knot $=$ 1 nmi/h $\doteq$ 6076 ft/h),
v_2	destroyer's speed in knots,
$a = v_2/v_1$,	ratio of destroyer's to carrier's speed,
T	time to rendezvous in hours,
$P = (x, y)$,	point of rendezvous,
$d = 2b$,	initial separation of ships in nautical miles.

In order to model the problem, assume that the ocean is a plane (flat surface) and superimpose on it an x, y coordinate system so that initially the carrier is at a point C, the destroyer is at a point D, and the distance between the ships is $d = 2b = |\overline{CD}|$. Also assume that both ships travel in a straight line, the aircraft carrier at a constant speed v_1, the destroyer at a constant speed v_2, and the ratio of the latter to the former is defined as $a \equiv v_2/v_1$.

Since the placement of the axes of our coordinate system is arbitrary, we can locate the y axis through the points C and D, with the origin at the midpoint of the line segment $\overline{CD}$, so that the coordinates are $C = (0, b)$ and $D = (0, -b)$. The point of rendezvous P has coordinates (x, y) and will be different for each course selected by the carrier. The angles θ_1 and θ_2 are the angles the paths of the ships make with the positive x axis (Figure 2.3).

If the ships rendezvous at the point P after a time T hours, then $|\overline{CP}| = v_1 T$ and $|\overline{DP}| = v_2 T = a v_1 T$, so that $|\overline{DP}| = a|\overline{CP}|$. Conversely, if $|\overline{DP}| = a|\overline{CP}|$ where $a = v_2/v_1$, then taking $T = |\overline{CP}|/v_1 = |\overline{DP}|/v_2$ gives

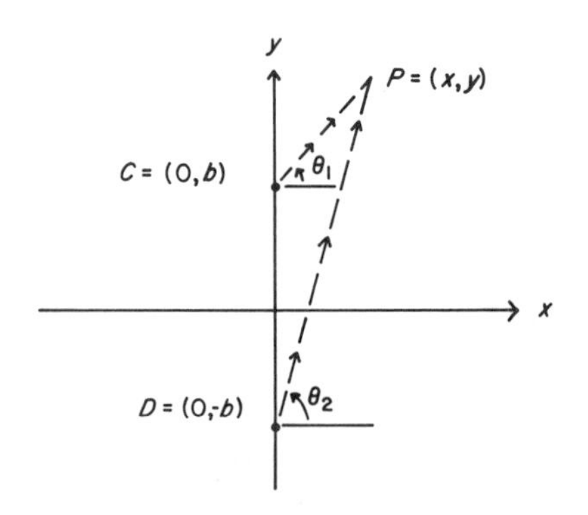

Figure 2.3. Rendezvous Paths.

$|\overline{DP}| = v_2 T$ and $|\overline{CP}| = v_1 T$, making it possible for the ships to rendezvous at P after T hours.

Thus we wish to find the locus of all points P such that

$$|\overline{DP}| = a|\overline{CP}|.$$

We rewrite this criterion with the following manipulations

$$|\overline{DP}|^2 = a^2|\overline{CP}|^2$$

$$x^2 + (y + b)^2 = a^2(x^2 + (y - b)^2) \tag{1}$$

$$(a^2 - 1)x^2 + (a^2 - 1)y^2 - 2(a^2 + 1)by + (a^2 - 1)b^2 = 0.$$

The last equation, the "rendezvous criterion", is the locus of all possible points of rendezvous and there are two cases to consider.

Case 1: $a = 1$. We then have as the rendezvous criterion

$$-4by = 0$$

whose solution is $y = 0$. Thus if the carrier's speed v_1 equals the destroyer's speed v_2, all possible points of rendezvous P have $y = 0$, i.e., they lie on the perpendicular bisector of the line segment $|\overline{CD}|$ (Figure 2.4).
The time to arrival at P is simply $T = |\overline{DP}|/v_2 = |\overline{CP}|/v_1$.

Case 2: $a \neq 1$. Dividing the rendezvous criterion by $a^2 - 1 \neq 0$, we get

$$x^2 + y^2 - 2\left[\frac{a^2 + 1}{a^2 - 1}\right]by + b^2 = 0,$$

and using completion of the square,

$$x^2 + \left(y - \left[\frac{a^2 + 1}{a^2 - 1}\right]b\right)^2 = \left[\frac{a^2 + 1}{a^2 - 1}\right]^2 b^2 - b^2 = \frac{4a^2b^2}{(a^2 - 1)^2},$$

or

$$x^2 + (y - h)^2 = R^2 \tag{2}$$

with

$$H \equiv \frac{a^2 + 1}{a^2 - 1}b, \qquad R \equiv \frac{2ab}{|a^2 - 1|}.$$

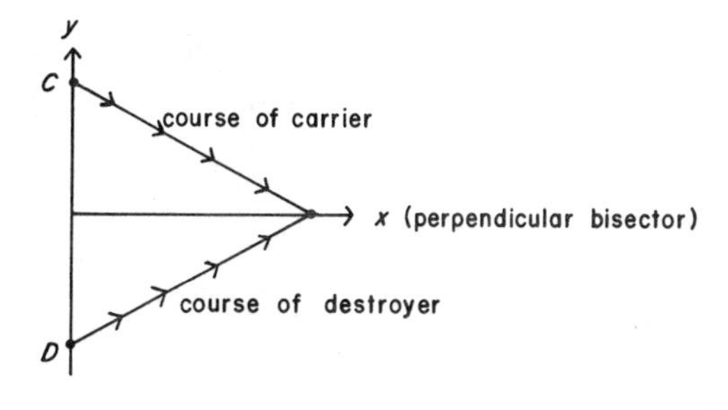

Figure 2.4. Rendezvous Paths if Speeds Are Equal.

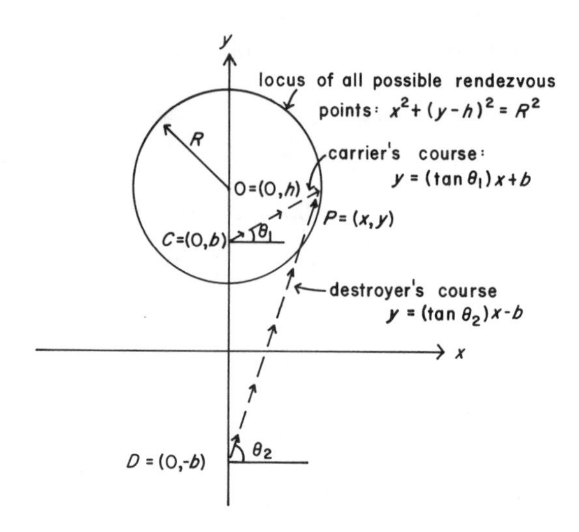

Figure 2.5. Rendezvous Paths when $v_1 < v_2$.

Thus the locus of all possible points P is the circle with center $\mathrm{O} = (0, h)$ and radius $R = 2ab/|a^2 - 1|$.

We are not yet done, for we would like to interpret equation (2). The number h is positive or negative depending on whether $a > 1$ ($v_2 > v_1$) or $a < 1$ ($v_2 < v_1$). Since $a^2 + 1 > |a^2 - 1|$, $|h| = ((a^2 + 1)/|a^2 - 1|)b > b$, so the center $\mathrm{O} = (0, h)$ lies on the line $\overline{CD}$ but does not lie between C and D. Moreover, $a \neq 1$ implies $(a - 1)^2 = a^2 - 2a + 1 > 0$ which implies $a^2 + 1 > 2a$, so that

$$|h| = \frac{(a^2 + 1)b}{|a^2 - 1|} > \frac{2a}{|a^2 - 1|}b = R.$$

So the circle lies entirely above (if $a > 1$) or entirely below (if $a < 1$) the x axis, which is the perpendicular bisector of $\overline{CD}$.

Of course, the latter fact is intuitively clear, since if $a > 1$ ($v_1 < v_2$), then the rendezvous points must be closer to the carrier than the destroyer and hence must lie on the carrier side of the perpendicular bisector of $\overline{CD}$. This situation with $v_1 < v_2$ ($a > 1$) is illustrated in Figure 2.5.

Note that if the carrier's speed v_1 were greater than the destroyer's speed v_2 (so $a < 1$), a rendezvous would be possible if and only if the carrier's course intersected the locus circle containing D. If the destroyer is faster, it can always steer a course to rendezvous, regardless of the carrier's course.

In summary, the mathematical procedure to determine the destroyer's course is to solve

$$x^2 + (y - h)^2 = R^2 \quad \text{(locus of rendezvous points)}$$

and

$$y = (\tan\theta_1)x + b \quad \text{(carrier's course)}$$

simultaneously for (x, y). Once $P = (x, y)$ is known, it can be substituted into

$$y = (\tan \theta_2)x - b$$

to determine θ_2, which specifies the destroyer's course.

EXERCISES

4. At 8 a.m., when the destroyer is 50 nmi south of the carrier, the carrier sets course $\theta_1 = 45°$, speed 20 knots, and orders the destroyer to rejoin as soon as possible at full speed (30 knots). What course does the destroyer steer to rendezvous? What is the estimated time to rendezvous? (Answers: $\theta_2 \doteq 28°9'$; $T = 4.06$ h.)

5. It is intuitively clear that if we interchange v_1 and v_2, then a is replaced by $a^* = 1/a$, and the old locus is replaced by a new locus in the x axis, since we have just interchanged the roles of the carrier and the destroyer. Verify this mathematically by examining the effect of the replacement $a \to a^* = 1/a$ in equation (2).

6. Write a computer program to calculate θ_2 and T based on given values of v_1, v_2, $d = 2b$, and θ_1. Do not forget the possibility that no rendezvous point may exist.

7. In this simple model, the carrier follows a straight line path $y = (\tan \theta_1)x + b$. Can you make any progress in reformulating the model if the carrier follows the parabolic path $y = x^2 + cx + b$ from $(0, b)$?

4. Submarine Detection, Long-Range Navigation

A tactic for locating submerged submarines with an aircraft involves placing a sonobuoy (an expendable air-dropped buoy having an underwater microphone suspended beneath it and a radio to transmit to the aircraft whatever sound is heard by the microphone) and a small explosive charge in the water a known distance z apart. The sonobuoy receiver in the aircraft is hooked up to the ink pen of a moving paper recording device; the speed of the moving paper is calibrated to the speed of sound in water. In this manner, the time delay between the receipt of the detonation at the buoy and the receipt of a possible echo from a sub can be measured and represents the quantity $1_1 + 1_2 - z$ as shown in Figure 2.6. The known quantity z is then added to obtain $1_1 + 1_2$. (Since the depth of the sub beneath the surface is small compared to the distances involved, we are essentially trying to locate a target lying in a plane just below the surface.) The locus of points representing the possible positions of the sub in this plane is an ellipse, since by definition an ellipse is the locus of all points the sum of whose distances from two fixed points (the foci) is a constant.

If the aircraft returns to the buoy and places another charge close to it, the time delay between detonation and an echo represents the distance 21_1 and the ellipse can be intersected by a circle centered at the buoy of radius

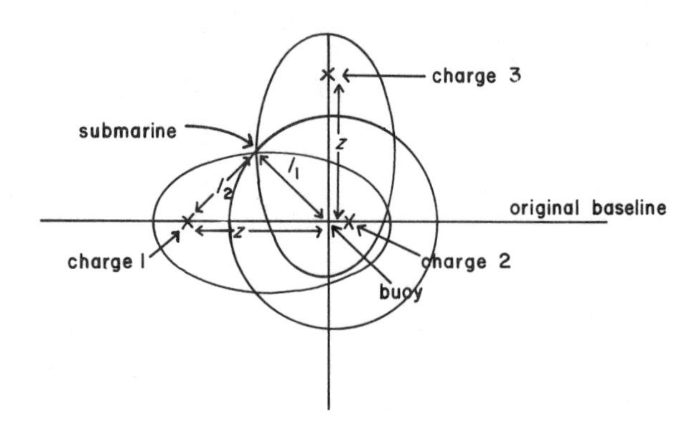

Figure 2.6. Scheme for Locating Submarine.

l_1. Since the sub lies on both the circle and the ellipse, only two locations are now possible. The ambiguity is resolved by placing another charge at a distance z from the buoy along a baseline perpendicular to the original baseline to obtain another ellipse which should intersect one of the two possible locations. The plotting, while now done by onboard computer, was initially done by dividers and a piece of string, then by a template of ellipses all with the same baseline z and having $l_1 + l_2$ plotted in multiples of 500 yds.

A system developed for long-range navigation (LORAN) makes use of a pair of radio transmitting stations located several hundred miles apart. The signals from these stations are synchronized so that the unit having a LORAN receiver can measure the time difference in microseconds between the receipt of the signal from the master station and that from a slave station.

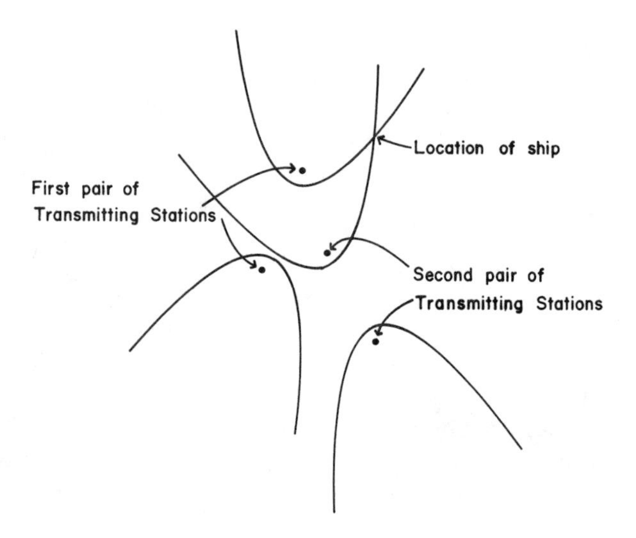

Figure 2.7. Scheme for Long-Range Navigation.

Knowing the speed of radio transmissions enables this time difference to be converted into a difference in distance from the two stations. Since by definition, a hyperbola is the locus of all points, the difference of whose distances from two fixed points is a positive constant, the locus of possible positions of the unit receiver lies on a hyperbola having the two transmitting stations as foci. In actuality, one's position is determined by locating oneself on one hyperbola from one pair of stations, then crossing this with another hyperbola from a second pair of transmitting stations (Figure 2.7). For ease of computation, LORAN charts are published with families of hyperbolas for each pair of transmitting stations labeled in microseconds. Each family is printed in different colors to avoid confusion.

Project. Try a computer simulation of one of these schemes, using standard formulas for ellipses, hyperbolas, and circles. You may have to study the effect of not having the hyperbolas aligned with the axes (rotations). You will need to master some computer techniques for finding roots of simultaneous equations.

5. Submarine Hunt

Problem 3

A hydrofoil patrol craft sights a surfaced enemy submarine and steers directly towards the submarine at its top speed of 60 knots. The submarine submerges when the patrol craft is 6 nmi away. It is known that the submarine will immediately proceed on an unknown straight course at its maximum submerged speed of 30 knots. What path should the patrol craft follow (at 60 knots) to *assure* that it eventually passes directly over the submarine? It is not necessary that the time to intercept be in some sense a minimum.

Symbols Used.

- r, θ polar coordinates in nautical miles and radians,
- s arclength in nautical miles (1 nmi $\doteq$ 6076 ft),
- T total time for complete search (worst case) in hours.

The statement of the problem contains no hint that polar coordinates would be a convenient tool in modeling the situation. However, polar coordinates in the form of "range" and "bearing" are familiar to naval officers and the path of the submarine is in a fixed direction (fixed bearing); hence we shall use a polar coordinate system.

Let A be the position of the patrol craft when the submarine submerges at B. Let B be the pole and BA be the polar axis for a system of polar coordinates. Since it is known that the submarine travels along a straight line at a known speed, the patrol craft can predict the distance r that the

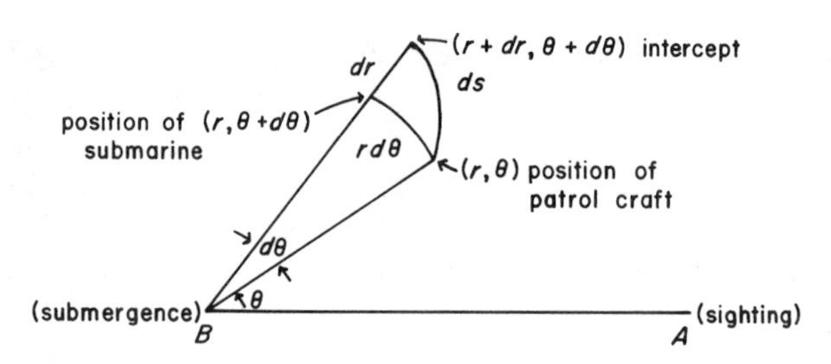

Figure 2.8. Positions of Submarine and Patrol Craft at Some Instant of Time, if Patrol Craft Has Slightly Misjudged Submarine's Direction of Travel.

submarine has traveled from the pole knowing only the elapsed time since the sighting. The only unknown of the submarine's path is the (constant) direction, which can be specified in terms of its polar angle. Suppose the patrol craft is at the point (r, θ) at a certain instant of time and the submarine has chosen to travel on a nearby path specified by the angle $\theta + d\theta$. Then the submarine lies at $(r, \theta + d\theta)$ at the same instant of time, where $d\theta$ is associated with the small error in direction that the patrol craft has made in locating the submarine. The patrol craft should now steer along its trajectory (which remains to be found) so as to reach the point $(r + dr, \theta + d\theta)$ at the same time as the submarine. (See Figure 2.8).

Since the patrol craft (pc) is twice as fast as the submarine, we have

$$(\text{speed of pc}) = ds/dt = 2dr/dt = 2(\text{speed of sub})$$

or

$$ds = 2dr. \tag{3}$$

However,

$$(ds)^2 = (dr)^2 + (rd\theta)^2, \tag{4}$$

which is a standard textbook result of substituting $x = r\cos\theta$, $y = r\sin\theta$ into $ds^2 = dx^2 + dy^2$ (see Figure 2.8). Substituting (3) into (4), we get

$$4(dr)^2 = (dr)^2 + (rd\theta)^2$$

and hence

$$\frac{dr}{r} = \frac{d\theta}{\sqrt{3}}. \tag{5}$$

Integrating both sides of the last equation

$$\ln|r| = \theta/\sqrt{3} + C,$$
$$|r| = \exp(\theta/\sqrt{3} + C),$$

or

$$r = A\exp(\theta/\sqrt{3}) \quad \text{where } A = \pm e^C. \tag{6}$$

This path is a logarithmic spiral. A logarithmic spiral $r = A\exp(a\theta)$ has the property that the angle between the polar radius and the tangent to the spiral is a constant $d = \arctan(1/a)$. In the above case, $d = \arctan(\sqrt{3}) = \pi/3 = 60°$.

Thus the resulting strategy is for the patrol craft to proceed in a straight line directly to some point $C = (r_0, \theta_0)$ of "possible intercept" (possible in the sense that the patrol craft, and the submarine will have the same distance r from the pole) and then follow the spiral $r = A\exp(\theta/\sqrt{3})$. The value of A is chosen so that (r_0, θ_0) satisfies $r_0 = A\exp(\theta_0/\sqrt{3})$, i.e., $A = r_0 \exp(-\theta_0/\sqrt{3})$ and

$$r = r_0 \exp((\theta - \theta_0)/\sqrt{3}). \tag{7}$$

EXAMPLE. One particular solution is for the patrol craft to proceed for 4 miles toward the point B of submergence, so that $C = (2, 0)$ and the submarine will have $r = 2$ since it travels at one-half the speed of the patrol craft (Figure 2.9). From C the patrol craft should follow the spiral $r = 2\exp(\theta/\sqrt{3})$.

EXERCISES

8. Show that the example solution is not really practical by examining the "instantaneous" turn that will be required of the patrol craft at C.

9. Verify that $C = (2\sqrt{3}, \pi/2)$ is a possible point of intercept and work out the details of the ensuing spiral. Is the turn required by the patrol craft at C more reasonable than in the example?

10. The submarine will be "caught" at some point during the $360°(2\pi)$ traversal on the spiral around B_0. In the worst case a complete search will consist of traversing $|\overline{AC}|$ plus the arc length around the complete spiral during which θ will undergo a net change of 2π. We can use the arc length integral

$$S = \int_{\theta_0}^{\theta_0+2\pi} \sqrt{\left(\frac{dr}{d\theta}\right)^2 + r^2}\, d\theta$$

to get the contribution from the spiral. Thus the time for a complete search will be

$$T = (|\overline{AC}| + S)/60.$$

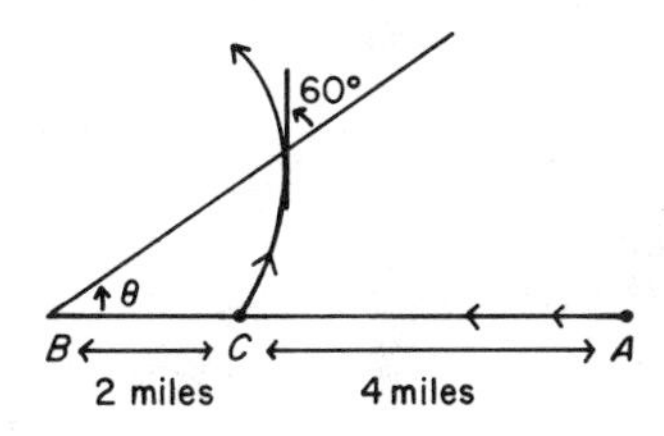

Figure 2.9. Path Followed by Patrol Craft in Example.

(a) Calculate the total time of traversal for the example. (Answer: 2 1/2 hours.)
(b) Do the same for Exercise 2. (Answer: 4 1/3 hours.)

11. It is perhaps easier to calculate the time for a complete search by computing how far the submarine has moved along its straight line path during the period that the patrol craft is making the search. At the end of a complete search the patrol craft will be at

$$r = r_0 \exp((\theta_0 + 2\pi) - \theta_0)/\sqrt{3})$$

and this is exactly the distance that the submarine has traversed at 30 knots, so

$$T = (r_0/30)\exp(2\pi/\sqrt{3}).$$

Repeat the calculations of Exercise 10(a) and (b) using this method.

12. Use the background gained in Exercise 11 to decide which choice of $C = (r_0, \theta_0)$ is "best", in the sense of T being smallest for that choice.

13. From Exercise 11, if we let $T(\theta)$ be the time to intercept at (r, θ), then

$$T(\theta) = (1/30)r_0 \exp(((\theta_0 + \theta) - \theta_0)/\sqrt{3}) = (r_0/30)\exp(\theta/\sqrt{3}).$$

If we assume that any course in the interval $0 \le \theta \le 2\pi$ is equally likely to be chosen by the submarine, then $f(\theta) = 1/2\pi$ is the (uniform) probability distribution function for the submarine's direction of travel. The expected time to intercept is

$$E(T) = \int_0^{2\pi} T(\theta)f(\theta)d\theta.$$

(a) Calculate the expected time to intercept for the strategy in the example. (Answer: 2/3 hour.)
(b) Calculate $E(T)$ for Exercise 2. (Answer: 1.16 hours.)
(c) Show that the strategy in the example is "best" in the sense of $E(T)$ being smallest.

14. As formulated above, the ratio (speed of pc/speed of sub) is two. Discuss what happens to the model if this ratio becomes larger or smaller.

15. Try to make some progress in discussing what happens if the submarine is not confined to move along a straight path θ = constant.

6. Satellite Surveillance

Problem 4

Almost two-thirds of the earth's surface is ocean and very little of this surface area can be monitored from land or ships. In order to better monitor the seas, the United States launched a spy satellite in circular orbit 900 nmi above the earth's surface. The satellite's wide-angle high-resolution camera returned to earth a picture of every point on earth in direct "line of sight" from the satellite. At any instant, how much area does the satellite spy on (Figure 2.10)? Also, what is the total area under the satellite's surveillance in one complete revolution around the earth?

Symbols Used.

R radius of earth = 3480 nmi (1 nmi = 6076 ft),
H height of satellite in nautical miles.

Consider the satellite of a point P, a distance $R + H$ from the center of the earth, and let T be a point of tangency of the line PT from P to the sphere. Let $\phi^* = \angle POT$. Since $\angle POT$ is a right angle, $\cos\phi^* = R/(R + H)$.

Consider the dark band of angle ϕ measured down from the line OP. A plane through P and O intersects this band as shown in Figure 2.11. The radius of this "circular" band is (approximately) $R\sin\phi$, so its circumference is $2\pi R\sin\phi$, and its width is $Rd\phi$, where $d\phi$ is the "small change" in ϕ encompassed by the band. Thus the band has area $dA = (2\pi R\sin\phi)Rd\phi = 2\pi R^2\sin\phi d\phi$. We must "add up" (or integrate) all the areas of such bands, as ϕ varies from 0 to ϕ^*. Thus

$$A = 2\pi R^2 \int_0^{\phi^*} \sin\phi d\phi = -2\pi R^2 \cos\phi \Big|_0^{\phi^*}$$

$$A = 2\pi R^2(\cos(0) - \cos\phi^*) = 2\pi R^2(1 - R/(R + H))$$

$$A = 2\pi R^2(H/(R + H)).$$

In our case $R = 3480$, $H = 900$ so,

$$A = 2(3480)^2(999/4380) \doteq 4.5 \times 10^6 \text{ nmi}^2.$$

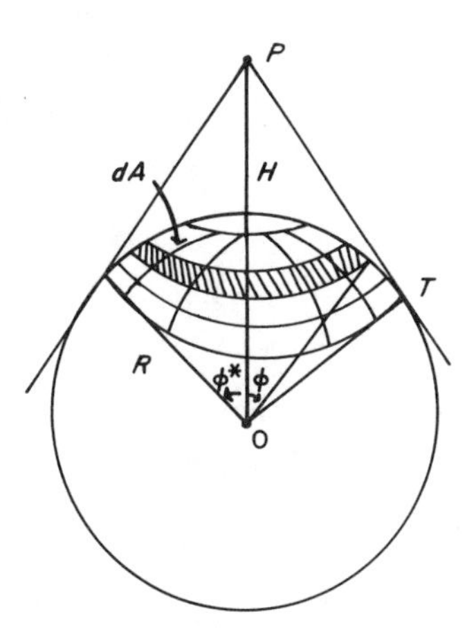

Figure 2.10. Area Seen by Satellite at Any Instant.

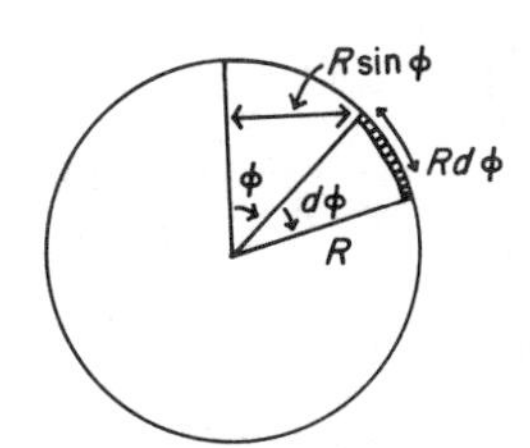

Figure 2.11. Variables for Area of Shaded Band in Figure 2.10.

To get the total area under the satellites's surveillance in one complete revolution of the earth, we introduce the angle θ which is measured from the ON axes rather than the OP axis (see Figure 2.12). The desired area is the lined region and the darkened band is a "circular" part of this region. Figure 2.13 is similar to Figure 2.11 and the arguments are similar to those above, so we have

$$dA = (2\pi R \sin\theta) R d\theta.$$

Hence

$$A = 2\pi R^2 \int_{\theta_1}^{\theta_2} \sin\theta d\theta = 2\pi R^2 (\cos\theta_1 - \cos\theta_2).$$

From Figure 2.12, $\theta_1 = \pi/2 - \phi^*$, $\theta_2 = \pi/2 + \phi^*$, so that

$$A = 2\pi R^2 (\cos(\pi/2 - \phi^*) - \cos(\pi/2 + \phi^*))$$

$$A = 2\pi R^2 (\sin\phi^* + \sin\phi^*) = 4\pi R^2 \sin\phi^*$$

$$A = 4\pi R^2 ((R + H)^2 - R^2)^{1/2}/(R + H)$$

$$A = 4\pi r^2 (2RH + H^2)^{1/2}/(R + H).$$

For our problem we get

$$A = 4\pi(3480)^2(2660/4380) \doteq 9.25 \times 10^7 \text{ nmi}^2.$$

Since $4\pi(3480)^2$ is the total area (in square nautical miles) of the earth, the satellite surveys $2660/4380 \doteq 0.6 = 60\%$ of the area of the earth during each orbit.

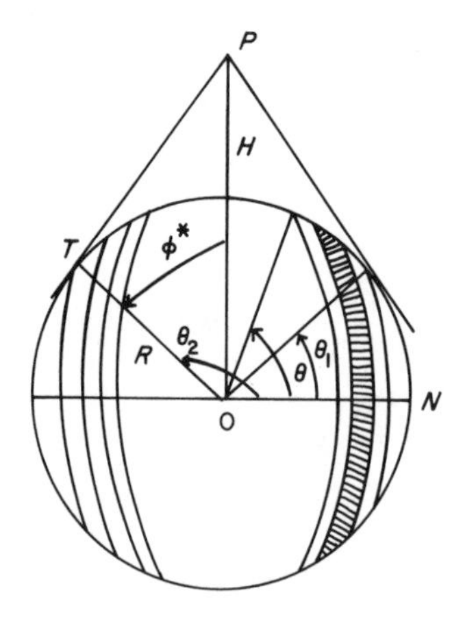

Figure 2.12. Area Seen by Satellite in One Revolution.

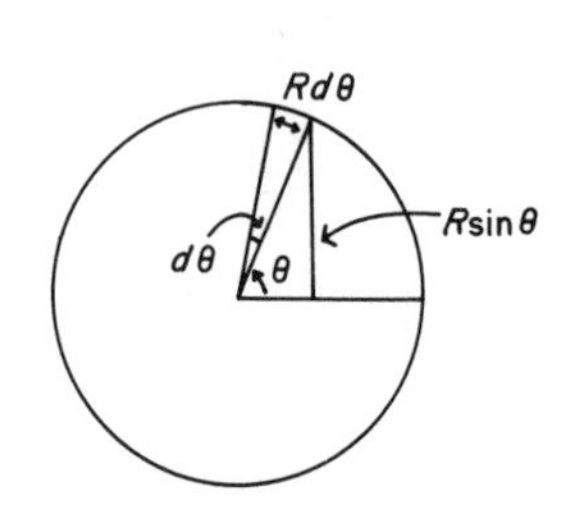

Figure 2.13. Variables for Area of Shaded Band in Figure 2.12.

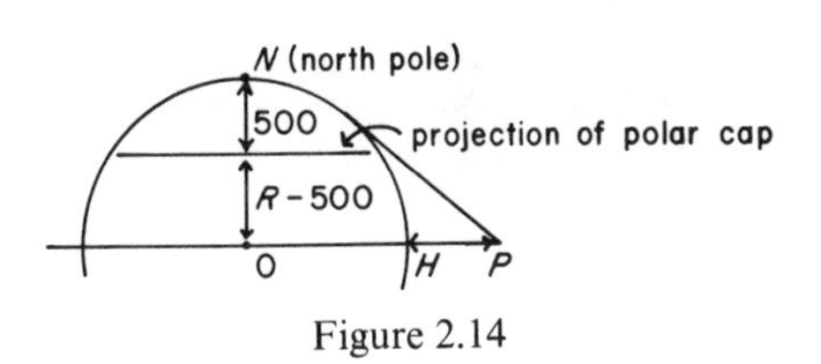

Figure 2.14

Exercises

16. Use the model in this section to calculate the area seen by an aviator at 10,000 ft. Compare your answer with Exercise 3 of Section 2.

17. At what height should a satellite be launched if it is to see 40% of the earth's surface in one revolution?

18. Assume we are interested in seeing all of the earth except the polar regions in one orbit of the earth. Assume that both polar caps are such that they subtend 500 nmi of the earth's radius (see Figure 2.14 which is a projection on the plane through O, N, and P). What should H be?

Notes for the Instructor

Objectives. Some applications of the calculus which could be used to supplement the usual textbook presentations are presented. These models could be read by students after the teacher has presented the prerequisite calculus topics. They would probably appeal to engineering and natural science majors more than to those in biology, social science, etc.

Prerequisites. These vary with the sections as follows.

Section 1, Distance to the Horizon: algebra and trigonometry.

Section 2, Rendezvous (with Fixed Speeds): Cartesian coordinates, equations of straight line and circle.

Section 3, Submarine Detection, Long-Range Navigation: definitions and equations of ellipse and hyperbola.

Section 4, Submarine Hunt: elementary integration, logarithm and exponential functions, polar coordinates, arc length (Exercise 3), probability distribution function (Exercise 6).

Section 5, Satellite Surveillance: integrals of a single variable.

Time. Each of the five models should occupy a one-hour class. The models are not dependent on each other, but a nice connection exists between Section 1, Exercise 3 and Section 5, Exercise 1.

Remark. This chapter comprises a collection of five loosely related models collected by a panel of officers at the United States Naval Academy. Techniques used are from the single variable differential and integral calculus.

CHAPTER 3

An Optimal Inventory Policy Model

Osvaldo Marrero*

1. Introduction

We are concerned here with a particular instance of an inventory model usually referred to in the literature as the Economic Lot-Size Model. Our objective is, roughly, to determine which is the best inventory policy one could adopt under certain conditions, with one objective being, of course, to ensure that costs incurred are kept at a minimum.

Our plan is to present an example of such a model in Section 2, where we obtain the desired result with the aid of techniques from differential calculus. This model considers shortages in inventory intolerable; the case where shortages are regarded as tolerable is dealt with in the Exercises. In Section 3 we present a particular numerical example to illustrate the discussion in Section 2. In Section 4 we present some concluding remarks.

The reader is expected to be acquainted with that part of differential calculus which deals with the determination of (local) extrema of functions. This is generally covered in the first course in calculus. To work through Exercise 6, the reader should be acquainted with the mechanics of computing partial derivatives of functions of two variables, as well as with the techniques for finding (local) extrema of such functions. No formal training in economics is required.

2. Statement and Analysis of the Problem

What is the best ordering policy a shoe retailer could adopt in order to keep his stock of goods at a satisfactory level?

* E. I. du Pont de Nemours & Company, Wilmington, DE 19898.

This question is much too vague to allow a careful mathematical analysis. We need to think carefully about the particular situation in order to formulate a more precise question. Of course, the application of the analysis that follows is not limited to shoe retailers; this analysis is just as relevant to any situation in which one is concerned with an inventory policy. We have chosen to consider a shoe retailing store in order to help the reader understand the ideas we want to discuss; once the reader has understood this particular example, it is a simple matter to apply a similar analysis to other appropriate situations.

Typically, goods travel from manufacturer to wholesaler to retailer, and finally to the consumer. Of course, the shoe retailer wants to obtain his inventory from a wholesaler at the lowest possible cost in order to minimize his expenses; the retailer also wants to maintain an adequate inventory of shoes in order to maximize his sales.

The retailer wants to purchase enough shoes from a wholesaler in order to ensure that "not too many" consumers fail to find what they want in the store, and at the same time, the retailer does not want to purchase so many shoes that they will be left over. That is, the retailer wishes to have enough inventory to satisfy demand but not more. Clearly, some shoe styles will sell faster than others, and therefore the retailer will want to reorder these faster selling shoes from a wholesaler at an appropriately faster rate. Thus, for simplicity, we will restrict our analysis to *one* particular shoe style.

Costs To be Considered. Three costs need to be considered (for *one* particular style of shoes):

(1) a *setup cost*, say K, charged at the time of ordering; this cost includes, for instance, the cost of the paper work involved in placing an order, record keeping, etc.;
(2) a *purchase cost*, say c dollars per item (pair of shoes); and
(3) an *inventory holding cost*, say h dollars, per item per unit of time, to cover the expenses of keeping the item in the store.

Two Assumptions. The retailer, of course, gathers and keeps data about sales over an appropriately long period of time. Again, to keep the mathematical model simple, we shall make two assumptions:

(1) the retailer sells items continuously to consumers at a known constant rate, say a; and
(2) when the retailer orders a quantity, say Q at a time, of pairs of shoes from a wholesaler, all Q items arrive at the retail store simultaneously when desired.

A Decision To Be Made. One decision must be made before proceeding further: will the retailer consider it tolerable to have a shortage of the particular shoe style we are considering in his store? (In this case we can reasonably assume that both alternatives—shortages are considered tolerable or shortages are considered intolerable—are feasible, and choosing one of these two alternatives is a personal matter to be decided by the retailer. Of course, this is not always the case: to think that a hospital would find it tolerable to have a shortage of anesthetics in its supply of goods is absurd.) We will assume that the retailer does not consider shortages tolerable; the other case will be dealt with in the exercises.

The Question Is More Precise Now. The initial question can now be stated more precisely: how often should the retailer place an order to a wholesaler for a particular shoe style, and how much should he order to ensure that the cost per unit time is a minimum? Recall that we are also assuming that shortages are not considered tolerable. We will take one month (30 days) as a unit of time.

Ordering Is a Cyclic Event. The retailer wishes to order from a wholesaler Q items at a time, and he expects to sell these items to consumers at the rate of a items per unit of time. Because shortages are not considered tolerable, it follows that the time between two consecutive orders is Q/a units of time; that is, the ordering of this particular shoe style is a cyclic event having cycle length Q/a, measured in months.

Cost per Unit Time. To obtain the cost per unit of time, we observe the following. The purchase cost per cycle is 0 dollars if $Q = 0$ (no shoes are ordered), or it is $K + cQ$ dollars if $Q > 0$ (Q pairs of shoes are ordered). To obtain the holding cost per cycle, note that the average inventory level during a cycle is $(Q + 0)/2 = Q/2$ items per unit of time and that the corresponding cost is $hQ/2$ dollars per unit of time. Since the cycle length is Q/a months, the holding cost per cycle is given by $(hQ/2)(Q/a) = hQ^2/2a$ dollars per cycle. It follows that the total cost per cycle is $K + cQ + hQ^2/2a$ dollars, and the total cost T per unit of time is

$$T = T(Q) = \frac{K + cQ + hQ^2/2a}{Q/a} = aK/Q + ac + hQ/2.$$

Minimum Value of Q. The value of Q, say Q^*, which minimizes T is (Exercise 1) $Q^* = \sqrt{2aK/h}$. In order to achieve the desired objective, the time t^* it takes between two consecutive orders is (that is, t^* is the optimal cycle length) $t^* = Q^*/a = \sqrt{2K/ah}$.

Conclusion. To achieve his objective, the retailer should order from a wholesaler every $\sqrt{2K/ah}$ months a quantity of $\sqrt{2aK/h}$ pairs of shoes of the particular style under consideration, where K is the setup cost of the order, h is the inventory holding cost per item per month, and a is the constant rate at which the items are sold by the retailer to the consumers.

3. A Numerical Example

Let us illustrate the foregoing with a particular numerical example. Suppose the particular shoe style we are considering is one specific type of women's shoes that is sold year round and is expected to continue to be in fashion for some time—long enough to warrant the consideration of the following analysis. The retailer estimates the setup cost incurred at the time of ordering to be \$20.00. Each such pair of shoes costs the retailer \$4.60 plus \$0.10 for delivery charges. The retailer regards the inventory holding cost to be \$0.84 per pair of such shoes per month. (Here's how the retailer arrived at what he considers his inventory holding cost: he sells each pair of such shoes for \$9.99; therefore, during each month, each such pair of shoes, he reasons, is worth $\$9.99/12 \doteq \$0.84$ per month to him.) Finally, the retailer expects to sell, on the average, 90 pairs of such shoes per month.

Initial Computation of Q^*. Here we must pay particular attention to the usual customs of the type of business being considered. Thus, in terms of the notation we used in Section 2, we now have $K = \$20.00$, $c = \$4.70$, $h = \$0.84$, and $a = 90$. It follows that the quantity Q^* of such pairs of shoes which should be ordered so as to minimize the total cost T per month is

$$Q^* = \sqrt{2aK/h} = \sqrt{2(90)(20)/0.84} = \sqrt{4285.71} = 65.47 \text{ pairs.}$$

Two problems must now be addressed. First, it is, of course, absurd to order 65.47 pairs of shoes; Q^* must be an integer. Second, the custom in the shoe retailing business is for a wholesaler to sell women's shoes to a retailer *only* by cases of either 18 or 36 pairs of shoes. For definiteness, let us assume the retailer may order the particular style of shoes under consideration from a wholesaler only by the case of 18 pairs.

Final Computation of Q^*. To resolve the two problems just mentioned, we first examine the behavior of the function $T(Q)$: on the feasible set $\{Q: 0 < Q < \infty\}$, the function $T(Q)$ is decreasing for $Q \in (0, \sqrt{2aK/h})$ and increasing for $Q \in (\sqrt{2aK/h}, \infty)$ (see Exercise 2). We have already seen that here $\sqrt{2aK/h} = 65.47$. Thus we need to compute $T(54)$ and $T(72)$, because 54 and 72 are those multiples of 18 which are closest to 65.47 and are, respectively, *less than* and *greater than* 65.47; one computes (Exercise 4) $T(54) = \$479.01$ per month and $T(72) = \$478.24$ per month. Therefore, the retailer should order $72 = Q^*$ pairs of such shoes.

Conclusion. Finally, in order to achieve the desired objective, the retailer should order 72 pairs of such shoes every $t^* = Q^*/a = 72/90 = 0.80$ month, that is, every 24 days (recall that, for us, "month" means 30 days).

4. Concluding Remarks

Other inventory models exist that we have not considered here. For instance, it is not unusual in various trades for a seller to allow for discounts according to volume of merchandise sold. In addition, the two models we considered are *deterministic*; i.e., the demand for a period is known. When this is not the case, the appropriate inventory model is a *stochastic* (or *probabilistic*) one, i.e., one in which the demand for a period is a random variable having a known probability distribution. Information about the models not considered here is available, for example, by Frederick S. Hillier and Gerald J. Lieberman in *Operations Research* (2d ed., San Francisco, Holden-Day, 1974, Chapter 11). In order to aid the reader who may wish to read further about inventory theory, our notation and terminology has been deliberately chosen to match as closely as possible that used by Hillier and Lieberman.

Exercises

1. Apply the techniques of differential calculus to show that if each of a, K, c, h, and Q is a positive real number, then $Q^* = \sqrt{2aK/h}$ is a (local) minimum for the function $T(Q) = aK/Q + ac + hQ/2$.

2. Plot the graph of the function $T = T(Q) = 1800/Q + 423 + 0.42Q$. This is the function we considered in the example given in Section 3. Observe that, by looking at the graph of this function T, one sees that it does attain a minimum value near the value 65.47 for Q. With this in mind, do you consider it necessary or worthwhile to go through Exercise 1 to find the minimum $Q^* = 65.47$? Comment about the relative advantages of graphical methods versus calculus techniques in determining (local) extrema of functions such as the one considered in this exercise.
3. Plot the graph of the inventory level Q versus time. Recall our assumptions: items are sold continuously at the rate of a items per unit of time, inventory ranges from 0 to Q items, and shortages are not considered tolerable.

4. Let $T(Q)$ be as in Exercise 1 and compute $T(54)$ and $T(72)$ to verify the values given in Section 3.

5. In Section 2 it was shown that the optimal cycle length t^* is given by $t^* = Q^*/a = \sqrt{2K/ah}$. Let $Q^* = 72$, $a = 90$, $K = 20$, and $h = 0.84$. Then compute $\sqrt{2K/ah}$, and note that this number is *not* equal to the value 0.80 obtained for t^* in the numerical example in Section 3. Comment about the difference in these numbers: is there an error in our analysis or computation?

6. This exercise considers the case where shortages in inventory are regarded as tolerable; for simplicity, the exercise is divided into several parts.

   We again consider one particular situation in order to help the reader understand the basic ideas. We consider a wine wholesaler in the U.S., and we are interested in the inventory of one particular wine the wholesaler imports from France. From past experience, the wholesaler knows that she must regard shortages as tolerable. A bad year in the vineyards abroad where this particular wine is made, difficulties in transporting the wine from the vineyard in France to her warehouse in the U.S., etc., or any combination of such situations could easily produce a shortage of this particular type of wine in the warehouse.

The wholesaler may buy this particular wine from the vineyard by cases of 12 bottles exclusively. However, she has found by experience that selling wine to retail dealers by the bottle is good business, as the retailers often want to mix the cases they buy with various wines. She therefore intends to keep a flexible policy. Thus one item in the wholesaler's inventory means one bottle of wine, and one unit of time will be one month (i.e., 30 days).

The three costs K, c, and h will be assumed to be the same as in the example of the shoe retailer in Section 2. In addition, shortages are considered tolerable; the wholesaler assigns a cost of p dollars for each demand of a bottle of wine that she cannot fill for one unit of time. Q and a will have a similar meaning as in the example of the shoe retailer in Section 2.

Let S denote the inventory of the particular French wine being considered on hand at the warehouse at the beginning of a cycle. Our objective now is to determine which values of S and Q will simultaneously yield the smallest total cost per unit of time for the wholesaler; this is the content of part (g) below. The several parts associated with this exercise are now considered.

(a) What is the *purchase cost per cycle*?

(b) What is the *holding cost per cycle*? (Note that the inventory level is positive for a time S/a, and the average inventory level during this time is $(S + 0)/2 = S/2$ bottles per month.)

(c) What is the *shortage cost per cycle*? (Note that shortages occur for a time $(Q - S)/a$, and the average amount of shortages during this time is $[0 + (Q - S)]/2 = (Q - S)/2$ bottles per month.)

(d) Add the results in (a), (b), and (c) to obtain the *total cost per cycle*.

(e) Use the result in (d) to write the *total cost T per unit of time*. (Recall that Q/a is one unit of time.)

(f) Observe that T is a function of two variables, S and Q. We wish to determine the (local) minimum of T. As a first step, compute $\partial T/\partial S$ and $\partial T/\partial Q$.

(g) Set each of $\partial T/\partial S$ and $\partial T/\partial Q$ computed in (f) equal to 0, and solve these equations simultaneously for S and Q. Let the solutions obtained be denoted by S^* and Q^*, respectively.

(h) Using the appropriate mathematical test (the one for functions of two variables analogous to the "second derivative test" for extrema of functions of one variable), verify that S^* and Q^* are in fact optimal.

(i) Compute the *optimal period length* $t^* = Q^*/a$ as a function of K, a, h, and p.

(j) Compute the *maximum shortage* $Q^* - S^*$ as a function of K, a, h, and p.

(k) Compute the *fraction of time that no shortage exists* $(S^*/a)/(Q^*/a)$ as a function of h and p.

(l) Draw the graph of the inventory level versus time. (Recall that the inventory on hand ranges from 0 to S; and, because shortages are considered tolerable, the amount Q ordered may exceed S. The inventory level is positive in cycles having length of time S/a.)

7. A numerical example. Repeat Exercise 6 using the following values: $K = \$40.00$; $c = \$6.80$ (this includes all transportation costs); $h = \$0.27$ (the wholesaler sells each bottle of the particular wine we are considering to retail dealers for \$9.99 and obtained this holding cost by reasoning similar to that of the shoe retailer in Section 3); $p = \$0.15$ (this is the subjective value assigned by the wholesaler); and $a = 426$ bottles per month. Keep in mind that the wholesaler may purchase this particular wine from the producer in France only by the case of 12 bottles each, and she sells it to wine retailers by the bottle (that is, although she always sells to retailers by the case, she allows retailers to mix the cases to suit their needs).

Notes for the Instructor

Objectives. We consider one example from each of two deterministic models in inventory theory. The two cases considered are 1) shortages are regarded as tolerable, and 2) shortages are regarded as not tolerable. The objective in each case is to determine the inventory level so as to minimize the associated costs.

Prerequisites. An acquaintance with that part of differential calculus which deals with the determination of (local) extrema of functions of one variable, and of two variables is necessary.

Remarks. This chapter is suitable for the following:
Courses in calculus—an application to economics, illustration of a case where one is interested in the determination of the (local) extrema of a function of one variable in one case, and of a function of two variables in the other case;
Courses in economics—an illustration of the application of calculus to inventory theory;
Courses in mathematical modeling or in applied mathematics.

CHAPTER 4

An Arithemetic Model of Gravity

Donald Greenspan*

1. Introduction

The immediate purpose of this chapter is to provide students and teachers of mathematics with a completely arithmetic formulation for the study of gravity. Shocking as it first may seem, *all* the results usually inferred by means of the calculus will be deduced using only the basic operations of addition, subtraction, multiplication, and division. A course in high school intermediate algebra, which may even be taken concurrently, is the only prerequisite for a complete understanding of all the ideas and results to be developed. For the reader who may wish, in addition, to explore arithmetic models of more complex physical forces, a concise survey and related references are provided in Sections 3 and 4, respectively.

Finally, a special warning must be added for those who may wish to use this chapter in a course in calculus. You must find your own answer to the inevitable question: "Why do we use calculus to solve problems about gravity when all we need is arithmetic?"

2. Gravity

2.0 Introduction

In developing an *arithmetic* model of gravity, how shall we choose mathematical formulas which will be right? In order to give an answer to this question, we first explore and analyze a simple experiment.

* Department of Mathematics, University of Texas at Arlington, Arlington, TX 76019.

Things fall. To discover the details of falling, consider the following experiment. Let us drop a solid metal ball, which we denote by P, from a height x_0 above ground and measure its height x above ground every Δt seconds as it falls. For example, if we had a camera whose shutter time was Δt, we could take a sequence of pictures at times $t_0 = 0$, $t_1 = 1(\Delta t)$, $t_2 = 2(\Delta t)$, $t_3 = 3(\Delta t)$, $\cdots$. From the knowledge of the initial height x_0, one could then determine the heights $x_1 = x(t_1)$, $x_2 = x(t_2)$, $x_3 = x(t_3)$, $\cdots$, from the photographs by ratio and proportion. For example, suppose $x_0 = 400$ is the height of a building from which the ball has been dropped. As shown in Figure 4.1, let an X axis be superimposed such that its origin is at the base of the building. If the top of the building is T, then determine from the photograph the ratio $|PO| : |TO|$. The actual height x of P is determined readily from

$$x : 400 = |PO| : |TO|.$$

In this manner, suppose that for $\Delta t = 1$ one finds, to the nearest foot,

$$\begin{aligned} x_0 &= 400 \\ x_1 &= 384 \\ x_2 &= 336 \\ x_3 &= 256 \\ x_4 &= 144. \end{aligned} \tag{1}$$

To analyze our data, let us first rewrite them in a fashion which exhibits clearly the *distance P has fallen*. Thus

$$\begin{aligned} x_0 &= 400 = 400 - 0 \\ x_1 &= 384 = 400 - 16 \\ x_2 &= 336 = 400 - 64 \\ x_3 &= 256 = 400 - 144 \\ x_4 &= 144 = 400 - 256. \end{aligned} \tag{2}$$

However, simple factoring now reveals that

$$\begin{aligned} x_0 &= 400 - 16(0)^2 \\ x_1 &= 400 - 16(1)^2 \\ x_2 &= 400 - 16(2)^2 \\ x_3 &= 400 - 16(3)^2 \\ x_4 &= 400 - 16(4)^2, \end{aligned} \tag{3}$$

so that

$$x_k = 400 - 16(t_k)^2, \qquad k = 0, 1, 2, 3, 4. \tag{4}$$

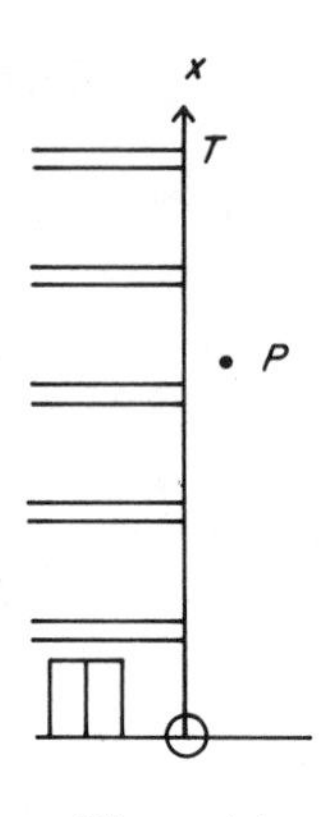

Figure 4.1

One might now be quite eager to assert that (4) is a *general* formula for the position of P at *any* time, not just at t_0, t_1, t_2, t_3, and t_4. This then requires further experimentation to accumulate more data. For example, from (4) one would *predict* for $k = 5$ that

$$x_5 = 400 - 16(5)^2 = 0,$$

which is indeed correct.

2.1 Velocity

In order to analyze the motion described in Section 2.0 in greater detail, it might be convenient to discuss next how fast the particle moved. In this connection, we will at present use the terms *velocity* and *speed* interchangeably, since the motion is one-dimensional only.

Our first problem is to devise some method by which we can determine P's velocities v_0, v_1, v_2, v_3, v_4 at the respective times t_0, t_1, t_2, t_3, t_4.

Since P was dropped from a position of rest, it is reasonable to assume that its initial velocity is zero, that is, $v_0 = 0$. Consider, next, v_1. The velocity v_1 at time t_1 must represent a measure of how fast P's height is changing with respect to time. Since P's height has changed $x_1 - x_0$ in the time $t_1 - t_0 = \Delta t$, let us define v_1 by the simple formula

$$\frac{v_1 + v_0}{2} = \frac{x_1 - x_0}{\Delta t}, \tag{5}$$

which is equivalent to

$$v_1 = -v_0 + \frac{2}{\Delta t}(x_1 - x_0). \tag{6}$$

Equation (5) is superior to various other possible formulas, like

$$v_1 = \frac{x_1 - x_0}{\Delta t}, \tag{7}$$

in two important ways. First, the left side of (5) is an averaging, or smoothing, formula for the velocities, and averaging, or smoothing, is of exceptional value in the analysis of approximate data. Second, (5) includes v_0. Never mind that v_0 was zero in our experiment, for we certainly did not have to *drop* P initially, we could actually have *thrown* it, in which case v_0 would not have been zero.

Having now defined v_1, it is consistent to define v_2 by

$$\frac{v_2 + v_1}{2} = \frac{x_2 - x_1}{\Delta t},$$

to define v_3 by

$$\frac{v_3 + v_2}{2} = \frac{x_3 - x_2}{\Delta t},$$

and so forth. Thus, in general, we define v_{k+1}, $k = 0, 1, 2, \cdots, n$, by

$$\frac{v_{k+1} + v_k}{2} = \frac{x_{k+1} - x_k}{\Delta t}, \qquad k = 0, 1, 2, \cdots, n, \tag{8}$$

which is, of course, equivalent to

$$v_{k+1} = -v_k + \frac{2}{\Delta t}(x_{k+1} - x_k). \tag{9}$$

Data (1), $\Delta t = 1$, $v_0 = 0$, and formula (9) now yield

$$\begin{aligned}
v_0 &= 0 \text{ ft/sec} \\
v_1 &= -v_0 + 2(x_1 - x_0)/(\Delta t) = 0 + 2(384 - 400)/1 = -32 \text{ ft/sec} \\
v_2 &= -v_1 + 2(x_2 - x_1)/(\Delta t) = 32 + 2(336 - 384)/1 = -64 \text{ ft/sec} \\
v_3 &= -v_2 + 2(x_3 - x_2)/(\Delta t) = 64 + 2(256 - 336)/1 = -96 \text{ ft/sec} \\
v_4 &= -v_3 + 2(x_4 - x_3)/(\Delta t) = 96 + 2(144 - 256)/1 = -128 \text{ ft/sec.}
\end{aligned} \tag{10}$$

2.2 Acceleration

To analyze the motion in our experiment in still greater detail, it may be of value to examine how fast the velocity is changing, which is called acceleration. For this purpose, let P's acceleration be a_0, a_1, a_2, a_3, and a_4 at the respective times t_0, t_1, t_2, t_3, and t_4.

With some thought, it becomes clear that we *do not know* a_0. We knew x_0 because it was given, and we decided that it was reasonable to assume $v_0 = 0$ because P was dropped from a position of rest. There simply does not seem to be any facet of the experiment which enables us to make a reasonable choice for a_0.

Since acceleration is to be a measure of how the velocity is changing with respect to time and since a_0 is not known, let us define a_0 by the very simple formula

$$a_0 = \frac{v_1 - v_0}{\Delta t}. \tag{11}$$

However, now that a_0 has been defined, it is consistent to define a_1, a_2, a_3, and so forth by

$$\begin{aligned} a_1 &= \frac{v_2 - v_1}{\Delta t} \\ a_2 &= \frac{v_3 - v_2}{\Delta t} \\ a_3 &= \frac{v_4 - v_3}{\Delta t}, \end{aligned} \tag{12}$$

and, in general, we define a_k, $k = 0, 1, 2, \cdots, n$, by

$$a_k = \frac{v_{k+1} - v_k}{\Delta t}, \qquad k = 0, 1, 2, \cdots, n. \tag{13}$$

For our experiment of Section 2.0, we have from (10) and (13)

$$\begin{aligned} a_0 &= (v_1 - v_0)/\Delta t = (-32 + 0)/1 = -32 \text{ ft/sec}^2 \\ a_1 &= (v_2 - v_1)/\Delta t = (-64 + 32)/1 = -32 \text{ ft/sec}^2 \\ a_2 &= (v_3 - v_2)/\Delta t = (-96 + 64)/1 = -32 \text{ ft/sec}^2 \\ a_3 &= (v_4 - v_3)/\Delta t = (-128 + 96)/1 = -32 \text{ ft/sec}^2. \end{aligned} \tag{14}$$

Thus (14) yields the remarkable result that, under the influence of gravity, P's acceleration at t_0, t_1, t_2, t_3 is constant, with the exact value -32 ft/sec^2.

2.3 A Mathematical Model

Let us see now if we can develop a *theory* for gravity. By this we mean that, assuming only the barest essentials, we will try to prove all the other results found thus far, and, perhaps, even more.

Let a particle P at height x_0 be dropped from a position of rest. For $\Delta t > 0$, let $t_k = k\Delta t$, $k = 0, 1, 2, \cdots, n$. At each time t_k, let the particle's position, velocity, and acceleration be x_k, v_k, and a_k, respectively. Assume that $v_0 = 0$ and that

$$\frac{v_{k+1} + v_k}{2} = \frac{x_{k+1} - x_k}{\Delta t}, \qquad k = 0, 1, 2, \cdots, n \tag{15}$$

$$a_k = \frac{v_{k+1} - v_k}{\Delta t}, \qquad k = 0, 1, 2, \cdots, n. \tag{16}$$

Finally, assume that

$$a_k = -32 \text{ ft/sec}^2, \qquad k = 0, 1, 2, \cdots, n. \tag{17}$$

We now wish to prove that the motion of P is exactly as was discovered in Sections 2.0–2.2. We will do this by showing how the above assumptions completely determine the motion of P from the fixed initial position x_0. The procedure, however, will require some observations and results in the use of summation symbols.

The symbol $\Sigma_{k=i}^{n} f(x_k)$, read "the summation from k equals i to k equals n of $f(x_k)$" means, let $k = i, i+1, i+2, \cdots, n$ in $f(x_k)$ and then add up all the resulting f's. Thus

$$\sum_{k=1}^{5} x_k = x_1 + x_2 + x_3 + x_4 + x_5$$

$$\sum_{k=0}^{5} [(-1)^k y_k] = y_0 - y_1 + y_2 - y_3 + y_4 - y_5.$$

The k in $\Sigma_{k=i}^{n} f(x_k)$ is called a "dummy index", since it can be changed to any other (unused) letter without changing the actual sum. For example,

$$\sum_{k=1}^{5} x_k^2 = x_1^2 + x_2^2 + x_3^2 + x_4^2 + x_5^2$$

is identical with

$$\sum_{j=1}^{5} x_j^2 = x_1^2 + x_2^2 + x_3^2 + x_4^2 + x_5^2.$$

To facilitate algebraic manipulation with summations, the following identities will be of exceptional value:

$$\sum_{k=i}^{n} [\alpha f_k + \beta g_k] = \alpha \sum_{k=i}^{n} f_k + \beta \sum_{k=i}^{n} g_k;\ \alpha, \beta \text{ constants} \tag{18}$$

$$\sum_{k=0}^{n-1} (x_{k+1} - x_k) = x_n - x_0 \tag{19}$$

$$\sum_{k=1}^{n} 1 = n. \tag{20}$$

The validity of (18) can be seen clearly from the following particular example:

$$\begin{aligned}\sum_{k=1}^{4} (3x_k - 4Y_k) &= (3x_1 - 4Y_1) + (3x_2 - 4Y_2) + (3x_3 - 4Y_3) + (3x_4 - 4Y_4)\\ &= 3(x_1 + x_2 + x_3 + x_4) - 4(Y_1 + Y_2 + Y_3 + Y_4)\\ &= 3\sum_{k=1}^{4} x_k - 4\sum_{k=1}^{4} Y_k.\end{aligned}$$

The validity of (19) can be seen clearly from the following particular example:

$$\begin{aligned}\sum_{k=0}^{4} (x_{k+1} - x_k) &= (x_1 - x_0) + (x_2 - x_1) + (x_3 - x_2) + (x_4 - x_3) \\ &\quad + (x_5 - x_4) \\ &= -x_0 + (x_1 - x_1) + (x_2 - x_2) + (x_3 - x_3) \\ &\quad + (x_4 - x_4) + x_5 \\ &= x_5 - x_0.\end{aligned}$$

The cancellation of all terms between x_5 and x_0, above, is called *telescoping* and (19) is called a telescopic sum. The validity of (20) follows readily, since

$$\sum_{k=1}^{n} 1 = \underbrace{1 + 1 + 1 + \cdots + 1}_{n \text{ terms}} = n.$$

Let us return now to (15)–(17). From (16) and (17), it follows that

$$v_{k+1} - v_k = -32\Delta t.$$

Summing both sides of this equation yields, for each integer $n \geq 1$,

$$\sum_{k=0}^{n-1} (v_{k+1} - v_k) = \sum_{k=0}^{n-1} (-32\Delta t), \qquad n = 1, 2, 3, \cdots,$$

or, by (18)–(20),

$$v_n - v_0 = -32n\Delta t.$$

Thus, since $v_0 = 0$ and $n\Delta t = t_n$, it follows that

$$v_n = -32t_n, \qquad n = 1, 2, 3, 4, \cdots \tag{21}$$

is the formula for the velocity of P at the time steps $t_1, t_2, t_3, \cdots$. For $\Delta t = 1$, in which case $t_0 = 0$, $t_1 = 1$, $t_2 = 2$, $t_3 = 3$, and $t_4 = 4$, formula (21) yields the exact values determined in (10). Observe also that these values have been deduced without any knowledge or use of x_0. That is, formula (21) is valid no matter what the initial height x_0 may be.

From (21), then,

$$\begin{aligned}\frac{v_{k+1} + v_k}{2} &= \frac{-32t_{k+1} - 32t_k}{2} \\ &= -16(t_{k+1} + t_k) \\ &= -16\Delta t[(k + 1) + k] \\ &= -16\Delta t(2k + 1),\end{aligned}$$

so that (15) can be rewritten as

$$\frac{x_{k+1} - x_k}{\Delta t} = -16\Delta t(2k + 1),$$

or, equivalently, as

$$x_{k+1} - x_k = -16(\Delta t)^2(2k + 1).$$

Summation of both sides of the latter equation then yields

$$x_n - x_0 = -16(\Delta t)^2 \sum_{k=0}^{n-1} (2k + 1), \qquad n = 1, 2, 3, 4, \cdots. \tag{22}$$

However, $\Sigma_{k=0}^{n-1}(2k + 1)$ is the sum of an arithmetic sequence of n terms whose first term is 1 and whose last term is $2n - 1$. This sum is then equal to n^2. Thus (22) simplifies to

$$x_n = x_0 - 16(n\Delta t)^2$$

or

$$x_n = x_0 - 16(t_n)^2, \qquad n = 1, 2, 3, 4, \cdots, \tag{23}$$

which is the formula for the position x_n of P at each time step t_n. For $x_0 = 400$ and $\Delta t = 1$, this formula yields exactly the same positions as those of (4).

3. A Survey of Other Arithmetic Models

Gravity is not the only force which can be studied using only arithmetic. In fact, all forces in Newtonian mechanics, classical molecular mechanics, and special relativistic mechanics can be so treated and all the conservation and symmetry laws continue to be valid. The price one pays for such immense mathematical simplicity is that, in general, one must use modern digital computers to solve related dynamic problems (see references [1], [3]), and these computer solutions differ from continuous ones by terms of order $(\Delta t)^3$ in both position and velocity [4].

Of special interest in both the Newtonian and classical molecular cases is that the formulations have resulted in new, heuristically appealing physical models of gravitation, oscillation, heat conduction, elasticity, shock wave generation, and laminar and turbulent fluid flow [1], [2]. Moreover, these new physical models allow for highly nonlinear behavior.

Exercises

1. Restate (1) in metric units. Restate (4) in metric units.

2. Write out each of the following sums.

   (a) $\sum_{k=0}^{5} y_k$.

   (b) $\sum_{k=1}^{6} 2y_k$.

(c) $\sum_{k=-1}^{5} (y_k^2)$.

(d) $\sum_{k=2}^{6} (-1)^k y_k^2$.

3. Use (18) to rewrite each of the following in an equivalent form.

(a) $\sum_{k=1}^{6} [7x_k - 9y_k]$.

(b) $\sum_{k=1}^{15} [7x_k^2 - 9e^k]$.

(c) $\sum_{k=0}^{20} 5 \sin x_k$.

(d) $\sum_{k=1}^{30} (-2k + 36k^2)$.

4. Show that each of the following is telescopic, and thereby simplify the sum.

(a) $\sum_{k=0}^{7} (y_{k+1} - y_k)$.

(b) $\sum_{k=-1}^{10} (y_{k+1}^2 - y_k^2)$.

(c) $\sum_{k=2}^{25} (y_{k+1}^3 - y_k^3)$.

5. Evaluate each of the following sums.

(a) $\sum_{k=1}^{5} 1$.

(b) $\sum_{k=0}^{5} 1$.

(c) $\sum_{k=-5}^{5} 1$.

6. Show that if a particle is under the influence of gravity and if $v_0 \neq 0$, then (15)–(17) imply

$$v_n = v_0 - 32t_n$$

$$x_n = x_0 + v_0 t_n - 16(t_n)^2.$$

7. A particle is thrown upward from the ground at 64 ft/sec. Using the results of Exercise 6, find how high the particle rises and how long it is in the air.

8. A particle is thrown upward from the ground at 640 ft/sec. Using the results of Exercise 6, find how high it rises and how long it is in the air.

9. A girl on the ground wishes to throw a particle exactly 144 ft into the air. With what initial velocity must she throw it?

References

[1] D. Greenspan, *Arithmetic Applied Mathematics*. Oxford: Pergamon, 1980.
[2] ——, *Computer-Oriented Mathematical Physics*. Oxford: Pergamon, 1981.
[3] R. A. LaBudde and D. Greenspan, "Discrete mechanics—A general treatment," *Computational Physics*, vol. 15, pp. 134–167, 1974.
[4] ——, "Energy and momentum conserving methods of arbitrary order for the numerical integration of equations of motion," Part I, *Numerische Mathematik*, vol. 25, pp. 323–346, 1976; Part II, *Numerische Mathematik*, vol. 26, pp. 1–16, 1976.

Notes for the Instructor

Objectives. The aim here is to provide teachers and students of mathematics with a completely arithmetic formulation of the study of gravity and to illustrate an alternate methodology for investigating the physical world.

Prerequisites. High school intermediate algebra.

Time. One or two lectures.

CHAPTER 5

Four-Way Stop or Traffic Lights? An Illustration of the Modeling Process

Edward W. Packel*

1. General Aspects of the Model

We compare the effectiveness of four-way stop signs with that of traffic lights in controlling traffic at an intersection. Though important considerations such as accident prevention may provide conclusive reasons for preferring one to the other, we limit ourselves to the consideration of the total (or average) delay time of all cars arriving at the intersection in a given time period as the criterion for effectiveness. The basic unit of time will be that required for one complete cycle of the traffic lights, henceforth to be called a *cycle*. For the purposes of estimating values of the various parameters to be defined, we may find it convenient to assume, for instance, that a cycle is one minute, but the general analysis will be carried out in terms of cycles.

As pictured in Figure 5.1, we assume an intersection of 2 two-way streets, each having one lane of traffic in each direction. The model will ignore pedestrians (a "fatal" error!) and will assume no left-hand turns in the case of traffic lights. We now define the key parameters which will figure in the development of the model.

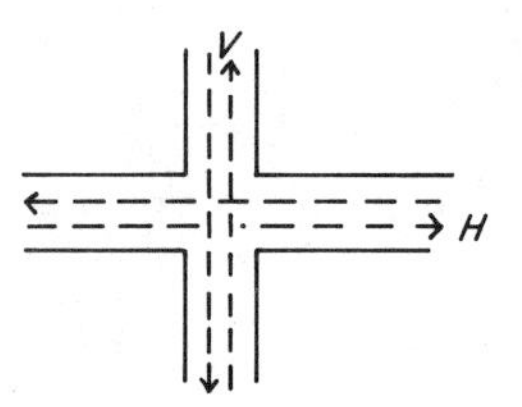

Figure 5.1. The Corner of H (Horizontal) and V (Vertical) Streets.

* Department of Mathematics, Lake Forest College, Lake Forest, IL 60045.

T	Total delay time (in cycles) of all cars arriving in a given cycle.
H	Number of cars/cycle arriving in the horizontal directions.
V	Number of cars/cycle arriving in the vertical directions.
B	$H + V$, number of cars/cycle arriving at the intersection in both directions.
S	Fraction of cycle required after starting up to regain normal speed (in the absence of other traffic) after being stopped (by light or stop sign).
C	Fraction of cycle required after starting up to clear the intersection for the next car (relevant only in stop sign case).
R	Fraction of cycle in which light is red in the horizontal direction.
$1 - R$	Fraction of cycle in which light is red in the vertical direction.

We emphasize that the total delay time T is the time lost in having to slow down or stop at the intersection. It will depend upon the other parameters as well as whether stop signs or traffic lights are present. It is this value we wish to minimize.

EXERCISES

1. Obtain an expression for the *average* delay time (in cycles) for each car arriving at the intersection in a given cycle. Then argue that if the number of cars/cycle is fixed, minimizing the average delay time is equivalent to minimizing the total delay time T.

2. Clearly, the red light times R and $1 - R$ in the model have a sum which corresponds to one full cycle of the traffic lights. From experience, suggest and discuss an aspect of traffic lights which threatens this assumption. Is the assumption valid?

3. Argue that, on the average, $H \cdot R$ cars will get caught by the light in the horizontal direction and that each of those caught will wait an average of $R/2$ cycles for the green light. Derive corresponding results for the vertical direction.

The speedup and clearing times S and C are perhaps the most questionable and least transparent parameters in the model. They are, however, important and will be reviewed in the next sections. For now, their definitions should be clear enough to see that generally $S > C$ and that typical values in seconds might be around six and three. Actual values, which could depend upon the nature of the intersection, speed limits, and aggressiveness of the drivers, could certainly be estimated by observation.

2. Traffic Light Analysis

In this section we obtain for the traffic light situation an expression for the total delay time T in terms of all other relevant parameters as defined in the previous section. Using this expression, we then determine the value of R

(red light percentage in the horizontal direction) which will minimize T. Note that R is the only parameter over which we have control, the others being fixed by the situation at hand.

We ignore any delay time (beyond the speedup time S) for a car which might arise from backing up of traffic at the light (but see Exercise 7) or from making left-hand turns. The model also ignores delay caused by the act of slowing down in both stop sign and traffic light situations, though the latter can be partially justified since the slow down time is essentially replaced by the stopping time as calculated below.

Assuming a uniform flow of traffic (an average of numerous trials using random flow would yield the same result) and using Exercise 3, the total delay time is determined by

$$T = \underbrace{\overbrace{S(H \cdot R + V(1 - R))}^{\#\text{ cars stopped by light}}}_{\text{total speed time}} + \underbrace{\overbrace{H \cdot R}^{\substack{\#\text{ cars stopped}\\ \text{in } H \text{ direction}}} \quad \overbrace{R/2}^{\text{ave. wait}} + V(1 - R)(1 - R)/2}_{\text{total stopping time}}$$

$$T = S(H \cdot R + V(1 - R)) + H \cdot R^2/2 + V(1 - R)^2/2. \tag{1}$$

Regarding T as a one variable function of R (a quadratic polynomial), a direct max–min calculation shows that T is minimized when

$$R = 1/B((1 + S)V - S \cdot H). \tag{2}$$

The desired expression for the minimum T is obtained by substituting (2) into equation (1). The resulting expression for T is

$$T_{\text{opt}} = \frac{(1 + 2S)^2 HV}{2B} - S^2 B/2,$$

(for this tricky computation see Exercise 8). Since we are interested in calculating R anyway, we leave the substitution for the computer program to perform in a later section.

EXERCISE

4. (a) (calculus prerequisite.) By treating T of equation (1) as $T(R)$ and differentiating, show that equation (2) does give the value of R which minimizes T. Be sure to prove that R is indeed a minimum rather than a maximum or an inflection point.

   (b) (No calculus required.) Write the expression in equation (1) in the form $T = a(R^2 + bR + c)$. Then complete the square to obtain $T = a((R + b/2)^2 + d)$. Finally, argue that T is minimized when $R = -b/2$, and by determining the appropriate value of b, check that equation (2) emerges.

It should be noted that our model has provided us with an unexpected bonus. In attempting to minimize T we obtain as a by-product the appropriate breakdown of the traffic light cycle into red and green. At this point it is good practice to see if our early results seem reasonable. Setting $S = 0$ in equation (2), we see that in the extreme case of zero speedup time, the light is red in the horizontal direction with a percentage $R = V/B$. Thus the percentage of time this light is red in a given direction equals the percentage of cars moving in the perpendicular direction, a result which seems intuitively reasonable.

Exercises

5. Prove that if the volume of traffic is equal in the horizontal and vertical directions, then $R = 0.5$ independent of the value of S. Discuss any implications of this result on the traffic light model.

This completes the traffic light component of the problem. Before proceeding to the stop signs, we consider in the following exercises two aspects of the model which require comment and perhaps strengthening.

6. The value of R in equation (2) must satisfy $0 \leq R \leq 1$ to be meaningful in terms of the model (why?).
   (a) Show that $0 \leq R \leq 1$ if and only if $S/(1 + S) \leq V/H \leq (1 + S)/S$.
   (b) If $S = 0.1$ and $H = 11$ what range of values for V will give meaningful R values in the model?
   (c) If $S = 0.1$, $H = 11$, and $V = 1$, what will the traffic light model predict and what is wrong with this result?

7. The description of the speedup time S assumes the absence of other traffic or (assuming all cars make the light) that all cars in the queue start accelerating simultaneously. More realistically, let us assume a lag time of $L = 0.01$ cycles between startup times of successive cars in the queue as the light turns green.
   (a) If $H = 15$, $R = 0.4$, and $L = 0.01$ show that cars in the horizontal direction contribute an additional lag time of $0.01 + 0.02 + 0.03 + 0.04 + 0.05 = 0.15$ to the delay time T.
   (b) Prove more generally that if $H \cdot R$ and $V(1 - R)$ are integers, then a total lag time of

$$L[H \cdot R(H \cdot R - 1)/2 + V(1 - R)(V(1 - R) - 1)/2]$$

   must be added to the expression for the total delay time T.
   (c) Use the results of (b), whether or not $H \cdot R$ and $V(1 - R)$ are integers, to obtain a more complex expression for T than appears in equation (1). Then minimize this expression as a function of R to obtain a corrected value of R. Compare the new R with that of equation (2) and discuss the net effect of incorporating lag time on the overall model for traffic lights.

8. Prove that, as claimed, substituting equation (2) into equation (1) gives

$$T_{\text{opt}} = \frac{(1 + 2S)^2 HV}{2B} - \frac{S^2 B}{2}.$$

Hint: First prove that $1 - R = 1/B[(1 + S)H - SV]$. The computation is tricky and *may* not be fun.

3. Stop Sign Analysis

The crucial parameter in stop sign situations is the clearing time C defined earlier as the time in cycles required upon starting up to clear the intersection for the next car. Thus if a car arrives at the intersection and then has to wait for four earlier arrivals to cross, this results in a delay of $4C$ cycles.

The stop sign model allows for right- and left-hand turns but ignores the fact that cars going in opposite directions on the same street can often negotiate the intersection simultaneously. It also ignores time lost in the required deceleration of each car in coming to a stop, but when other cars are already present in the intersection this time is absorbed in the clearing times. In the absence of other cars many people seem to stop rather abruptly if at all!

We consider separately two assumption about traffic flow at the four-way stop intersection. The first case of uniform flow is clearly unrealistic but will be useful for the method used and the simplicity of the solution.

Case I: Uniform Flow

We assume that B cars arrive at the intersection in a uniform manner—precisely one car every $1/B$ cycle. Under this assumption we can readily see that if $C \leq 1/B$ no backing up of cars takes place at the intersection, and the total delay time per cycle is given by

$$T = S \cdot B \quad \text{(each car loses } S \text{ cycles in speedup time).}$$

If $C > 1/B$ there will be some backing up of traffic. Then, in addition to the S cycle speedup time for each car, we obtain the following progression. The first car has no backup wait, the second must wait $(C - 1/B)$ cycles (the difference between its arrival and the first car's clearance), and generally the Kth car accumulates $(K - 1)(C - 1/B)$ cycles of backup wait. Thus in the case $C > 1/B$ of uniform flow,

$$\begin{aligned} T &= \underbrace{S \cdot B}_{\text{speedup time}} + \underbrace{(C - 1/B) + 2(C - 1/B) + \cdots + (B - 1)(C - 1/B)}_{\text{backup time}} \\ &= S \cdot B + (C - 1/B)[1 + 2 + \cdots + (B - 1)] \\ &= S \cdot B + (C - 1/B)[B(B - 1)/2] \quad \text{(summing the arithmetic progression).} \end{aligned}$$

In summary, we have for uniform flow:

$$T = \begin{cases} S \cdot B, & \text{if } C \leq 1/B \\ S \cdot B + (C - 1/B)B(B - 1)/2, & \text{if } C > 1/B. \end{cases} \tag{3}$$

EXERCISES

9. Assume uniform flow.
   (a) Given $S = 0.15$, $H = 6$, $V = 4$, and $C = 0.08$, what value of T does the stop sign model give?
   (b) Replace $C = 0.08$ by $C = 0.12$ leaving other values as in (a). Compute T for the stop signs in two ways. First, set up a table of arrival and departure times for each of the 10 cars and second, use formula (3). Your answers should agree!

10. Let $S = 0.15$, $H = 6$, $V = 4$, and assume uniform flow. Compute the traffic light value of T for this case. Considering parts (a) and (b) of Exercise 9 separately, would stop signs or traffic lights be preferred for each of these cases? Explain your reasoning.

Having warmed up with the uniform case, we now consider the more realistic case of random flow.

Case II: Random Flow

We assume that B cars/cycle arrive at the intersection as determined by B numbers generated randomly between 0 and 1 and then arranged in increasing order.

Let $A(K)$, $K = 1, 2, \cdots, B$, denote the arrival times in increasing order as obtained in the random manner indicated above. As was the case previously, our assumptions lead to a total speedup time/cycle of $S \cdot B$ to be augmented by a backup time which will depend on B (but not on the relative contributions of H and V). In fact, as long as we make sure to have a car crossing the intersection whenever it is clear and a car is available to cross, we need not concern ourselves with the nature of the queueing at the intersection or the complicated psychological and behavioral questions which govern such situations. Indeed these factors, as a result of our simplifying assumptions, will not affect the total delay time.

We first present a simple numerical example to clarify what follows. Let $B = 6$, $C = 0.1$, and let the arrival times be given by 0.12, 0.15, 0.27, 0.48, 0.56, and 0.88. Then we obtain the following table.

Car # (K)	Arrival Time ($A(K)$)	Departure Time	Backup Wait ($W(K)$)
1	0.12	0.12	0
2	0.15	0.22	0.07
3	0.27	0.32	0.05
4	0.48	0.48	0
5	0.56	0.58	0.02
6	0.88	0.88	0

Note that the backup wait is simply departure minus arrival, and the departure times are obtained either as immediate departure (no backup) or by adding $C = 0.1$ to the previous departure time (backup).

The backup time can be determined inductively without a table by means of the following algorithm:

$$\begin{cases} W(1) = 0, & \text{the first car has no backup wait} \\ W(K+1) = \text{MAX}\{0, W(K) + C - (A(K+1) - A(K))\}, & \\ \qquad\qquad\qquad\qquad K = 1, 2, \cdots, B-1. & \end{cases} \tag{4}$$

(The $(K+1)$st car to arrive either has no backup wait or else must wait however long the Kth car waited plus C cycles for clearing minus the difference in their arrival times.)

Thus the total backup time (per cycle) is just

$$W(1) + W(2) + \cdots + W(B) = \sum_{K=1}^{B} W(K),$$

and the total delay time in the random case at a four-way stop is

$$T = S \cdot B + \sum_{K=1}^{B} W(K). \tag{5}$$

Exercise

11. (a) Letting $B = 6$, $C = 0.1$ as in the numerical example, show that no backup wait occurs under the assumption of uniform flow.
    (b) Give an informal argument that for fixed values of B and C, the T value for random flow is always greater than or equal to the T value for uniform flow. Does this agree with your intuition?
    (c) Letting $B = 6$, $S = 0.15$, $C = 0.12$, and using the arrival times given in the numerical example, compute T using formulas (4) and (5).

We conclude this section by mentioning two shortcomings of the stop sign models. An immediate problem occurs when traffic flow is heavy—in particular, when C approaches or exceeds $1/B$. The difficulty is that our model computes the backup delay associated with a given cycle but ignores the fact that backup will then be present at the very start of the next cycle. This could happen (for random flow) even when $C < 1/B$, provided arrivals cluster towards the end of one cycle and the beginning of the next. In defense of our model, we should note that situations with C close to or exceeding $1/B$ are basically those with serious potential for congestion anyway, and stop signs would be inferior to a two-way stop on the side street, traffic lights, or in the case of high volume in both directions, an underpass.

To cope with this difficulty in the stop sign model when $C < 1/B$, one might compute delay time over a large number of cycles instead of stopping after a single cycle. By taking an average, a more realistic value of T could then be obtained for comparison with the traffic light delay (see Exercise 14).

In its present state, we can only state that the model is less trustworthy at high volumes of traffic then at lower ones.

A second shortcoming of the model is that expression (5) for T is probabilistic in nature, having no closed form solution and depending on the particular random numbers generated. While this indeterminacy is an intriguing and realistic aspect of the model, calling for and illustrating computer simulation, a closed form equation such as that of (1) or (3) would be desirable both practically and theoretically. One could then try to compare the values for T in (1) and (5) analytically, possibly deriving a threshold relationship among H, V, S, and C where stop signs and traffic lights would be equally good. One standard way to approach a closed form solution is to impose a probability distribution on the arrival time gaps and use theoretical means to obtain an expected backup time in closed form. This more mathematically advanced approach can be found by the interested reader in references [2]–[4].

The dependence of backup times upon the particular numbers generated is easier to deal with and we do so in the computer program of the next section. We simply allow for repeated trials of the random number generation process and then take an average of these trials to obtain a more stable backup time.

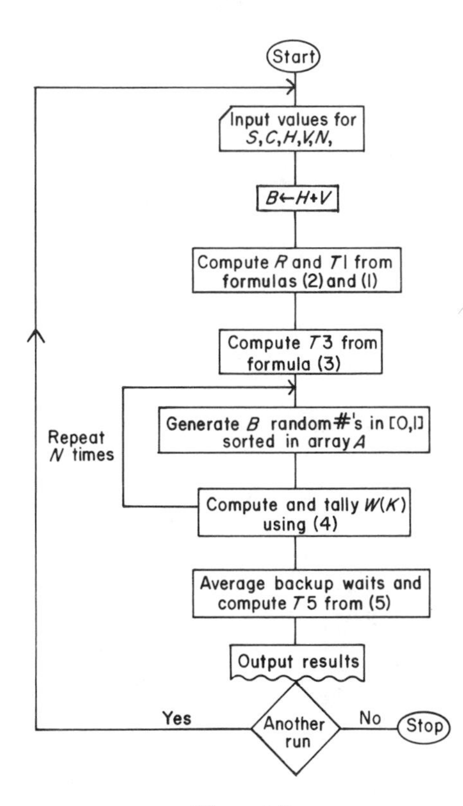

Figure 5.2

4. Computer Implementation

We now combine the results of the last two sections in a BASIC program so that delay times can be compared. The flowchart (Figure 5.2) summarizes the program, which proceeds in essentially the same order as the results were derived. Variable names coincide with those used previously, except that the three total delay times are called $T1$ (lights), $T3$ (signs, uniform), and $T5$ (signs, random). The integer N is the number of repeated trials for the stop sign simulation as described in the last paragraph of Section 3.

```
10  PRINT 'SPEEDUP TIME S IN CYCLES'
11  INPUT S
20  PRINT 'CLEARING TIME C IN CYCLES'
21  INPUT C
30  PRINT 'NUMBER OF CARS/CYCLE H IN HORIZONTAL
    DIRECTION'
31  INPUT H
40  PRINT 'NUMBER OF CARS/CYCLE V IN VERTICAL
    DIRECTION'
41  INPUT V
50  PRINT 'NUMBER OF RANDOM TRIALS N'
51  INPUT N
55  PRINT 'DO YOU WANT RANDOM ARRIVAL TIMES PRINTED
    (YES OR NO)'
56  INPUT P$
60  LET B = V+H
65  R = ((1 + S)*V-S*H)/B                     }
66  T8 = S*(H*R+V*(1-R))                      } Computes T1
67  T9 = .5*(H*R*R+V*(1-R)*(1-R))             }
68  T1 = T8 + T9                              }
70  F = 0                                     }
71  IF C > 1/B THEN F = 1                     } Computes T3
72  T3 = S*B+F*.5*B*(B-1)*(C-1/B)             }
74  S1 = 0                                                  }
75  FOR J = 1 TO N                                          }
80  FOR I = 1 TO B     }                                    }
85  A(I) = RND(0)      } generates random                   }
90  NEXT I             } numbers                            }
100 FOR Y = 1 TO B-1                   }                    }
110 FOR Z = Y+1 TO B                   }                    }
120 IF A(Y) < = A(Z) THEN 160          }                    }
130 A = A(Y)                           } sort into          }
140 A(Y) = A(Z)                        } increasing         } Computes T5
150 A(Z) = A                           } order              }
160 NEXT Z                             }                    }
170 NEXT Y                             }                    }
180 IF P$ = 'NO' THEN 200              }                    }
184 FOR K = 1 TO B                                          }
186 PRINT A(K):                                             }
188 NEXT K                                                  }
195 PRINT                                                   }
```

```
200 S2 = 0                                       }
210 W(1) = 0                                     }
220 FOR K = 1 TO B-1                             }
230 W = W(K) + C-(A(K+1)-A(K))                   }
240 W(K+1) = (W+ABS(W))*.5                       }
250 S2 = S2 + W(K+1)                             } Computes T5
260 NEXT K                                       }
270 S1 = S1 + S2                                 }
280 NEXT J                                       }
290 T5 = S*B+S1/N                                }
300 PRINT                                        }
350 PRINT 'PERCENTAGE OF TIME LIGHT IS RED IN
    HORIZONTAL DIRECTION = ';R
354 PRINT
356 PRINT 'TOTAL WAITING TIMES IN CYCLES'
360 PRINT 'T-LIGHTS =    '; T1
361 PRINT 'T-UNIFORM =    '; T3
362 PRINT 'T-RANDOM =    '; T5
365 PRINT
380 PRINT 'DO YOU WANT TO INPUT NEW PARAMETER VALUES
    (YES OR NO)'
385 INPUT A$
387 PRINT '----------------------------------'
390 IF A$ = 'YES' THEN 10
400 END
```

Various sample runs of the program follow. The output provides further information on the reasonability of the model as discussed in the next section. Note that a delay time for the uniform flow stop sign case ($T3$) is output for information purposes even though it has no bearing on the final decision on the form of traffic control. An option is included in the program for outputting the sequences of random arrival times and it is interesting to note just how unequally spaced these random times are (in contrast to our unrealistic uniform traffic flow case).

```
-----------------------------------------
SPEEDUP TIME S IN CYCLES                    RUN 1
!.1
CLEARING TIME C IN CYCLES
!.05
NUMBER OF CARS/CYCLE H IN HORIZONTAL DIRECTION
!3
NUMBER OF CARS/CYCLE V IN VERTICAL DIRECTION
!3
NUMBER OF RANDOM TRIALS N
!2
DO YOU WANT RANDOM ARRIVAL TIMES PRINTED (YES OR ON)
!YES
.0615234 .212372 .478729 .645844 .699951 .906616
.162598 .52466 .566284 .574066 .873627 .887421

PERCENTAGE OF TIME LIGHT IS RED IN HORIZONTAL
DIRECTION = .5
```

```
TOTAL WAITING ITEMS IN CYCLES
T-LIGHTS = 1.05
T-UNIFORM = .6
T-RANDOM = .650394

-------------------------------------------
SPEEDUP TIME S IN CYCLES                RUN 2
!.1
CLEARING TIME C IN CYCLES
!.05
NUMBER OF CARS/CYCLE H IN HORIZONTAL DIRECTION
!4
NUMBER OF CARS/CYCLE V IN VERTICAL DIRECTION
!2
NUMBER OF RANDOM TRIALS N
!1
DO YOU WANT RANDOM ARRIVAL TIMES PRINTED (YES OR NO)
!YES
.310547 .311737 .386963 .460571 .559662 .816376

PERCENTAGE OF TIME LIGHT IS RED IN HORIZONTAL
DIRECTION = .3

TOTAL WAITING TIMES IN CYCLES
T-LIGHTS = .93
T-UNIFORM = .6
T-RANDOM = .672394

-------------------------------------------
SPEEDUP TIME S IN CYCLES                RUN 3
!.1
CLEARING TIME C IN CYCLES
!.05
NUMBER OF CARS/CYCLE H IN HORIZONTAL DIRECTION
!5
NUMBER OF CARS/CYCLE V IN VERTICAL DIRECTION
!1
NUMBER OF RANDOM TRIALS N
!2
DO YOU WANT RANDOM ARRIVAL TIMES PRINTED (YES OR NO)
!YES
.00537109 .251679 .321259 .352387 .651245 .651611
.0765381 .202484 .207245 .342407 .603607 .74707

PERCENTAGE OF TIME LIGHT IS RED IN HORIZONTAL
DIRECTION = .1

TOTAL WAITING TIMES IN CYCLES
T-LIGHTS = .57
T-UNIFORM = .6
T-RANDOM = .656872

DO YOU WANT TO INPUT NEW PARAMETER VALUES (YES OR NO)
!YES
```

```
-------------------------------------------
SPEEDUP TIME S IN CYCLES                    RUN 4
!.1
CLEARING TIME C IN CYCLES
!.05
NUMBER OF CARS/CYCLE HH IN HORIZONTAL DIRECTION
!4
NUMBER OF CARS/CYCLE V IN VERTICAL DIRECTION
!4
NUMBER OF RANDOM TRIALS N
!1
DO YOU WANT RANDOM ARRIVAL TIMES PRINTED (YES OR NO)
!YES
.0356445 .222138 .229248 .291138 .423706 .678436 .693207
.961517

PERCENTAGE OF TIME LIGHT IS RED IN HORIZONTAL
DIRECTION = .5

TOTAL WAITING TIMES IN CYCLES
T-LIGHTS = 1.4
T-UNIFORM = .8
T-RANDOM = .909119
-------------------------------------------
SPEEDUP TIME S IN CYCLES                    RUN 5
!.1
CLEARING TIME C IN CYCLES
!.05
NUMBER OF CARS/CYCLE H IN HORIZONTAL DIRECTION
!6
NUMBER OF CARS/CYCLE V IN VERTICAL DIRECTION
!6
NUMBER OF RANDOM TRIALS N
!1
DO YOU WANT RANDOM ARRIVAL TIMES PRINTED (YES OR NO)
!NO

PERCENTAGE OF TIME LIGHT IS RED IN HORIZONTAL
DIRECTION = .5

TOTAL WAITING TIMES IN CYCLES
T-LIGHTS = 2.1
T-UNIFORM = 1.2
T-RANDOM = 1.3626
-------------------------------------------
SETUP TIME S IN CYCLES                      RUN 6
!.1
CLEARING TIME C IN CYCLES
!.05
NUMBER OF CARS/CYCLE H IN HORIZONTAL DIRECTION
!8
NUMBER OF CARS/CYCLE V IN VERTICAL DIRECTION
!8
```

```
NUMBER OF RANDOM TRIALS N
!2
DO YOU WANT RANDOM ARRIVAL TIMES PRINTED (YES OR NO)
!NO

PERCENTAGE OF TIME LIGHT IS RED IN HORIZONTAL
DIRECTION = .5

TOTAL WAITING TIMES IN CYCLES
T-LIGHTS = 2.8
T-UNIFORM = 1.6
T-RANDOM = 2.26541

------------------------------------------
SPEEDUP TIME S IN CYCLES                    RUN 7
!.1
CLEARING TIME C IN CYCLES
!.05
NUMBER OF CARS/CYCLE H IN HORIZONTAL DIRECTION
!10
NUMBER OF CARS/CYCLE V IN VERTICAL DIRECTION
!10
NUMBER OF RANDOM TRIALS N
!16
DO YOU WANT RANDOM ARRIVAL TIMES PRINTED (YES OR NO)
!NO

PERCENTAGE OF TIME LIGHT IS RED IN HORIZONTAL
DIRECTION = .5

TOTAL WAITING TIMES IN CYCLES
T-LIGHTS = 3.5
T-UNIFORM = 2
T-RANDOM = 3.59331

------------------------------------------
SPEEDUP TIME S IN CYCLES                    RUN 8
!.1
CLEARING TIME C IN CYCLES
!.05
NUMBER OF CARS/CYCLE H IN HORIZONTAL DIRECTION
!12
NUMBER OF CARS/CYCLE V IN VERTICAL DIRECTION
!12
NUMBER OF RANDOM TRIALS N
!5
DO YOU WANT RANDOM ARRIVAL TIMES PRINTED (YES OR NO)
!NO

PERCENTAGE OF TIME LIGHT IS RED IN HORIZONTAL
DIRECTION = .5

TOTAL WAITING TIMES IN CYCLES
T-LIGHTS = 4.2
```

```
T-UNIFORM = 4.7
T-RANDOM = 5.87775

------------------------------------------------
SPEEDUP TIME S IN CYCLES                    RUN 9
!.1
CLEARING TIME C IN CYCLES
!.05
NUMBER OF CARS/CYCLE H IN HORIZONTAL DIRECTION
!16
NUMBER OF CARS/CYCLE V IN VERTICAL DIRECTION
!16
NUMBER OF RANDOM TRIALS N
!4
DO YOU WANT RANDOM ARRIVAL TIMES PRINTED (YES OR NO)
!NO

PERCENTAGE OF TIME LIGHT IS RED IN HORIZONTAL
DIRECTION = .5

TOTAL WAITING TIMES IN CYCLES
T-LIGHTS = 5.6
T-UNIFORM = 12.5
T-RANDOM = 12.8477

------------------------------------------------
SPEEDUP TIME S IN CYCLES                    RUN 10
!.1
CLEARING TIME C IN CYCLES
!.05
NUMBER OF CARS/CYCLE H IN HORIZONTAL DIRECTION
!12
NUMBER OF CARS/CYCLE V IN VERTICAL DIRECTION
!8
NUMBER OF RANDOM TRIALS N
!3
DO YOU WANT RANDOM ARRIVAL TIMES PRINTED (YES OR NO)
!NO

PERCENTAGE OF TIME LIGHT IS RED IN HORIZONTAL
DIRECTION = .38

TOTAL WAITING TIMES IN CYCLES
T-LIGHTS = 3.356
T-UNIFORM = 2
T-RANDOM = 3.1613

DO YOU WANT TO INPUT NEW PARAMETER VALUES (YES OR NO)
!YES

------------------------------------------------
SPEEDUP TIME S IN CYCLES                    RUN 11
!.1
CLEARING TIME C IN CYCLES
!.05
```

```
NUMBER OF CARS/CYCLE H IN HORIZONTAL DIRECTION
!12
NUMBER OF CARS/CYCLE V IN VERTICAL DIRECTION
!8
NUMBER OF RANDOM TRIALS N
!16

DO YOU WANT RANDOM ARRIVAL TIMES PRINTED (YES OR NO)
!NO

PERCENTAGE OF TIME LIGHT IS RED IN HORIZONTAL
DIRECTION = .38

TOTAL WAITING TIMES IN CYCLES
T-LIGHTS = 3.356
T-UNIFORM = 2
T-RANDOM = 3.46718

------------------------------------------
SPEEDUP TIME S IN CYCLES                  RUN 12
!.1
CLEARING TIME C IN CYCLES
!.05
NUMBER OF CARS/CYCLE H IN HORIZONTAL DIRECTION
!13
NUMBER OF CARS/CYCLE V IN VERTICAL DIRECTION
!7
NUMBER OF RANDOM TRIALS N
!5
DO YOU WANT RANDOM ARRIVAL TIMES PRINTED (YES OR NO)
!NO

PERCENTAGE OF TIME LIGHT IS RED IN HORIZONTAL
DIRECTION = .32

TOTAL WAITING TIMES IN CYCLE
T-LIGHTS = 3.176
T-UNIFORM = 2
T-RANDOM = 3.37983

------------------------------------------
SPEEDUP TIME S IN CYCLES                  RUN 13
!.1
CLEARING TIME C IN CYCLES
!.05
NUMBER OF CARS/CYCLE H IN HORIZONTAL DIRECTION
!9
NUMBER OF CARS/CYCLE V IN VERTICAL DIRECTION
!3
NUMBER OF RANDOM TRIALS N
!2
DO YOU WANT RANDOM ARRIVAL TIMES PRINTED (YES OR NO)
!NO
```

```
PERCENTAGE OF TIME LIGHT IS RED IN HORIZONTAL
DIRECTION = .2

TOTAL WAITING TIMES IN CYCLES
T-LIGHTS = 1.56
T-UNIFORM = 1.2
T-RANDOM = 1.55942

-----------------------------------------
SPEEDUP TIME S IN CYCLES                    RUN 14
!.08
CLEARING TIME C IN CYCLES
!.05
NUMBER OF CARS/CYCLE H IN HORIZONTAL DIRECTION
!9
NUMBER OF CARS/CYCLE V IN VERTICAL DIRECTION
!3
NUMBER OF RANDOM TRIALS N
!2
DO YOU WANT RANDOM ARRIVAL TIMES PRINTED (YES OR NO)
!NO

PERCENTAGE OF TIME LIGHT IS RED IN HORIZONTAL
DIRECTION = .21

TOTAL WAITING TIMES IN CYCLES
T-LIGHTS = 1.4754
T-UNIFORM = .96
T-RANDOM = 1.3274

-----------------------------------------
SPEEDUP TIME S IN CYCLES                    RUN 15
!1
CLEARING TIME C IN CYCLES
!.07
NUMBER OF CARS/CYCLE H IN HORIZONTAL DIRECTION
!9
NUMBER OF CARS/CYCLE V IN VERTICAL DIRECTION
!3
NUMBER OF RANDOM TRIALS N
!3
DO YOU WANT RANDOM ARRIVAL TIMES PRINTED (YES OR NO)
!NO

PERCENTAGE OF TIME LIGHT IS RED IN HORIZONTAL
DIRECTION = .2

TOTAL WAITING TIMES IN CYCLES
T-LIGHTS = 1.56
T-UNIFORM = 1.2
T-RANDOM = 1.64985
-----------------------------------------
```

EXERCISES

12. • Incorporate results of Exercise 7 in the computer program and using sample runs discuss the extent to which the results are different.

13. • Working with fixed values of S and C of your own thoughtful choosing, try to find various pairs of H and V values for which the model is essentially indifferent between traffic lights and stop signs (random case).

14. • Review the problem of excessive backup delay discussed in the fourth and third from final paragraphs of Section 3. Then add to the program an option which allows the backup delay calculation to run over any number of cycles (with H and V constant) and averages the results.

5. Conclusions

A look at the various red light percentages R computed from formula (1) suggests that the lights in the busier direction should be red in a percentage slightly less than the percentage of cars moving in the perpendicular direction (see runs 2, 3, and 10). This result appears in the literature as a rule of thumb for simple signalized intersections, thus providing encouragement for the predictions of our model.

As seen by comparing runs 1, 4, 5, 6, 7, 8, and 9, the preference for stop signs when traffic is light gives way to a preference for lights as traffic gets heavy. The threshold values depend on C and D too, of course, (see runs 13, 14, 15) in a manner which agrees with intuition. By comparing runs 7, 11, and 12 we see further that for a fixed overall volume of traffic B, traffic lights are more efficient when traffic is unevenly divided, again a sensible result.

The discrepancy between T-RANDOM values in runs 10 and 11 indicates that repeated trials of the arrival time generation process are necessary for stabilization. Ideally, the number N of repetitions should be large for consistent results. This serves as a simple illustration of the use and value of *Monte Carlo simulation*, where a probability or theoretical value is estimated by repeated random number generation.

In summary, the model seems to yield results which agree with our expectations; and the model and its program enable us to treat the original question in quantitative fashion provided we obtain reasonable values for C and D.

As with any first effort on a model, many refinements can be made, assumptions questioned, and results tested. The model presented here was constructed using only elementary mathematical techniques and in the absence of preliminary empirical data. For refined and considerably more technical approaches, consult the references (a very small subset of the extensive traffic theory literature). The journal *Transportation Science* might also be scanned as a source of traffic models based on real data and of a considerably more complex nature.

Independent of the literautre, it is important to realize that model building is a highly individualistic, open-ended, and creative endeavor. While mathematical skills are essential (the more the better), the modeler also needs to bring a personal touch well grounded in observation and experience (the more the better) to the task at hand. Listed below are some open-ended projects which may serve to sharpen the technical and creative skills of the interested model builder.

Projects

(1) Carefully choose a four-way stop intersection and obtain experimental values for C and S there. Then attempt to measure total delay time (say, per minute) at the intersection with various traffic volumes and compare your results with those predicted by the model. Then pick a traffic light intersection and determine S and R there. Is the value of R and the total delay time/cycle close to that suggested by our model?

(2) It has been suggested somewhat facetiously (though some people drive as if they believe it) that the safest and most efficient form of traffic control at an intersection would be to require all drivers to cross the intersection at speeds in excees of 100 mph. Set up a model for and analyze this novel approach. Then discuss its advantages and drawbacks using results from your model and common sense.

(3) A third and very common alternative to stop signs or lights consists of an uninterrupted traffic flow on the main street and a two-way stop on the side street. Model this situation with appropriate simplifying assumptions as well as you can, incorporate the model into our computer program, and run the program to find out and discuss the new results. Note: This third situation has been considered extensively in the literature. After completing the first model, see [4] for a mathematical approach.

References

[1] A. French, "Capacities of one-way and two-eay streets with signals and with stop signs," *Highway Research Board*, vol. 112, 1956. (This reference, most closely related to what we have done, includes some actual data on the various control possibilities.)

[2] D. L. Gerlough, and F. Barnes, *Poisson and Other Distributions in Traffic*. Eno Foundation, 1971. (Introduces and applies the Poisson and exponential distribution to statistical models of traffic volume and arrival gaps.)

[3] F. A. Haight, *Mathematical Theories of Traffic Flow*. New York: Academic

Press, 1963. (A standard introductory work in the field touching upon all the main aspects of traffic flow.)

[4] J. H. Kell, "Analyzing vehicular delay at intersections through simulations," *Highway Research Board*, vol. 356, 1962. (Combines a statistical model with simulation techniques in considering the case of an unrestricted main street with stop signs at a cross street.)

[5] M. Wohl and B. V. Martin, *Traffic System Analysis for Engineers and Planners*. New York: McGraw-Hill, 1967. (Treats stop sign questions in Chapter 13, traffic light questions of green time, cycle length, etc., in Chapter 14, and simulation methods in Chapter 15.)

Notes for the Instructor

Objectives. To construct and analyze a simplified model of traffic flow at an intersection, to illustrate the general modeling process, and to provide a natural example of computer simulation.

Prerequisites. High school algebra, sum of an arithmetic progression, flow charting, BASIC programming, elementary calculus (at single, avoidable spots).

Time. Two to four hours.

Courses. This chapter is appropriate in the following courses: mathematics appreciation, finite mathematics, modeling or simulation, and computer programming.

Remarks. This model takes the traffic flow question of whether four-way stop signs or lights are appropriate at a given intersection and provides answers in quantitative terms by means of a mathematical model. The model developed and the short flowchart and BASIC program which implement it provide a simple but worthwhile illustration of modeling and simulation requiring from two to four hours of class time.

Interspersed throughout most sections are a number of Exercises which call for validation of stated results, provide numerical examples, and suggest further analysis and extension of the model. These exercises are intended to be straightforward and readily amenable to student solution. The exercises marked by a bullet (•) may be of a more challenging and time-consuming nature. The final section concludes with some ambitious and truly open-ended suggestions for the strongly motivated reader.

One vital aspect of modeling is the need to simplify the problem at hand, and the reader will soon become aware of the oversimplified nature of the model. In several instances, shortcomings of the model are purposefully ignored until a later exercise or paragraph. It is hoped that the student may catch these oversights as they occur and be prepared to discuss them.

Great activity and a vast literature exist in the area of traffic flow modeling and simulation. Much of the work begins with laboriously collected data and follows with both empirical and mathematically deduced formulas to shape the overall model. In our example we rely more on a mental picture of the situation to be modeled, though data collection would be appropriate to supply actual parameter values for the particular intersection. Nevertheless, the inductive process of selecting key variables and their relationships and the deductive process of doing the mathematics are both neatly illustrated by the model.

Happily, the results predicted by the model seem quite reasonable. To be seriously tested and sharpened, the results could be compared with empirical data collected by teams of stopwatch (and tape measure?)—wielding students. Finally, the instructor might wish to point out that the model could be implemented by placing traffic sensors around the intersection and interfacing them with a microcomputer which could then control or convert to blinking red (stop signs) the lights at the intersection depending on the traffic flow. This idea is already being used in selected traffic situations and our model serves as an illustration of how this process might work.

CHAPTER 6

A Model for Municipal Street Sweeping Operations[1]

A. C. Tucker* and L. Bodin**

1. Introduction

An enormous variety of human activities can be the subject of mathematical modeling. To optimize the performance of an auto assembly line, the process of putting two screws into an auto chassis may be broken into, say, 25 carefully defined steps and then an order and timing of these steps is determined to make the process as easy, error-free, and quick as possible. As another example, in a recent criminal trial in New York, the defense attorneys relied heavily on computer-generated profiles to pick a jury most favorably disposed towards the defendants.

When an activity entails substantial costs in time, money, or people, then the precise analysis produced by a mathematical model is often invaluable. Of course such analysis itself requires time, money, and personnel. While most branches of the federal government and all large corporations have used mathematical models to plan their operations for many years, state and municipal operations are just now beginning to do so. Recently, New York City has taken the lead in using mathematical modeling to improve urban services and cut their costs. In this module, we shall examine a model that was developed to route and schedule street sweepers.

A few words are in order to explain the special difficulties inherent in

[1] The work of the first author on this chapter was partially supported by National Science Foundation Grant GP-PO33568-X00.

* Department of Applied Mathematics and Statistics, State University of New York, Stony Brook, NY 11794.

** School of Business, University of Maryland, College Park, MD 20742.

building good mathematical models for municipal services. It can take up to a year and thousands of dollars just to collect the data (which often varies seasonally) needed in the model. Few municipal governments feel they can justify such long-term investments when they are faced with immediate and more pressing problems in the community. Often the task to be modeled is performed by workers belonging to a strong union which resists any apparent attempt by City Hall to tell the men in the streets how to do their job. On the mathematical side, a major difficulty is that mathematical models need a precise formulation of the problem and its constraints. However, many constraints in urban problems are hard to quantify as they involve administration (i.e., bureaucracy), politics, unions, and the people who perform the service, plus many practical difficulties of which, as the unions rightly claim, only the workers are aware. Moreover, the more precisely one defines the problem, the more unwieldy the model becomes. Thus to make a nice mathematical solution possible, one often has to make idealized assumptions. Yet the results that come out of the mathematical analysis must be realistic and workable. For example, when things go wrong or special events occur, a district dispatcher using the model should be able to get the job done as well as he could with the old ways.

Some attempts to optimize services have been in practice for many years. For example, garbage trucks do not just go out into an area and go down streets however they wish. The routes have been planned by capable people who looked carefully at maps of the areas to be covered. One of the first objectives in any computer analysis of this situation would be to see how closely the hand-drawn routes approximate the optimized mathematical solutions.

We have chosen to discuss the problem of efficient street sweeping because its analysis contains all the major features typical of a large-scale routing or scheduling problem, yet the mathematical techniques required are comparatively simple, and for small problems the calculations can be done by hand. As in many routing problems, some parts of the sweeping problem can be handled very precisely, while other parts require simplification, idealization, and even guess work. This problem also has its share of extra administrative and practical constraints.

The model described here was developed for the New York City Department of Sanitation by L. Bodin and S. Kursh in the Urban Science Program at the State University of New York at Stony Brook. The New York City Sanitation Department has a $200,000,000 annual budget of which $10,000,000 goes to street sweeping. The computerized sweeper routing based on this model, if implemented city-wide, is expected to save close to $1,000,000. The model was used in a part of the District of Columbia where it cut costs by over 20%.

Section 2 describes our sweeping problem and Section 3 presents the basic mathematical model. Section 4 develops the analysis of the model. Section 5 gives the algorithms that are required in our analysis. Section 6 contains a summary of the steps in the analysis, and Section 7 gives a detailed example. Section 8 has some comments about computer implementation of the model. Section 9 presents some final comments and extensions. Section 10 contains a set of exercises. While reading the lengthy Section 4, some readers may want to refer to the summary in Section 6 and to the example in Section 7.

2. Statement of the Problem

Our problem is to design an efficient set of routes for sweeping the streets in some city. While we speak of "sweeping a street," actually only the sides of a street along the curb are to be swept. This task is performed by vehicles called "mechanical brooms" or "brooms" for short.

At first the problem seems to be trivial: send a broom up and down the length of one street, north-south or east-west, and have it repeat this on the next parallel street; and partition the city up into sets of north-south and east-west streets such that each set can be covered during the time of a single broom's period on the streets (2–4 hours, depending on union work rules). However, for the broom to sweep along a curb, no parked cars must be present. Thus we quickly see that we are at the mercy of parking regulations. Many smaller cities with well-disciplined citizenry can institute parking regulations designed to coincide with the type of broom routes mentioned above. However, in a city like New York City, simple consistent parking regulations,[2] such as allowing cars to be parked on alternate sides of the street on alternate days, are essential in most business districts. In addition, major arterial routes cannot be swept during rush hours. In residential and manufacturing districts, where the full parking capacity of the streets is needed much of the time, special regulations are needed. One-way streets and areas where streets do not have a common north-south and east-west alignment cause further complications. Finally, a city is usually divided up into administrative districts and broom routes may not cross district lines. The character of our problem has now changed from trivial to extremely complex, and the large and disparate set of constraints seems overwhelming.

To close this section, we note one additional constraint. When among one-way streets, broom routes should try to avoid turns where the curb to be swept will switch from the left to right side of the vehicle or vice versa (this requires the driver to get out and reposition the brooms on the other side of the vehicle). To a lesser extent, they should also avoid U-turns or left turns.

[2] Even if one could make a good case for changing certain regulations, the changeover cost might be prohibitive. It costs about $2000 per mile to install new parking signs, and about $1000 per mile to change them. New York City has 11,000 miles of streets.

3. The Mathematical Model

The sweeping problem's set of diverse constraints all but forces us to seek an abstract model. We need a general structure that can incorporate all this information without getting bogged down by its complexity. This structure is a directed graph. A *directed graph* $G = (N, E)$ consists of a set N of *nodes* and a set E of *edges*, each directed from one node to another node. References [1]–[3] discuss the theory of graphs and their applications. (In the next section we also encounter undirected graphs, whose edges simply link pairs of nodes without direction.) An edge represents *one side* of a street (in one block) (see Figure 6.1). In this problem we will assign a "length" to each edge, namely, the length of time it takes a mechanical broom to traverse the edge. In Figure 6.1(b), the lengths are written beside each edge. A node will usually represent a street corner. An edge e_j from node x to node y can be written as $e_j = (x, y)$. Indeed, it turns out to be most convenient to represent a graph

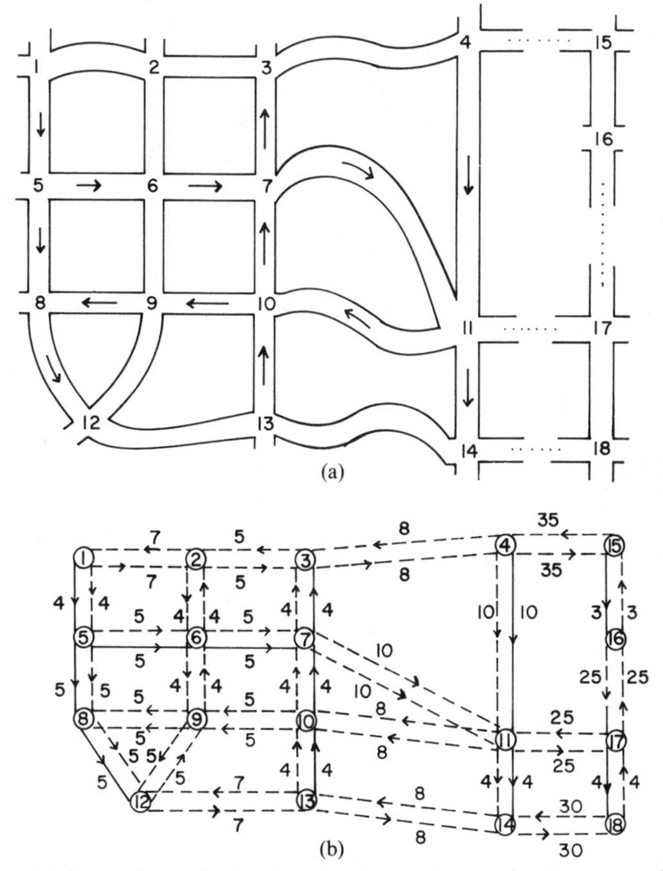

Figure 6.1. (a) Streets in a District; Streets without Arrows Are Two-way. (b) Graph for the Streets in (a). Solid Edges Make Up Subgraph of Streets with No-parking 8 a.m.–9 a.m.

as a set of nodes and a set of directed connections between pairs of nodes. (The connection (x, y) would be listed twice if a one-way street lies between x to y.) Hence our edge problem will be presented in a node-based model. This is natural since the nodes, not the edges, play the active role in our problem; the decision about which edge to sweep next is made at a node.

Next let us consider the time constraints imposed by parking regulations. These constraints turn out to be a blessing in disguise, for they force us to break the original large problem into many short, manageable problems. That is, for each period of the day or week, and for each administrative district, we form the subgraph of edges on which parking regulations (and rush-hour constraints) permit sweeping (Fig. 6.1). Since any given edge should not appear in more than one subgraph, we occasionally must impose artificial constraints (to determine the unique subgraph in which to put an edge that represents, say, a side of a street where parking is always banned) to reduce the period when an edge can be swept.

Our problem can now be given a mathematical formulation. For each of the subgraphs generated in the foregoing, find a minimal set of routes (i.e., minimize the number of routes) that collectively cover all the edges in the subgraph. Lengths (times for traversal) have been assigned to the edges and each route cannot require more time than the length of the no-parking period. In addition, the route should be as free of U-turns and other undesirable turns as possible. The primary difficulty arises from the fact that a street sweeper frequently must raise up its broom and travel some distance along streets that are not to be swept (or were swept already) to get to a new sweeping site. (Typically, a sweeper travels twice as fast when the broom is up making the effective length of streets half as long.) Some students might want to see how few hour-long routes they need to sweep all the solid edges in Figure 6.1b.

The problem of instituting parking regulations in a district with no regulations can be done simultaneously with the minimal routing (see Exercise 16). We shall mention additional questions in Section 9 and in the Exercises, such as linking together routes in successive short periods.

We can now restrict our attention to the problem of finding a minimal set of routes (of a given length) covering the edges of a directed graph. The solution to this problem would then be applied to each of the subgraphs representing streets to be swept in a given district during a given period.

4. An Analysis of the Mathematical Problem

Two different approaches exist to the problem of seeking a minimal, or in practice, near minimal, number of routes to cover the edges of a directed graph. We can either divide the graph up into subgraphs of a size small enough that each subgraph can be covered by a single route (of a given length), or we can find one extended route covering all the edges in the graph and then break it up into appropriate-size routes. These methods are called "cluster first–route second" and "route first–cluster second", respectively. We shall use the term "tour" for an extended route covering all edges. The route first method is much more tractable because an exact mathematical solution exists to the minimal tour problem. However, as we see in the example in Section 7, the subsequent clustering need not lead to a minimal set of routes. On the other hand, when we try to cluster first, cutting the graph into the right subgraphs, we must rely essentially on guesswork and quickly drawn (inefficient) possible routes. This causes us to cluster into conservatively small subgraphs. Another major problem is that after several "nice" subgraphs have been cut out, the remaining edges may be scattered about in many small sections. Some other not so obvious drawbacks are associated with the cluster first method. The net effect is that the route first approach is easier mathematically and gives better solutions for most graphs. Note that the cluster first method does have one merit that is important to administrators: the routes are generally contained in a nice, compact region and different routes do not cross. This makes it easy to determine whose route contains a missed (unswept) street without recourse to a detailed map marking out each route. The cluster first method is discussed in [6].

We now develop the route first–cluster second approach. Our task will be to obtain a minimal tour that covers all the edges of a given graph. The subsequent job of breaking up that tour into feasible routes is treated later. Notice how much we have been able to simplify the complex problem we had in Section 2.

For convenience, let us assume that the tour must be a circuit. (Formally, a *circuit* is a sequence of edges $(e_1, e_2, \cdots, e_n)$ such that the end node of e_i is the start of e_{i+1} and e_n ends at the start of e_1.) This assumption means that there will be many possible ways to break up this tour later into subtours.

While in general any circuit (minimal or not) covering all edges will pass some edges twice, in some graphs a circuit that traverses each edge of the graph exactly once can be obtained. A circuit covering all edges once (and all nodes) is called a *Euler circuit* (or Euler tour). Clearly, any Euler circuit is a shortest possible tour, since by the definition of our problem, we must traverse each edge at least once. The conditions for the existence of a Euler circuit, therefore, will tell us how a shortest-tour construction must procede even when a Euler circuit does not exist.

There is an obvious condition for the existence of a Euler circuit, namely, that as many edges be directed into each node as are directed out of it, since a circuit arrives at a node as many times as it leaves a node. Let us define the *inner* (*outer*) *degree* of a node to be the number of edges directed into (out of) the node. A graph is *connected* if a set of edges exists, *disregarding their directions*, that links together any given pair of nodes. Clearly, graphs with Euler circuits are connected.

Theorem 1. *A directed graph has a Euler circuit if and only if the graph is connected and for each node, the inner degree equals the outer degree.*

PROOF. We have alread observed that graphs with Euler circuits satisfy these two conditions. Let us have a graph satisfying the two conditions. At each node, we arbitrarily pair off and tie together the inward and outward edges. Thus we have tied all the edges together into long strings which, having no ends, must be circuits. We can repeatedly combine any two circuits with a common node (by repairing the edges (on the two circuits) incident to the common node). By the connectedness of G, the result cannot be two or more vertex-disjoint circuits. Thus we obtain a single circuit, an Euler circuit. □

Note that the construction in the above proof would allow us to pair inward and outward edges at each node in a manner designed to minimize unwanted turns (U-turns, etc.) at each node.

Now let us consider the problem of finding a minimal tour covering all edges in an arbitrary connected graph. We handle disconnected graphs by solving our problem for each connected part, called *component*, of the graph. (Later in this section we describe how to tie these solutions together.) The difficulty in arbitrary graphs arises at nodes whose outer degree does not equal the inner degree. At such corners, a vehicle will lift up its broom and drive over to some other corner where the broom will be lowered and sweeping will recommence. The time spent with the broom up is called *deadheading time*. This is what must be minimized (the time needed to sweep is predetermined). Define the *degree* $d(x)$ of a node x to be the outer degree of x minus the inner degree. If a node x has $d(x) < 0$, i.e., an excess of inward degrees, then our tour must eventually deadhead on an edge when it leaves x. A similar difficulty arises if $d(x) > 0$.

We can draw a graph of the route of a tour. See the example in Figure 6.2. Here we represent the added deadheading edges as dashed edges. These deadheading edges are drawn from the larger graph of all streets in the city. The dashed edges are duplicates of existing (no parking) edges when the broom

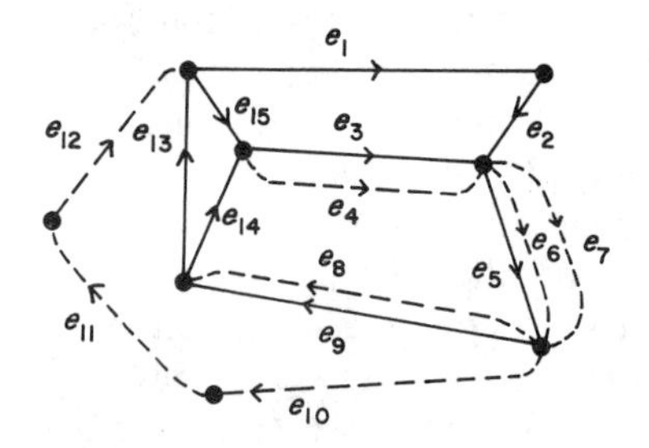

Figure 6.2. Directed Graph of Edges To Be Swept (Solid Edges) with Additional Dead-heading Edges (Dashed Edges) Needed for a Complete Tour. A Possible Tour is $(e_1, e_2, e_5, e_9, e_{14}, e_4, e_6, e_8, e_{13}, e_{15}, e_3, e_7, e_{10}, e_{11}, e_{12})$.

deadheads down a street that must be swept. In other places, the dashed edges represent sides of streets where parking is not banned. Observe that *the added deadheading edges necessarily expand the original connected graph into a graph having a Euler circuit*, that is, by Theorem 1, into a connected graph where each node x has $d(x) = 0$. So the problem of finding a minimal tour becomes the problem of finding a minimal-length set of edges whose addition to the original graph balances the inward and outward degrees of each node.

Theorem 2. *Let G be a directed graph with a length assigned to each edge, and let H be a larger graph containing G. Let A be a minimal-length collection of edges (a single edge may be counted several times in A) drawn from the graph H such that the addition of the edges of A to G makes $d(x) = 0$ for each node x in the new graph. We assume such a set A exists. Then A may be partitioned into paths (a consecutive sequence of edges) from nodes of negative degree to nodes of positive degree. If $\deg(x) = -k$ (or $+k$) in G, then k of the paths start (end) at x.*

PROOF. Let G^* be the graph generated by just the edges in A (with k copies of an edge that is counted k times in A). The only nodes of G^* with nonzero degree will be the nodes of G with nonzero degrees, since the sum of the degree of a node in G and the degree in G^* is zero. Let us arbitrarily pair off, as in the proof of Theorem 1, inward and outward edges at each node of G^* as much as possible. Thus if $d(x) = -3$ in G^*, three inward edges at x would remain unpaired. We again get a set of long strings of edges except that now some strings have beginnings and ends. The beginnings are necessarily nodes of positive degree (negative degree in G) and the ends, nodes of negative degree (positive in G). If any string formed a circuit, we could drop that set of edges from A without changing the degrees of any nodes. So by the minimality of A, all the strings are paths from nodes of negative degree in G to nodes of positive degree in G. □

Theorem 2 tells us how to minimize the deadheading time. We must look at all ways of pairing off negative nodes with positive nodes with deadheading paths (a negative or positive node x would occur in $|d(x)|$ deadheading paths) and then pick the set of pairings that minimizes the total deadheading time, i.e., minimizes the sum of the lengths of the shortest paths between the paired negative and positive nodes. Recall that when deadheading, a mechanical broom normally travels twice as fast and so the total deadheading time is actually half the sum of the lengths. When the minimal set of deadheading pairings is found, we add the dashed deadheading edges to our graph. Now the resulting graph has zero-degree nodes, and we can apply the method in Theorem 1. Note that when we are building a Euler tour in Theorem 1, we could create a deadheading path from the negative node x to a node of zero degree. For example, this happens in the tour in Figure 6.2 when the positions of e_8 and e_9 are exchanged, for once the minimal set of dashed edges has been appended, we can trace out a Euler circuit in many different ways. Any such Euler circuit will be a minimal deadheading route for a mechanical broom.

To solve the minimal pairing problem, we need a matrix giving the lengths of shortest paths between ith negative node x_i and jth positive node y_j, for all i, j. (Of course the routes of the shortest paths are also needed.) If the graph is disconnected, we should solve the pairing problem for the whole graph at once since an optimal pairing often links nodes in different components (this situation occurs in the example in Section 7). Note that Theorem 2 does not require G to be connected. In the next section, we give an algorithm for finding the shortest paths between pairs of nodes in a graph. Remember when looking for the shortest path, we need not restrict ourselves to the graph at hand, the graph of no-parking streets at a certain time. Rather, we should look at the graph for all the streets in that district of the city. We note that for large problems, a geographical estimate can be used to find the shortest paths between the negative and positive nodes. The computer program used in the New York City project knew the coordinates of each node (corner), and by adding the differences of the x coordinates and of the y coordinates of two nodes, it derived what is often called the "Manhattan" distance. Multiplying this number by an appropriate constant yields a good estimate of the travel time between the two nodes. After a minimal pairing is found from these distances, we go back to our shortest-path algorithm to find the specific deadheading edges (and actual length) of a shortest path between each two paired nodes.

Let us now reformulate the problem of finding a minimal set of pairings between the negative and positive nodes. We have a matrix A with a row for each negative node and a column for each positive node. Entry a_{ij} is the "cost" (length of shortest path) in going from the ith negative node x_i to the jth positive node y_j. On the left side of the matrix beside row i, we enter the "supply" at x_i, $b_i = |d(x_i)|$, and on the top of the matrix above, column j, we enter the "demand" at y_j, $c_j = d(y_j)$. Our problem in this setting is to minimize the total cost of "moving" the supplies from the x_i to meet the

	40	80	20
60	3	6	7
50	8	2	4
30	5	4	1

Figure 6.3. Typical Data for a Transportation Problem.

Turn	Weight
Straight ahead	0
Right turn	1
Left turn	4
U turn	8
[†]Switch sides of street	10 additional
Raise or lower broom	5 additional

[†] Only applies when sweeping

Figure 6.4

demands of the y_j. Problems of this sort arise frequently in operations research and are called *transportation problems*. See the example in Figure 6.3, which could have come from the problem of finding a minimal-cost schedule for shipping grain from three different grain elevators, having supplies of 60, 50 and 30 tons, respectively, to three mills with demands of 40,80, and 20, respectively. Note that we always need $\Sigma b_i = \Sigma c_j$. (If supplies exceed demand, we can add an artificial column to balance this equation.) We give an algorithm for the transportation problem in the next section.

The solution of the transportation problem gives a minimal set of pairings; that is, it tells us which paths of deadheading edges to add. Let G' be G plus these additional deadheading edges. We now build a Euler circuit for G'. If G' is not connected, we build a Euler circuit for each component. Our next step is the pairing of inward and outward edges at each node needed in the construction in Theorem 1. We do this pairing in a manner designed to minimize undesirable turns. We assign a weight to each possible type of pairing in Figure 6.4. An extra weight is given to switching from dashed to undashed edges (or vice versa), because one wants to raise or lower the broom as infrequently as possible. (One shortcoming with this model is that in the final Euler circuit no way exists to force a deadheading path that starts from right-side sweeping to terminate at an edge requiring right-side sweeping. For this reason, we do not assign a side-of-street to deadheading edges. No changing-side-of-street penalty can occur with them.) At each node, we seek a pairing of the nodes's inward and outward edges in G' that minimizes the sum of the weights. To put this problem in a standard form, we make a matrix $W^{(k)}$ for the kth node v_k with a row for each inward edge to v_k and a column for each outward edge from v_k. Entry w_{ij} in $W^{(k)}$ is the weight for pairing the ith inward edge with the jth outward edge. The problem of a minimal pairing of the row elements with the column elements is called the *assignment problem*. The assignment problem is actually a "degenerate" transportation problem with row supplies and column demands all equal to one. In practice, a node usually has at most two inward edges and thus at most two pairings exist, so one computes the weights of each pairing and picks the lesser one. (Another possible approach to minimizing undesirable turns is to minimize the largest weight that occurs in the pairing—this is called a *bottleneck problem*—but this approach often gives bad pairings.)

After the assignment problem has been solved at each node, we form circuits as in the proof of Theorem 1. We combine these circuits together as in that proof. The re-pairing needed to combine pairs of circuits may occasionally form bad turns. However, in practice few of the circuits exist and so it is not important to find an optimal way to paste them together (in a typical 1000-edge street-sweeping graph, only 5–8 circuits occur; in small graphs, often little or no choice arises of where to paste).

Finally, we consider how to handle the case of a disconnected graph. Assume we have Euler circuits (forming minimal-length tours) for each *component* of G. We pick an arbitrary component G_1 and start the prescribed tour of G_1 at a chosen node x_0. Then at some x_1 in G_1, we cross over to some node x_2 in component G_2' and now begin the tour of G_2. If we do not leave G_2 for another component, then we complete the tour of G_2 and go from x_2 back to x_1 and resume the tour of G_1. (Note that if the tour on G' deadheads from x_1 to some node x_1', then we would do better to return directly to x_1'; such short-cuts are difficult to systematize and are ignored at this stage, but in the final stage of our procedure obvious shortcuts can be inserted by hand.) We shall require that the tour of G_2, which may be interrupted by side tours to other components, should always be completed before we ever come back to G_1; further, once the G_2 tour is completed, we must return directly to node x_1 in G_1. See the example in Figure 6.5(a). These requirements apply to all components.

We can define a component graph H, an undirected graph, with a node for each component of G' and an *undirected* edge between node x_i and x_j in H if our tour crosses between components G_i and G_j. See Figure 6.5. We have implicitly required that H contain no circuits. A connected graph with no circuits is called a *tree* (that is what it looks like). If the length of an edge (x_i, x_j) in the component graph is the shortest distance of a round-trip (paths in both directions) between some node in G_i and some node in G_j, then an optimal H would minimize the sum of the edges. Such a tree is called a *minimal spanning tree*. Since it is usually true, we assume that a minimal spanning tree of components minimizes the amount of deadheading time needed to link together the tours of the individual components (however, some cases exist where circuits in H may give an improved solution; one case occurs when the same node in G_i is the closest node of G_i to several nearby components; see the counterexample in Figure 6.6). In the next section, we give an algorithm for finding a minimal spanning tree in an undirected graph. To find the distance between the closest nodes in each pair of components, we again use the approximate lengths supplied by the Manhattan distance (based on the difference of the coordinates). With this measure, it is not too hard to sort through (by hand or programmed heuristics) the many possible pairs of nodes to find the closest pair of nodes for a given pair of components.

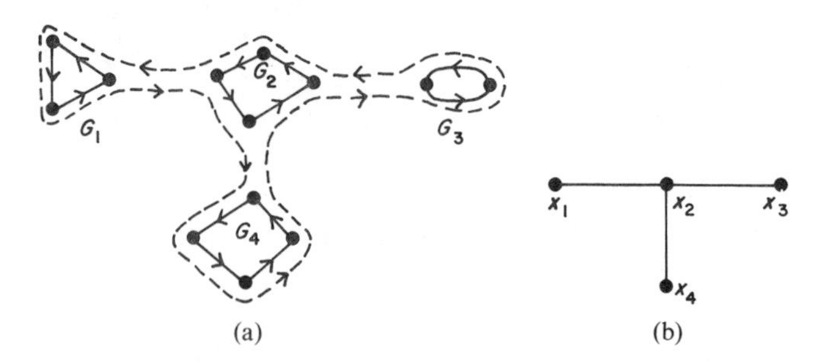

Figure 6.5. (a) Tour Linking Components. (b) Associated Component Graph.

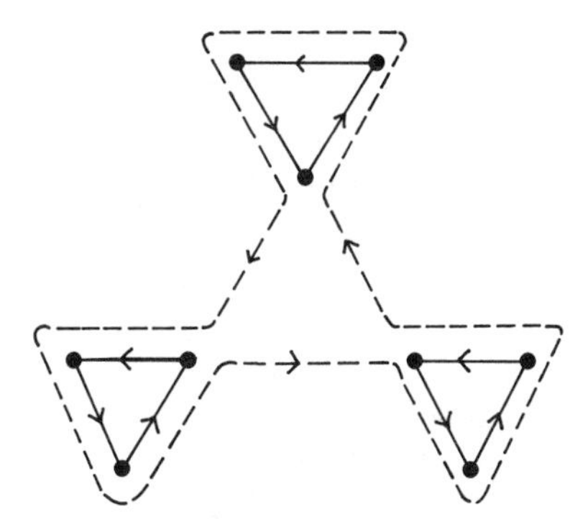

Figure 6.6. A Graph in Which the Minimal Tour Does Not Form a Tree among the Components.

As before, once the minimal spanning tree is found and we know which pairs of nodes between different components are to be linked, then we turn to our "shortest path" algorithm to find actual shortest paths *in each direction* between these pairs of nodes. Using these shortest paths, we join the tours of the individual components of G' into a tour of all of G'—a tour covering all edges of the original graph G.

The only remaining step is breaking up the grand tour into routes of a length that do not exceed the length of the no-parking period associated with the given graph. We start at an arbitrary node x_0 and follow the tour as far as we can within the time period. Suppose the first route ends at node y_0. The next route starts at y_0 and continues as far as possible. The process is repeated until the whole tour has been covered and we have returned to x_0. *Note that if a route started (or ended) with a stretch of deadheading time, that part of the tour can be omitted.* How much deadheading we can omit depends on which node was chosen as x_0. To try other possibilities, we pick the node following x_0 on the first route and let it be the starting point for generating successive routes. Then we try the node after that as the start. We repeat this process until we reach y_0, the end of the original first route. Since one of the routes must begin somewhere in the stretch between x_0 and y_0, we have thus checked all possible ways of generated routes. The starting point that resulted in the fewest routes is the one we use.

Note that while the above single-tour solution was optimal (except for possible shortcuts or circuits in the process of linking the components of G'), the multiple-route solution is not guaranteed to be optimal. This is because not all positive or negative nodes need be paired off in a multiple-route solution (a route could end or start at such nodes). The use of a minimal spanning tree to join components presumes that the tour has to return to the component at which it started. This is not necessary in a multiple-route solution. The example in Section 7 illustrates this difficulty. However, in connected graphs, the solutions of our model are usually quite close to optimal. In the routing system used in New York City, the (final) stage of breaking up the grand tour into routes was done by hand. In addition to finding possible improvements in disconnected graphs, hand routing allows for practical adjustments (and incorporating the possible shortcuts when linking together components of G') and permits breaking up the tour in a way that started or ended many of the routes close to the district garage.

This completes the analysis of the broom routing problem in one district during one no-parking period. Of course, in a real problem this must be repeated for all districts and all periods. In the next section, we present the three special algorithms we need to find shortest paths, to solve a transportation problem, and to find a minimal spanning tree. Section 6 contains a summary of the preceding analysis.

5. Minimizing Algorithms

In the preceding section, we showed how to break the broom routing problem into several parts; we identified places where particular minimization subproblems arose; and we showed how to piece the solutions of those subproblems together to get an efficient solution to the original routing problem. In this section, we present algorithms to solve the different minimization subproblems. Three such minimization problems arose and all are basic problems of operations research. Actually, a fourth problem, the assignment problem, also arose, but we chose to convert it into a transportation problem. A slightly faster algorithm specially for the assignment problem also exists. All these algorithms are discussed at greater length in standard operations research texts such as [4], [5].

5.0 Shortest Path Algorithm

The algorithm we shall discuss is due to Dijkstra and can be used to find shortest paths from any given node a to any other node z, or from the node a to all other nodes. Recall that the analysis developed in the preceding section needs such an algorithm to find shortest paths between various pairs of negative and positive nodes and between closest nodes in two neighboring components.

Shortest Path Algorithm. We need to determine the shortest paths from a given node a to each node. Let $k(e)$ denote the length of edge e. Let the variable m denote a "distance counter." For increasing values of m, the algorithm labels nodes whose (minimal) distance from node a is m.

(1) Set $m = 1$ and label node a with $(-, 0)$, where the "–" represents a blank.
(2) Check each edge $e = (p, q)$ from some labeled node p to some unlabeled node q. Suppose p's labels are $(r, s(p))$. If $s(p) + k(e) = m$, label q with (p, m).
(3) If all nodes are not yet labeled, increment m by 1 and go to step (2). Otherwise, go to step (4). (If we are only interested in a shortest path to z, then we go to step (4) when z is labeled.)
(4) For any node y, a shortest path from a to y has length $s(y)$, the second part of the label of y, and such a path may be found by backtracking from y (using the first part of the labels) as will be described.

A brief discussion of the idea behind this algorithm and an example follow.

Observe that instead of concentrating on the distances to specific nodes, this algorithm solves the questions: how far can we get in 1 unit, how far in 2 units, in 3 units, . . . , in m units, . . . ? Formal verification of this algorithm requires an induction proof (based on the number of labeled vertices). The key idea is that to find a shortest path from a to z we must first find shortest paths from a to the "intervening" nodes. Suppose that $x_1, x_2, \cdots, x_k$ are the set of nodes with edges going to node y, that for each x_i we found a shortest path P_i from a to x_i with length s_i, and that $k_i = k(x_i, y)$ is the length of the edge from x_i to y. Since a shortest path to y must pass through one of the x_i, the length of the shortest path from a to y equals $\min_i(s_i + k_i)$. Moreover, if x_i is a minimizing node, then P_i followed by edge (x_i, y) is a shortest a-y path. To find this shortest path, we shall not need to obtain the distance from a to all nodes adjacent to y, since only those x_i that are closer to a than y are of interest.

Let $P_n = (x_1, x_2, \cdots, x_n)$ be a shortest path from x_1 to x_n. Then $P_n = P_{n-1} + (x_{n-1}, x_n)$, where $P_{n-1} = (x_1, x_2, \cdots, x_{n-1})$ is a shortest path to x_{n-1}. Similarly $P_{n-1} = P_{n-2} + (x_{n-2}, x_{n-1})$ and so on. Then to record a shortest path P_i to x_i, all we need to store (as the first part of a label in the above algorithm) is the name of the next-to-last node on the path; namely, x_{i-1}. To get the node preceding x_{i-1} on P_i, we go to the first stored node for P_{i-1}, that is, the next-to-last node on P_{i-1} which is x_{i-2}. By continuing this backtracking process, we can recover P_i.

The algorithm given above has one significant inefficiency: in step (2), if all sums $s(p) + k(e)$ have values of at least $m' > m$, then the distance counter m should be increased immediately to m'.

Example 1 (Shortest Path Problems). A newly married couple, upon finding that they are incompatible, wants to find a shortest path from point N (Niagara Falls) to point R (Reno) in the road network shown in Figure 6.7. We apply the shortest path algorithm. First N is labeled $(-, 0)$. For $m = 1$,

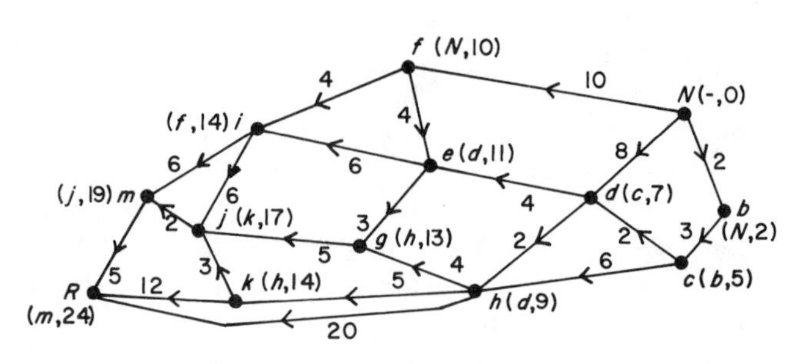

Figure 6.7

no new labeling can be done (we check edges (N, b), (N, d) and (N, f)). For $m = 2$, $s(N) + k(N, b) = 0 + 2 = 2$, and we label b with $(N, 2)$. For $m = 3, 4$, no new labeling can be done. For $m = 5$, $s(b) + k(b, c) = 2 + 3 = 5$, and we label c with $(b, 5)$. We continue to obtain the labeling shown in Figure 6.7. Backtracking from R, we find the shortest path to be $(N, b, c, d, h, k, j, m, R)$ with length 24.

If we want simultaneously to find shortest distances between all pairs of nodes (without directly finding all the associated shortest paths), we can use Floyd's simple algorithm. Let matrix D have entry $d_{ij} = \infty$ (or a very large number) if there is no edge from the ith node to the jth node; or else $d_{ij} =$ the length of the edge from the ith node to the jth node. Then Floyd's algorithm is most easily stated by giving the following Fortran code.

```
DO 1 K = 1, N
DO 1 I = 1, N
DO 1 J = 1, N
IF D(I,K) + D(K,J) < D(I,J) THEN
   D(I,J) = D(I,K) + D(K,J)
1 CONTINUE
```

When finished, d_{ij} will be the shortest distance from the ith node to the jth node.

5.1 Minimal Spanning Tree Algorithm

We present two minimal spanning tree algorithms for connected, undirected n-node graphs. Note that any spanning tree of a connected, undirected n-node graph has $n - 1$ edges (Exercise 30a).

Kruskal's Algorithm. Repeat the following step until the set S has $n - 1$ edges (initially S is empty): add to S the shortest edge that does not form a circuit with edges already in S.

Prim's Algorithm. Repeat the following step until the tree T has $n - 1$ edges: add to T the shortest edge between a node in T and a node not in T (initially pick any edge of shortest length).

In both algorithms, when a tie occurs for the shortest edge to be added, any of the tied edges may be chosen. Showing that Kruskal's algorithm does indeed form a spanning tree is left as an exercise. Note that Prim's algorithm is intuitively quicker because there are fewer edges to check in each iteration and no worry about forming circuits. Note also that Prim's algorithm is very similar to the shortest path algorithm: if we consider nodes in T as labeled nodes, then both algorithms repeatedly search all edges from a labeled node to an unlabeled node (although the search is for different purposes). Indeed, the edges used to label new nodes in step (2) of the path algorithm form a directed spanning tree (Exercises 29).

The difficult part in the minimal spanning tree problem obviously is proving the minimality of the two algorithms. We give the proof for Prim's algorithm and leave Kruskal's as an exercise.

Theorem 3. *Prim's algorithm yields a minimal spanning tree.*

PROOF. Suppose $T = \{e_1, e_2, \cdots, e_{n-1}\}$ is the spanning tree constructed by Prim's algorithm, with the edges indexed in order of their inclusion into T, and T' is a minimal spanning tree chosen to have as many edges in common with T as possible. We shall prove that $T = T'$. Assume $T \neq T'$, and let $e_k = (a, b)$, chosen on the kth round of the algorithm, be the first edge of T (having smallest index) that is not in T'. Let $P = (e'_1, e'_2, \cdots, e'_n)$ be the (unique) path in T' from a to b (in T, the path from a to b is simply e_k). If every edge on P is shorter than e_k, then on the kth and later rounds, Prim's algorithm would have incorporated successive edges of P before considering the longer e_k. (A technical note: the algorithm could choose e'_1 on the kth round without fear of forming a circuit since edges chosen up to the kth round plus the edge e'_1 are all in the (circuit-free) tree T'.) If P has an edge with the same length as e_k, we remove this edge from T' and replace it by e_k. It is not hard to show that the new T' is still a spanning tree which has the same minimal length and which has one more edge in common with T, this contradicts the choice of the original T'. If P has an edge with greater length than e_k, we remove it from T' and replace it by e_k to get a shorted spanning tree; this contradicts the minimality of T'. □

EXAMPLE 2 (Minimal Spanning Tree). We seek a minimal spanning tree for the network in Figure 6.8. Both algorithms start with a shortest edge. There are three edges of length 1: (a, f), (l, q), and (r, w). Suppose we pick (a, f). If we follow Prim's algorithm, the next edge we would add is (a, b) of length 2, then (f, g) of length 4, then (g, l), then (l, q), then (l, m), etc. The next-to-last addition would be either (m, n) or (o, t), both of length 5 (suppose

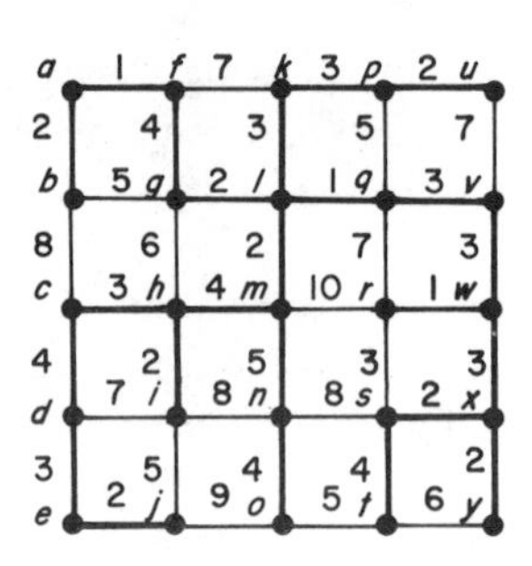

Figure 6.8

we choose (m, n)), and either one would be followed by (n, o). The choice of (m, n) or (o, t) brings out the fact that *minimal spanning trees are not unique*. The final tree is indicated with darkened lines. On the other hand, if we follow Kruskal's algorithm, we first include all three edges of length 1. Next we would add all the edges of length 2: (a, b), (e, j), (g, l), (h, i), (l, m), (p, u), (s, x), (x, y). Next we would add almost all the edges of length 3: (c, h), (d, e), (k, l), (k, p), (q, v), (r, s), (v, w), but not (w, x) unless (r, s) were omitted (if both were present we would get a circuit containing these two edges together with edges of shorter length (r, w) and (s, x)). Next we would add all the edges of length 4 and finally either (m, n) or (o, t) to obtain the same minimal spanning tree(s) produced by Prim's algorithm. This similarity is no coincidence.

5.2 The Transportation Problem

We are given an $n \times m$ matrix A of transportation costs, i.e., a_{ij} = cost of shipping one unit of our commodity from origin 0_i to depot D_j, along with supply and demand vectors B and C, i.e., b_i = supply at 0_i, c_j is demand at D_j. We assume $\Sigma b_i = \Sigma c_j$. The goal is to minimize the bill (total cost) for shipping the commodities to the depots. Remember that although the a_{ij} were lengths of time in the sweeping problem (and b_i and c_j were nonzero degrees of nodes), we can treat the minimal pairing of nonzero degree nodes as a transportation problem.

Let x_{ij} be the number of units sent from 0_i to D_j. Then our problem is to minimize

$$\text{total costs} = \sum_{i,j} a_{ij} x_{ij}$$

subject to the constraints

$$\begin{aligned} &\text{(supplies)} \quad \sum_{j=1}^{m} x_{ij} = b_1, \qquad i = 1, 2, \cdots, n, \\ &\text{(demands)} \quad \sum_{i=1}^{n} x_{ij} = c_j, \qquad j = 1, 2, \cdots, m. \end{aligned} \qquad (*)$$

In addition, we want x_{ij} to be nonnegative integers. It suffices to require that x_{ij} be nonnegative: for all optimal solutions then they turn out to be integral (when B and C are integral). In this algebraic setting, the transportation problem has the form of what is called a *linear program*. While there are good algorithms for solving linear programs, we shall present an elegant algorithm of Hitchcock's that is much faster than general linear programming methods. See [4] for a more detailed discussion of this algorithm.

Our algorithm has 4 steps. First we seek an initial solution of positive x_{ij} that satisfy the constraints (*). We want to use as few positive x_{ij} as possible. We shall see that only $(n + m - 1)$ positive variables are needed. Our method for obtaining such a solution is best presented through an example. For the transportation problem given in Figure 6.9 (the costs are written in the top right box in each entry; the other numbers will be explained later), we choose x_{11} as large as possible. The first origin constraint implies $x_{11} \leqq b_1$ and the first depot constraint implies $x_{11} \leqq c_1$. So we set $x_{11} = \min(b_1, c_1)$. In this case $x_{11} = \min(40, 30) = 30$. To try to use up the rest of the supplies of 0_1, we set $x_{12} = \min(c_2, b_1 - x_{11}) = \min(40, 10) = 10$. Now we try to fill the rest of the demand at D_2 with x_{22}. We continue in this fashion, setting $x_{22} = 30$, $x_{23} = 20$, and $x_{33} = 50$. Each new x_{ij} exhausts some supply or fills out some demand. Apparently, $n + m$ positive x's will be needed. However, because $\Sigma b_i = \Sigma c_j$, when we use the last x_{ij} we must simultaneously satisfy both the last origin and last depot constraint. Thus $n + m - 1$ positive x's are needed. This technique for finding a solution to the constraints (*) is called the northwest-corner rule, because we zig-zag our way across the matrix from the northwest corner. See [4] for a further explanation of the northwest-corner rule and for modifications in the "degenerate" case where fewer than $n + m - 1$ positive x's are needed.

The next step involves writing down a solution to a "shadow" problem. In that problem, an outside shipper wants to determine two sets of prices; first, v_j, the price at which he will sell a unit of the commodity at D_j; and

		Depot			
u_i		30	40	70	
O		7	3	5	
r	40	30	10	(0)	$u_1 = 0$
i		6	1	3	
g	50	(1)	30	20	$u_2 = 2$
i		8	2	6	
n	50	(0)	(-2)	50	$u_3 = -1$
		$v_1 = 7$	$v_2 = 3$	$v_3 = 5$	

Figure 6.9

second, u_j, the price at which he will buy from us a unit at 0_i. Then $v_j - u_i$ is the cost to us of shipping a unit from 0_i to D_j. To do business with us, his transportation costs should not exceed the cost of our solution above. Since the shipper wants to make as much money as possible, he picks his costs to be exactly equal to ours for each x_{ij} used in the above solution. Thus his constraints are (i) $v_1 - u_1 = 7$; (ii) $v_2 - u_1 = 3$; (iii) $v_2 - u_2 = 1$; (iv) $v_3 - u_2 = 3$; and (v) $v_3 - u_3 = 6$. Since the prices are relative; i.e., only the differences are important, we can arbitrarily set $u_1 = 0$ (we always do this). Then (i) and (ii) determine $v_1 = 7$ and $v_2 = 3$. Then (iii) determines $u_2 = v_2 - 1 = 2$, and subsequently $v_3 = 3 + u_2 = 5$ and $u_3 = v_3 - 6 = -1$. Note that because we had only five equations (in general, $n + m - 1$ equations) in six unknowns (in general, $n + m$ unknowns), we had to determine one unknown arbitrarily. (In a "degenerate" case with less than $n + m - 1$ equations, more unknowns are set equal to 0; see [4].) With these prices, it now costs us as much to sell all our supplies to the shipper at the origins and buy the required amounts from him at the depots as it costs to use the solution obtained above.

The third step involves the choice of a new x_{ij} to be used to lower the cost of the existing solution. To beat the cost of our first solution satisfying the constraints (*), it suffices to get a solution that beats the cost of the outside shipper (since his total cost was our total cost). A natural first step is to compare our costs a_{ij} with the shipper's cost $v_j - u_i$ at each entry where x_{ij} is currently zero. In each such entry in Figure 6.9, we write $a_{ij} - (v_j - v_i)$ in parenthesis. We see that this difference is negative for entry (3, 2). That means we save \$2 for each unit we ship internally from O_1 to D_2 instead of with the shipper (\$2 versus $3 - (-1) =$ \$4). Then we alter the current solution so as to permit x_{32} to become positive. In general, we seek to increase the x_{ij} where $a_{ij} - (v_j - u_i)$ is most negative. If none of the differences is negative, then there is no way to beat the shipper's cost (thus our own cost), and the current solution is minimal.

The fourth step determines the best improvement possible when we make the new x_{ij} positive. As we increase the new x_{ij}, in this case, x_{32}, we must balance the constraints (*) by changing other x_{ij}. We shall only be permitted to change the values of *positive* x_{ij} *used in the current solution*. The reason is that from step three we know that using other (currently zero) variables would either raise the cost of the new solution or not lower it as much. As we increase x_{32}, we must compensate by decreasing x_{22} and x_{33}. To compensate for the latter changes, we increase x_{23}. So $\Delta x_{22} = \Delta x_{33} = -\Delta x_{32}$ and $\Delta x_{23} = \Delta x_{32}$. Observe that a unit increase in x_{32} decreases the cost of the solution by $a_{22} + a_{33} - a_{32} - a_{23} = 1 + 6 - 2 - 3 = 2$—this checks with the "predicted" savings in step three. The best improvement using x_{32} will obviously come from increasing x_{32} as much as possible. That is, we increase x_{32} until x_{33} or x_{22} becomes 0. So we can set $x_{32} = 30$ with $x_{33} = 20$, $x_{22} = 0$, and $x_{23} = 50$. The other x_{ij} are unchanged. Note that we apparently could have decreased x_{12} instead of x_{22} to balance the increase of

(a)

	30	40	70	
40	7 / 30	3 / 10	5 / (−2)	$u_1 = 0$
50	6 / (3)	1 / (2)	3 / 50	$u_2 = 4$
50	8 / (2)	2 / 30	6 / 20	$u_3 = 1$
	$v_1 = 7$	$v_2 = 3$	$v_3 = 7$	

(b)

	30	40	70	
40	7 / 30	3 / (2)	5 / 10	$u_1 = 0$
50	6 / (1)	1 / (2)	3 / 50	$u_2 = 2$
50	8 / (0)	2 / 40	6 / 10	$u_3 = -1$
	$v_1 = 7$	$v_2 = 1$	$v_3 = 5$	

Figure 6.10

x_{32}. Then we would need to increase x_{11} to balance the decrease in x_{12}, but now there is no way to balance this increase in x_{11}, since $x_{12} = x_{13} = 0$. As it turns out, one unique way always exists to compensate for an increase in a previously unused variable. If the way is not obvious, we can find it by a systematic search. For further details, see [4].

Now we have the solution shown in Figure 6.10(a). We repeat steps two, three, and four, and get the solution in Figure 6.10(b). Now in step three, no entry appears that is less than the shipper's cost, and so we have an optimal solution.

6. Summary of Procedure

The analysis developed in Section 4 can be divided into four general steps. First, we add a minimal-length set of edges so that the resulting graph G' has a Euler circuit in each of its components. Second, we pair inward and outward edges at each node so as to minimize unwanted turns and use these pairings to construct a Euler circuit for each component of G'. Third, if G' is not connected, then we link these circuits together in a grand tour. Fourth, we allow for various practical considerations as we manually break up the grand tour into routes of limited length. We assume the given graph G is contained in the larger graph H.

1. Append edges to G to get a graph G' with Euler circuits in each of its components.

(a) Obtain the matrix of approximate shortest distances in H between negative nodes x_i and positive nodes y_j of G. See pages 83–84. (If relatively few nodes are involved, the minimal path algorithm is used to find shortest distances; see pages 88–90.

(b) Solve the transportation problem with the matrix of step 1(a) and with row supplies $-d(x_i)$ and column demands $d(y_j)$. See pages 84–85 and 92–95.

(c) Using the minimal path algorithm, find the shortest paths in the larger graph H between the negative and positive nodes paired in step 1(b). See pages 88–90.

(d) Append dashed edges, duplicate and new, to G so that there are k dashed copies of an edge if that edge occurred on k of the paths in step 1(c). Call the new graph G'. See pages 83–85.

2. Build Euler circuits in each component of G'.

(a) Match up inward and outward edges at each node in G' by "inspection" or using the assignment problem approach. Use the weights in Figure 6.4. See pages 84–85.

(b) Form the circuits arising from the match-ups in step 2(a), and paste these circuits together to get a Euler circuit in each component of G'. See pages 82 Proof of Theorem 1 and 85–86.

3. Link together the components of G'—only if G' is not connected.

(a) Find the (approximate) shortest round-trip distances between the components of G'. See pages 85–87.

(b) Using the distances in step 3(a), find a minimal spanning tree for the component graph. See pages 89–92.

(c) Find the shortest paths (in both directions) joining the closest pair of nodes in each pair of components linked in the minimal spanning tree. See pages 85–87.

(d) Use these shortest paths to unite the tours of each component of G' to get the desired grand tour which covers all edges of G with minimal length.

4. Breaking up the grand tour.

(a) Break up the grand tour into subtours of feasible length. At this point, the grand tour may be modified to allow for various constraints and other considerations. See pages 86–88.

7. An Example

Our procedure will now be illustrated with an example based on Figure 6.1. The solid-lined edges in Figure 6.1(b) represent sides of streets on which parking is banned from 8:00 a.m. to 9:00 a.m. (the time period for our example). The graph G of the solid edges alone is given in Figure 6.11. The side of a street represented by an edge is indicated by the position of the edge relative to its end nodes in Figure 6.1b. The dashed edges in Figure 6.1(b) represent the other streets in the district which may be used in dead-

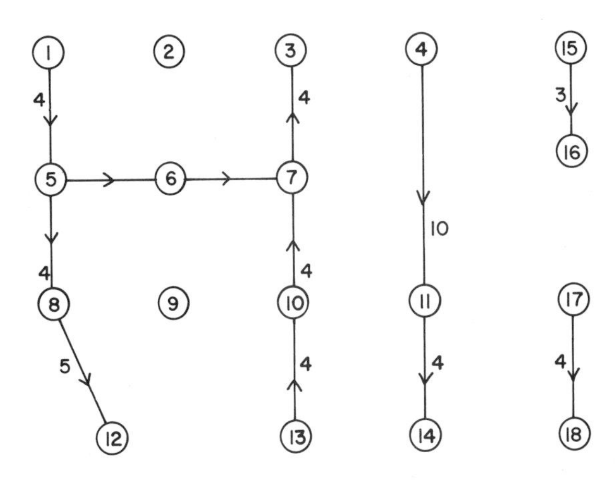

Figure 6.11

heading. The time, in minutes, to sweep each edge is indicated in Figure 6.1(b) and 6.11. We assume that the time needed to deadhead an edge is one half of the time needed to sweep it.

Our analysis will proceed as outlined in Section 6. Most of the computations are left as exercises, and the results are given without explanation or are obtained by inspection and heuristics.

Step 1. We pick out the nodes of negative and positive degree which will be the origins and depots, respectively, of the transportation problem. The origins are nodes 3, 7, 12, 14, 16, 18. The depots are nodes 1, 4, 5, 13, 15, 17. The supply or demand at each of these nodes is one. (The unit supplies and demands will make the transportation problem "degenerate.") We must now compute the distance between each origin and depot. In the computerized analysis described in Section 4, the Manhattan distance was calculated from the nodes' coordinates. Because the graph in Figure 6.1(b) is relatively small, the coordinates of nodes have been omitted. The exact distances are readily obtained (if desired, coordinates can easily be invented). Most can be gotten by inspection and other distances can be ignored, as mentioned below. Figure 6.12 presents the matrix of relevant distances (students are asked to compute these distances in Exercise 39). For simplicity, the distances are given in sweeping time, not deadheading time; times on deadheading edges will be divided by two later. Only a moment's thought is necessary to see that nodes 16 and 15 and nodes 18 and 17 should be paired in an optimal solution to our transportation problem. Thus we can restrict the problem to the remaining four origins and four depots. The matrix in Figure 6.12 reflects this restriction.

One optimal solution to this transportation problem pairs node 3 with node 4, 7 with 1, 12 with 5, 14 with 13, as well as 16 with 15 and 18 with 17. (Other optimal solutions exist.) The sum of the lengths of the paths in this

	Depots			
Origins	1	4	5	13
3	12	8	16	25
7	16	12	20	22
12	20	26	24	7
14	32	28	36	8

Figure 6.12

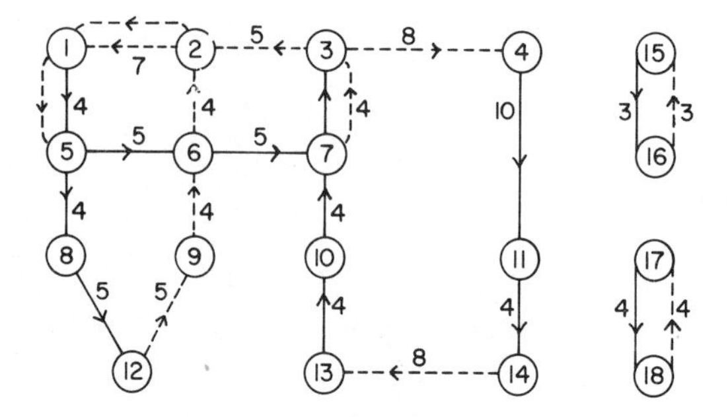

Figure 6.13

pairing is 63 minutes, or, at deadheading speed, 31 1/2 minutes. The deadheading edges, that is, the edges of the shortest paths between the paired nodes, are now added to the graph in Figure 6.11 to produce the graph G' in Figure 6.13. We easily check that every node in G' does have zero degree as desired. This completes step 1.

Let us note in passing some interesting properties of G': (i) G' has three components while G had four components; (ii) edge (2, 1) is deadheaded twice; (iii) edges (1, 5) and (7, 3) are swept and deadheaded in the same direction.

Step 2. Each node in G' is now examined, and the inward–outward edge assignment problem is solved. Only nodes 1, 2, 3, 5, 6, and 7 have two inward edges and at each of these an optimal assignment is obvious. The table in Figure 6.14(a) lists a set of optimal edge pairings at the above six nodes. Since only the nodes have names, the pairings are written as a triplet node sequence. We shall underline the parts of a sequence in Figure 6.14 that contain deadheading edges (these appear as italic numerals in the sequences within the text). From these pairings, we get a set of circuits described in the proof of Theorem 1. They are listed in Figure 6.14(b). Note that, as is typical in most practical problems, there is little changing sides of a street. Moreover, there is no choice about how often or where to change sides—it must happen at nodes 4 and 14. Finally, we join the first and second circuits in Figure 6.14(b); whether we join them at node 7 or 3 does not matter, for like circuits result. This completes step 2.

Step 3. We need to find the shortest distances between the three components of G'. Let us call the large component C_1, the nodes 15–16 component C_2, and the nodes 17–18 component C_3. The closest nodes linking C_1 and C_2 are nodes 4 and 15 with distance 35. The closest nodes linking C_2 and C_3 are 16 and 17 with distance 25. The closest nodes linking C_1 and C_3 are 11 and 17 with distance 25. The minimal spanning tree among components is indicated with darkened edges in Figure 6.15. Using the corresponding edges, we connect the components to get the grand tour t: 1–5–8–*12–9–6–2–1–5*–6–7–*3–4–11–17–18–17–16–15–16–17–11–14–13*–10–7–*3–2–1*. The tour T is shown in Figure 6.16. It will take 137 1/2 minutes to complete this tour, 56 minutes of sweeping and 1/2 (163) = 81 1/2 minutes of deadheading (remember that deadheading edges are traversed at twice the sweeping speed). This completes step 3.

Step 4. At first, it seems obvious that we cannot break the tour T into two 60-minute subtours, since T runs 137 1/2 minutes. We have many ways to break T into three feasible subtours. However, it might be possible to break T so that long deadheading stretches around node 17 fall at the start or end of the subtours and thus can be dropped from the subtours. Clearly, this is the only way we might be able to use only two subtours. Unfortunately, it is not quite possible to do this (the reader should verify this for himself). One

Node	Pairings
1	2–1–5, 2–1–5
2	6–2–1, 3–2–1
3	7–3–2, 7–3–4
5	1–5–8, 1–5–6
6	5–6–7, 9–6–2
7	6–7–3, 10–7–3

(a)

Circuits

1–5–8–12–9–6–2–1–5–6–7–3–2–1
7–3–4–11–14–13–10–7
15–16–15
17–18–17

Combination of first two circuits

1–5–8–12–9–6–2–1–5–6–7–3–4–11–14–13–10–7–3–2–1

(b)

Figure 6.14

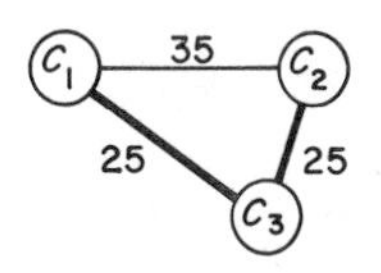

Figure 6.15

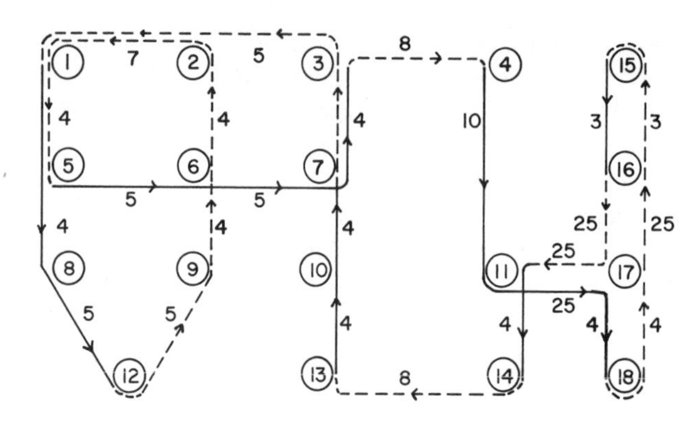

Figure 6.16. Minimal Tour Covering All Edges of Graph in Figure 6.11.

way around this difficulty is to consider other ways to link together the components of G'. Two other ways of linking components exist which, while not giving a minimal-length grand tour, do allow us to break the tour at long deadheading stretches so as to obtain two feasible subtours. If we combine components C_1 and C_3 of G' between nodes 14 and 18 instead of between 11 and 17, then the resulting grand tour would be T^*: 1–5–8–*12–9–6–2–1–5*–6–7–*3–4*–11–*14–18–17–16–15–16–17–18–14–13*–10–*7–3–2–1*. This can be broken into subtours T_1 : 1–5–8–*12–9–6–2–1–5*–6–7–*3–4*–11–14, and T_2: 15–*16–17–18–14–13*–10–7 of lengths 57 and 46 1/2 minutes, respectively. Note that by deadheading edge (1, 5) the first time we traverse it (instead of the second time) in T_1, we get a further deadheading edge at the start of T_1. Thus we get T_1': 5–8–*12–9–6–1*–5–6–7–*3–4*–11–14 of length 56 minutes. Together T_1' and T_2 take 102 1/2 minutes with only 45 1/2 minutes of deadheading (T had 81 1/2 minutes of deadheading). If we link component C_1 of G' with component C_2 instead of C_3, we again get a grand tour containing two feasible subtours (the reader should find these feasible subtorurs). This concludes step 4.

8. Computer Implementation

While graphs are easy to represent (on paper) to the human eye, they are more difficult to represent for use inside a computer. As noted at the start of Section 2, we should represent edges as ordered pairs of nodes. Then the question is how to represent nodes in a computer. The usual answer is as numbers. If a graph has n nodes, we use some scheme to assign each node a different number between 1 and n. Then we can read into the computer the set of edges—ordered pairs of nodes—for streets in a given district. We would also need secondary information about each edge: its length, side of street, and no-parking period. If we are going to estimate distances between nodes with the Manhattan distance, we need to read in the coordinates of each node. The coordinate information can also be used to check for errors in the node list as follows. A relatively small bound usually exists for the distance between any two nodes joined by an edge. We program the computer to check that the distance between the ends of each edge is within this bound.

Inside the computer, we should reorganize the edges in terms of their end nodes. Thus for each node we want a list of incoming and outgoing edges. This information is just what we need in the shortest path and assignment problem calculations. The secondary information about each edge can be stored elsewhere. When a specific time period for sweeping is chosen, we can flag all the nodes and edges in the graph for that period. We still need to retain the other edges for possible use in deadheading.

Now students are ready to program the four steps in our sweeping procedure. A program for the whole procedure is a major undertaking and so some parts might be fudged. A programmer must resolve certain loose ends. The main ones are as follows.

(1) How does one incorporate deadheading edges in a graph on a computer?
(2) How does one store and represent circuits? How do you link them together?
(3) Heuristics to find nearest pairs of nodes for pairs of components of G'.
(4) Sometime you need to determine how many components the graph G' (with deadheading edges added) has. When? How? Hint: The "when" and "how" are naturally related.

Note that for step 4, the program should print out the grand tour for manual breaking up and also print out a minimal set(s) of feasible subtours found by the method at the end of section 4.

9. Summary and Extensions

In this module, we have presented a detailed mathematical procedure for routing mechanical brooms during a given period of time in a given district of a large city. A present, only a few cities in the country have the personnel and funds needed to utilize and maintain a computer program implementing this procedure. Only two cities, New York City and Washington, D.C., are now working with the procedure. Further, only in a fairly large city can the procedure improve over hand-drawn routes on the number of brooms needed in a given period in a given district. However, the model's analysis would be useful to a large city in several ways.

Even if the mathematically obtained set of subtours for a graph is not smaller than the present set, the individual tours would probably be shorter (with less deadheading time). Thus vehicles would be used less. The model's set of subtours serve as a good standard against which the manually generated routing can be rated. Most importantly, with such a model, one can quickly check out the results of various proposed changes in a city's parking regulations. (In New York City, almost every week sees some minor change in parking regulations in some of the districts.) The model can also be used to examine how changing the boundaries of the districts could result in a city-wide reduction in the number of brooms needed at peak demand. (We must note that the methodology for solving this city-wide problem is in a primitive stage; further, administrators at present feel that the complications involved in changing district lines would not be worth the possible savings.)

Recall that many constraints exist that are difficult to quantify in urban routing problems. In the current problem we have incorporated all major constraints. Yet the precision of our model must not make us forget the uncertainties of day-to-day operations. For example, the time required to go down an edge (street) can only be an average time, varying with traffic congestion. The final step in which we manually check out the breaking of the grand tour into subtours permits us some practical adjustments to compensate for any overprecision of the mathematical analysis. We also should remember, as we saw in Section 7, that the set of subtours our procedure gives may not be minimal (but in practice so far, the computerized sets of tours have always been as good as hand drawn sets and usually have at least 20% less deadheading).

While we have included all major constraints of the original problem, the analysis in this model can be extended in many ways. One could seek a more systematic method of pasting together the circuits into a Euler circuit in each component of G', a method that minimizes unwanted turns. One could examine various ways of linking the components of G' in search of a grand tour in which feasible subtours can be made to start and end at the long deadheading stretches between components, as we did in the example in Section 7. One could consider the problem of linking up the routes of one period with the routes of the next period (perhaps, even design routes in one period to shorten this linking stage). Or one could incorporate into the model the time required to move from or to the district garage when dealing with sweeping periods at the start or end of a work period. If we permit a vehicle to travel along a curb in the opposite direction of the traffic flow when the vehicle is sweepinging (few major cities permit this), then the edges of our graph become undirected (but deadheading would implicitly be directed). Most of our procedure would require substantial changes in this case.

In our model, we have pushed the mathematical analysis almost past the "state of the art" in urban science. Still, the resulting procedure has been used to achieve cost savings. We chose this model because it does show the state of the art and because it uses a wide array of basic concepts and algorithms of operations research. Another model [8] developed at Stony Brook for New York City Department of Sanitation to optimize manpower scheduling contained less interesting mathematics but resulted in an annual savings of about $10,000,000 per year in New York City (the model even played a principal role in union contract negotiations).

Exercises

These exercises are divided into three sections. Exercises 1–10 present a variety of graph modeling. Exercises 11–38 deal with graph-theoretic concepts, variations, and extensions of our analysis, and the theory behind the algorithms. Exercises 38–53 are numerical exercises.

1. In a football season, each pair of teams in a football league plays each other once. We assume that the result of each game is a win for one of the teams (no ties).
   (a) Describe how a directed graph can be used to represent the outcomes (who won) of each game in a season.
   (b) Suppose A, B, C, D are the teams in the league and that A beats B, D; B beats C, D; C beats A; and D beats C. Draw the associated graph.
   (c) A ranking of teams is a list of the teams in which the ith team in the list beat the $(i + 1)$st team. Give a graph-theoretic interpretation in the associated graph of a ranking.
   (d) Using the graph-theoretic interpretation of a ranking, find a ranking for the league in (b) by looking at the graph.

2. Consider a group of individuals who pass information (rumors) among themselves.
   (a) Describe how an undirected graph can be used to describe the "lines of communication" in this group, i.e., the various pairs of people who talk together.
   (b) Draw the graph for a group of five people in which each person talks with exactly two other people in the group.

3. Suppose that we need to match a set X of objects, say boys, with elements in another set Y of objects, say girls, and that each object x in X can only be paired with elements in the subset $N(x)$ in Y.
   (a) Describe how an undirected graph can be used to show which are possible pairs in a matching.
   (b) Suppose Bill likes Ann, Diana, and Lolita; Fred likes Ann, Carol, and Lolita; John likes Carol and Lolita; and Harry likes Diana and Lolita. Draw the associated graph and find a matching that assigns each boy to a different girl.
   (c) A graph such as in (a) is called bipartite—that is, the nodes divide into two parts, and all edges go between, rather than within, the two parts. Show that an undirected graph is bipartite if and only if all its circuits are of even length.

4. A map of several countries is often colored so that countries with a common border (not just a common corner but a border of positive length) are drawn with different colors.
   (a) Describe how an undirected graph can be used to indicate which pairs of countries on the map have a common border.
   (b) Treat the network in Figure 6.7 as a map—the edges are the borders. Draw the associated graph of bordering countries.
   (c) Restate the condition for coloring countries of a map in terms of coloring nodes in the associated graph.
   (d) What is the minimum number of colors needed to color the nodes in the graph of bordering countries in (b)?
   (e) Draw a graph that could *not* represent a set of bordering countries on map. Explain.
   (f) Draw two graphs such that each cannot be colored with just three colors and such that each could be three-colored after *any* one node was removed.

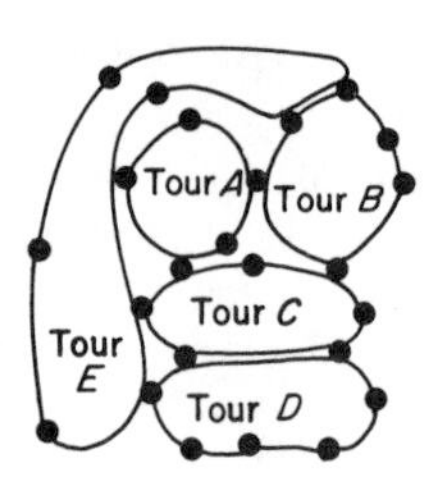

Figure 6.17

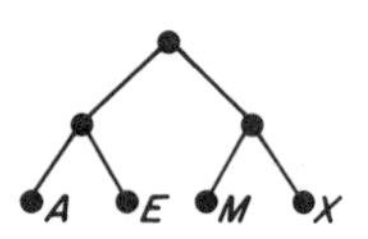

Figure 6.18

5. A state legislature has many committees. Certain senior legislators are on several committees. Thus the memberships of the different committees overlap.
   (a) Describe how a graph can be used to represent which committees have overlapping membership.
   (b) Committee A has legislators 1, 3, 5, 6; B has 2, 4, 8, 10; C has 1, 7, 9; D has 2, 5, 8; E has 2, 4, 10; and F has 11, 12, 13. Draw the associated graph described in (a).

6. A collection of garbage truck routes is drawn for a 2-day period so that each pickup site is visited by one or two routes, depending on whether the site needs daily or every-other-day service. A sample set of routes is shown in Figure 6.17. We want to partition the routes between the first and second day of the period so that no daily site is visited both times on the same day.
   (a) Describe how a graph can be used to indicate whether or not any pair of routes can be assigned to the same day.
   (b) Draw the graph for the set of routes in Figure 6.17.
   (c) What condition must this associated graph satisfy to guarantee that the routes can be partitioned as required? Does the graph in (b) satisfy this condition?
   (d) Restate the condition in (c) in terms of a coloring of the nodes in the associated graph.

7. In a computer, a search to identify an unknown word, or for simplicity, an unknown letter, is performed as follows. The computer can test whether the letter is larger (in alphabetical order) than a given letter, and one uses a scheme of such tests to identify the unknown letter. For example, if the unknown letter is one of A, E, M, X, then an efficient scheme would first test if the letter is greater than E; then one would ask if it is greater than A if the first answer were no, and if it is greater than M if the first answer was yes. This scheme can be represented with a graph called a binary search tree with the branching process which narrows down the identification of the unknown letter. See example in Figure 6.18.
   (a) Draw the graph associated with the testing scheme that asks is the unknown letter greater than A, then greater than E, then M, then X.
   (b) If each possible value of the unknown letter has a given probability of occurring, we can calculate the average number of tests needed to identify an unknown letter. If each of A, E, M, X has probability 1/4, what is the average number of tests for a letter in the scheme (graph) in (a)?
   (c) What graph minimizes the average number of tests when each letter has probability 1/4? Prove your answer carefully.
   (d) Suppose the probabilities are A: 3/10, B: 5/10, M: 1/10, X: 1/10. Now which graph minimizes the average number of tests?

8. Suppose we have an $n \times m$ rectangular chess board (instead of the standard 8×8 board) and we wonder whether a sequence of permissible steps exists by which a knight can go from one given square to another given square.
   (a) Describe how to draw an associated graph such that a path in the graph would represent a sequence of permissible knight moves.
   (b) Draw this graph for a 3×3 chess board.
   (c) Is there a sequence of permissible knight moves between every pair of squares on a 3×3 board? Explain in terms of the graph.

9. Some famous experiments about the structure of genes involved the following data. Many small segments along a certain gene had been identified and from the experiments one knew which segments overlapped. The general question was, When combined together did these segments form a long string (linear structure) or would the combined structure have circuits, branches, etc.? In particular, one wanted to know if the overlap could possibly come from a linear structure or could it only be generated by a more complex structure.
   (a) Describe how to draw an associated graph that would represent the overlap information.
   (b) From the overlap information in Figure 6.19 (an "x" indicates overlap) draw the associated graph.
   (c) Draw some graphs which could *not* arise from overlapping segments on a line.
   (d) Could the graph in b) arise from overlapping segments on a line?

	1	2	3	4	5	6
1	x	x				x
2	x	x		x		
3			x	x	x	
4		x	x	x	x	x
5			x	x	x	
6	x			x		x

Figure 6.19

10. In psychophysics, one attempts to define measures for such concepts as how appetizing various meals appear to a given person. Here one wants a preference function f that assigns to each possible meal x a value $f(x)$ such that the person prefers meal x to meal y if and only if $f(x) > f(y) + 1$ The "+1" factor allows for the fact that it is hard for a person to discriminate between two meals that are different yet similar, e.g., half a pie versus 49/100th's of a pie. Conversely, a person would be indifferent between x and y if and only if $|f(x) - f(y)| \leqq 1$. We can define an indifference graph to represent which meals are indistinguishable to the person.
    (a) How might a preference function be defined in terms of the indifference graph?
    (b) Suppose the matrix in Figure 6.19 tells which pairs of six meals are indifferent. Draw the indifference graph and find a preference function for that graph (if one exists).
    (c) Draw two different four-node indifference graphs that have no preference functions.

11. Suppose we let nodes, instead of edges, represent sides of a street in a graph model for the sweeping problem.
    (a) Describe how the edges of this other graph should be defined. Where do the street "lengths" go?
    (b) Draw this other graph for the road network in Figure 6.1(a).
    (c) Instead of finding a minimal set of routes covering all edges, what do we want in this other graph?

12. Suppose instead of an Euler circuit, we wanted an Euler path—a path crossing every edge just once but not ending where it started. State and prove Theorem 1 for Euler paths in a directed graph.

13. Consider a connected network of *two-way* streets.
    (a) Prove that a tour (a circuit) exists that traverses each edge once in each direction.
    (b) Pretend that all streets in Figure 6.1(a) are *two-way*. Now draw a tour of the type in (a).
    (c) When tracing out a tour in (b), let S_x denote the street just traversed as we arrive at corner x the first time. Prove that if one follows the rule, "When at corner x, do not leave that corner along street S_x unless it is the only remaining possibility," then you will always trace the tour in one step (no pasting circuits as in Theorem 1).

14. In an undirected graph, an Euler circuit is a circuit that traverses each edge once (in some direction, not both directions). State and prove Theorem 1 for undirected graphs.

15. What is the sum of the degrees of all the nodes in a directed graph? Prove your result rigorously.

16. If a directed graph has exactly two nodes of nonzero degree, prove that a path from one to the other exists.

17. If k is the sum of the degrees of all positive nodes in a directed connected graph G, prove that a set of k paths exists such that each edge of G is on exactly one of these paths.

18. To get a tour covering all edges of a directed graph, we needed to add deadheading edges in order to make each node have zero degree.
    (a) What equivalent condition would deadheading edges have to satisfy if one were working with an undirected graph?
    (b) Restate and prove Theorem 2 for a minimal set of deadheading edges in an undirected graph which satisfy the condition in (a).

19. Suppose that instead of a minimal-length circuit covering all edges, one wanted a minimal-length path covering all edges in a directed connected graph (assume the graph does not have an Euler circuit). Thus deadheading edges are added to get a graph with an Euler path. (See Exercise 12.)
    (a) Restate and prove the appropriate form of Theorem 2.
    (b) How should the transportation problem be modified to get the set of paths required by the revised form of Theorem 2 in (a)?
    (c) Suppose that one wishes to find a minimal-length set of K (or fewer) paths covering all edges. Restate (but do not prove) the appropriate form of Theorem 2 and modify the transportation problem accordingly.

20. Prove that if an undirected graph with n nodes is a tree, then it has $n - 1$ edges.

21. Prove that if an undirected graph is a tree, then a unique path exists between any given pair of nodes.

22. An arborescence is the directed version of a tree: it is a tree when we ignore the direction of the edges, and further all its edges are directed out from a given node called the root. See the example in Figure 6.20. An inverted arborescence is like an arborescence except that all edges are now directed towards the root. A spanning (inverted) arborescence of a graph G is a subgraph that is an (inverted) arborescence and which contains all nodes of G.

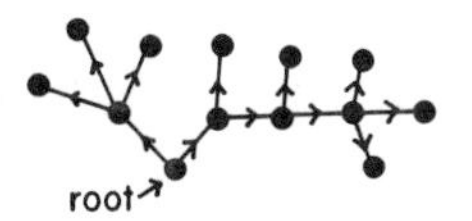

Figure 6.20

(a) Prove that if a directed graph satisfies the conditions of Theorem 1 then it has a spanning arborescence and inverted arborescence rooted at any given node. Hint: Use part of the Euler circuit.

(b) Prove that if paths exist in both directions between every pair of nodes in a directed graph, then a spanning arborescence exists rooted at any given node.

23. Suppose that we have a directed graph satisfying the conditions of Theorem 1 and that we are given a spanning inverted arborescence A (see Exercise 22) rooted at node x. Show that if we start tracing a path from x and only follow an edge of A when no other unused edge exist by which to leave the node we are currently at, then we will trace out a Euler circuit in one pass (with no pasting as in Theorem 1's proof).

24. Give a formal proof that the shortest path algorithm finds the shortest path from a to every other node in the graph.

25. Prove that Floyd's shortest path algorithm finds shortest paths.

26. (a) Restate the shortest path algorithm for undirected graphs.
(b) Ignoring the directions of the edges in Figure 6.7, find the shortest path from N to m.

27. Alter the shortest path algorithm to find shortest path from each node to a node z in a directed graph.

28. Speed up the shortest path algorithm by searching in step 2 among all edges from a labeled node p to an unlabeled node q for an edge that minimizes $d(p) + k(p, q)$. Explain how this would work.

29. Prove that in the shortest path algorithm, the edges used in step 2 to label new nodes form a spanning arborescence (see Exercise 22).

30. (a) Prove that Kruskal's algorithm gives a minimal spanning tree.
(b) Prove that the spanning tree of a connected, undirected n-node graph has $n - 1$ edges.

31. Prove the following lemma used in the middle of the proof of Prim's algorithm. If T and T' are spanning trees, if $e_k = (a, b)$ is an edge in T but not in T', and if $P = (e'_1, e'_2, \cdots, e'_m)$ is a path in T' from a to b with e'_j being an edge on P not in T, then removing e_k from T and replacing it by e'_j will yield another spanning tree.

32. Modify Kruskal's algorithm so that it finds a minimal spanning tree that contains a prescribed edge e. Prove your modification works.

33. Modify Kruskal's algorithm so that it finds a maximal spanning tree.

34. Assume that if the edges of the undirected connected graph G are ordered, then when a tie occurs in Kruskal's algorithm for the next edge to be added, the edge

of smaller index is chosen. Recall that G often has several minimal spanning trees. Prove that the edges can be ordered so that Kruskal's algorithm, with the above tie-breaking rule, will yield any given spanning tree.

35. Describe how any transportation problem with integer supplies and demands can be converted into a transportation problem with supplies and demands all equal to one.

36. A transportation problem with unit supplies and demands can be converted into a matching problem in a bipartite graph (see Exercise 3) in which one seeks a matching which minimizes the sum of the lengths of the edges in the matching.
    (a) Describe this conversion.
    (b) Prove that the intermediate and final solutions obtained in our transportation problem algorithm all correspond to spanning trees in the associated bipartite graph of a).

37. Suppose that we have generated sets of routes in a given district for periods 1 and 2. Four routes exist in each period. Suppose x_i is the node where the ith route in period 1 ends and y_j is the node where the jth route in period 2 begins. One wants to pair up the routes of periods 1 and 2 so as to minimize the sum of the distances between the ends of the period 1 routes and the beginning of the period 2 routes. Pose this problem in a form which can be solved by one of the algorithms introduced in this module.

38. Suppose we are working in a district with no parking regulations (parking is never banned) and that we can introduce any regulations we want, but parking can only be banned for one hour a day on any street. Describe how, after solving the sweeping problem for a whole day (say, an 8-hour period) and getting a collection of 8-hour routes that cover all edges, we can then set up "nice" (compatible) parking regulations.

39. (a) Verify the distances in the matrix in Figure 6.12.
    (b) Use the transportation problem algorithm to solve the node pairing problem for nodes in Figure 6.12.

40. Find shortest paths in Figure 6.7 from f to R and from f to h.

41. Direct the edges in Figure 6.8 by the alphabetical order of the nodes, e.g., edge (h, m), not (m, h). Find the shortest paths from b to t and from a to y.

42. Find shortest paths in Figure 6.1(b), using all edges
    (a) from 1 to 14,
    (b) from 14 to 1,
    (c) from 5 to 15.

43. Program Floyd's algorithm for Figure 6.7 or manually use this algorithm for the subgraph in Figure 6.7 involving just nodes N, b, c, d, e, f, g.

44. Ignoring the directions of edges, find a minimal spanning tree for Figure 6.7. Is it unique?

45. Ignoring the directions of edges, find a minimal spanning tree for the whole graph (dashed and solid edges) in Figure 6.1 (b). Is it unique?

46. Find a spanning tree for Figure 6.8 which includes the edges (b, c) and (b, g) and which is of minimal length. (See Exercise 32.)

47. Find a maximal-length spanning tree for Figure 6.8 (See Exercise 33.)

48. Solve the following transportation problems.

(a)

	60	30	30
40	6	3	4
40	3	1	7
40	5	3	7

(b)

	50	30	90
70	5	2	6
40	3	1	8
60	7	4	9

(c)

	10	3	2	3
5	6	4	3	2
9	5	2	4	4
4	4	2	2	4

49. In the end of the example in Section 7, we claimed that a set of two feasible subtours can be obtained if one links component C_1 with C_2, instead of C_3, when building the grand tour. Find this set of two feasible subtours.

50. Consider the set of dashed edges in Figure 6.1(b) involving just nodes 1, 2, 3, 4, 5, 6, 7.
    (a) Generate a minimal-length circuit covering all these edges following the procedure outlined in Section 6.

51. Consider the set of dashed edges in Figure 6.1(b) involving just nodes 1, 2, 3, 5, 6, 7, 8, 9, 10, 12, 13.
    (a) Repeat part (a) in Exercise 50 for these edges.
    (b) Assuming a one-hour period, use your result in (a) as a basis for finding a minimal set of feasible tours covering all these edges.

52. Consider the set of all dashed edges in Figure 6.1(b).
    (a) Repeat part (a) in Exercise 50 for these edges.
    (b) Assuming a *two-hour* period, use your result in (a) as a basis for finding a minimal set of feasible tours covering all dashed edges.

53. Consider the graph in Figure 6.7. Suppose we add extra dashed edges (R, N) of length 25, (m, b) of length 20, and (f, N) of length 15.
    (a) Repeat part (a) in Exercise 50 for the original solid edges in Figure 6.7.
    (b) Assuming a *two-hour* period, use your result in (a) as a basis for finding a minimal set of feasible tours covering all solid edges.

References

[1] F. Busacker and T. Saaty, *Finite Graphs and Their Applications*. New York: McGraw-Hill, 1966. A good basic text in graph theory that covers a lot of ground; the second half is a survey of applications.
[2] O. Ore, *Graphs and Their Uses*. New York: Random House, 1963. A short, introduction to graph theory written for able high school students (and lower division undergraduates); of the many interesting applications, the logical puzzles (Sections 2.4 and 6.1) are especially enjoyable.
[3] R. Wilson, *Introduction to Graph Theory*. New York: Academic Press, 1972. A concise, well-written undergraduate text.
[4] F. Hillier and G. Lieberman, *Introduction to Operations Research*. San Francisco:

Holden-Day, 1967. The classic undergraduate operations research text; the algorithms used in this module are discussed in greater detail in Chapters 6 and 7.

[5] H. Wagner, *Principles of Operations Research*. Englewood Cliffs, NJ: Prentice-Hall, 1969. The other standard operations research text; is a bit more advanced than [4]; algorithms used in this module are in Chapter 6.

[6] L. Bodin, "A taxometric structure for vehicle routing and scheduling problems," *J. Computers and the Urban Society*, vol. 1, pp. 11–29, 1975. This paper presents the cluster-first method and some node-covering problems (in this module we did edge covering) as well as a primitive version of the procedure presented in this module.

[7] E. Beltrami and L. Bodin, "Networks and vehicle routing for municipal waste collection," *Networks*, vol. 4, pp. 65–94, 1973. This paper describes two other garbage routing problems as well as a primitive version of the procedure presented in this module.

[8] J. Meckling, "Chart day problem: A case study in successful innovation," *J. Urban Policy*, vol. 2, no. 2, Nov. 1974. This paper, written by a New York City Sanitation Department administrator, describes some mathematical analysis which did not contain as interesting modeling as in this module but which resulted in an *annual* savings to the city of about $10,000,000.

[9] J. Edmonds and E. Johnson, "Matching, Euler tours, and the Chinese postman," *Mathematical Programming*, vol. 5, pp. 88–124, 1973. This paper presents a way to build a Euler circuit in a directed graph all at once (without first getting several circuits that must be joined together as we did in Theorem 1); their method is based on a directed spanning tree and is quite intuitive.

Notes for the Instructor

A major reason for the presentation of this particular problem is the natural way it lends itself to an open-ended teaching style, in which the students themselves can develop, with the guidance of an instructor, the mathematical model and an analysis of the problem in terms of this model. Because the analysis contains several basically independent steps, one can present some of the steps in a lecture and leave others to the class. Two of the three minimization algorithms required in the solution, the shortest path and minimal spanning-tree algorithms could be "discovered" by the class.

If an open-ended approach is used, guiding the students into the "route first–cluster second" method is important. Also, students should not worry about disconnected networks or the turns made at each corner until the other problems have been analyzed. Beyond this, the students can progress as they wish. Several different proofs exist for Theorems 1 and 2. (When students later are studying the problem of turns at each corner, they will discover the proof of Theorem 1 given in this text.) If students write a computer program for street sweeping, they will find several minor but challenging loose ends in the model (some are indicated in Section 8). Even though these problems often can be ignored in actual practice, some class discussion concerning heuristic solutions to these problems is worth including, if time permits.

Note that even in an open-ended approach, the instructor should consider distributing sections of the text after the students have developed the particular analysis.

If the material is presented in a normal lecture method, it is advisable to insert examples after each stage of the analysis in Section 4. In addition to the extended example in Section 7, several of the exercises in Section 10 can serve as examples. The instructor may also choose to introduce the algorithms when needed rather than waiting until the analysis is completed. With students who have a limited exposure to graphs, the instructor should consider introducing other types of problems that are naturally modeled with graphs. The exercises have several examples of graph modeling, and more may be found in [1], [2].

CHAPTER 7

Finite Covering Problems

Ronald E. Prather*

1. Introduction

In this chapter we conduct a survey of the occurrence of finite covering problems in graph theory, operations research, and miscellaneous fields, seeking to unite such instances within a common framework. We then offer a unified approach to the solution of such problems. Elements of a general theory are presented, and two exhaustive solution methodologies are given, those we call the branching method and the algebraic method. The latter leads rather naturally to an introduction to distributive lattices and their application to discrete decision problems. This report should serve equally well as an introductory survey for readers wishing to become acquainted with the origins and the mathematical treatment of covering problems and as a classroom module for supplementing courses in mathematical modeling, applications of graph theory, discrete mathematical structures, and the like.

Only a few modest prerequisites are assumed of the reader. S/He should be familiar with the standard set theory notation and with elementary algebraic manipulation. An introduction to graph theory notions is desirable but is not necessary. The idea of a graph is so easily motivated and its definition so quickly presented that in the interest of completeness, it seemed advisable that we include these at the outset. An early undergraduate student should have little difficulty in following the major portion of this chapter. Some additional mathematical maturity will probably be necessary, however, for gaining a full understanding of some of the later, more abstract portions and for acquiring the abilities needed to handle some of the more difficult exercises. With this in mind, the instructor should be able to employ the

* Department of Mathematics, University of Denver, Denver, CO 80210.

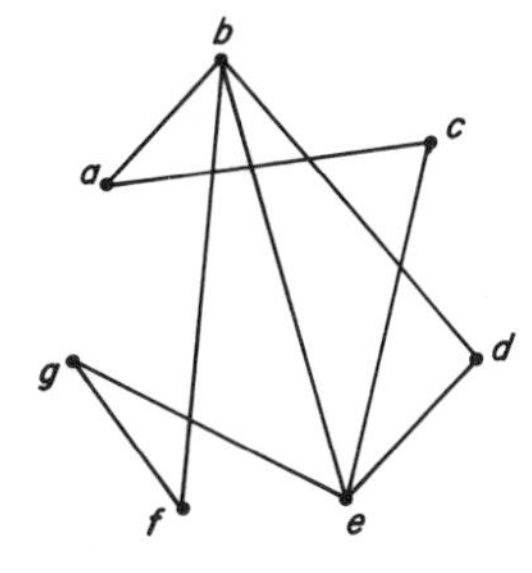

Figure 7.1

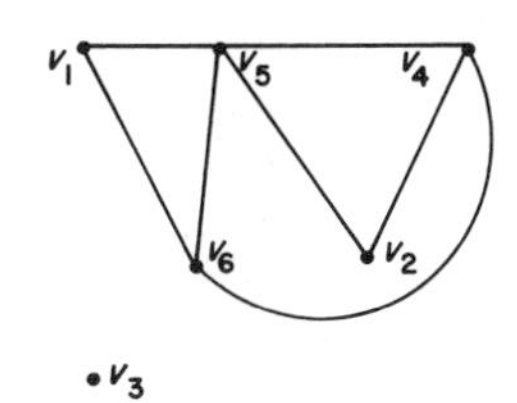

Figure 7.2

portions of the chapter consistent with the mathematical maturity of the class. This is one reason why we present two distinct solution techniques: the "branching method," for its intuitive appeal, and the "algebraic method" with its more theoretical foundation.

A fair number of exercises are presented throughout the paper. Most of these are routine illustrations of the text material, but a few are aimed at the mathematically inclined student of at least intermediate maturity, perhaps requiring proofs for the questions raised. In some cases, the problem will require computer programming skills, particularly experience with non-numerical algorithms—combinatorial computing, if you like. Such problems are easily identified by the instructor and will be chosen or not depending on the intent.

The chapter is reasonably self-contained, by design, and should require no outside reference on the part of an instructor. Those few references we do cite are given in the interest of completeness and to provide the more inquisitive student with a guide to some of the available literature and background on the subject. In this connection, we might mention that this literature is quite fragmented and in itself was one of the reasons for our writing this report.

2. Graph-Theoretic Origins

Most readers will know that a *graph* $G = (V, E)$ consists of a set V of "vertices" and a collection E of "edges," these being considered as unordered pairs $\{v, w\}$ of distinct vertices of the graph. A pictorial representation of a graph is given in Figure 7.1. In this case, $V = \{a, b, c, d, e, f, g\}$, and E is the collection

$$E = \{\{a, b\}, \{a, c\}, \{b, d\}, \{b, e\}, \{b, f\}, \{c, e\}, \{d, e\}, \{e, g\}, \{f, g\}\}.$$

For the graph of Figure 7.2 we have $V = \{v_1, v_2, v_3, v_4, v_5, v_6\}$ and

$$E = \{\{v_1, v_5\}, \{v_1, v_6\}, \{v_2, v_4\}, \{v_2, v_5\}, \{v_4, v_5\}, \{v_4, v_6\}, \{v_5, v_6\}\}.$$

Note that in these pictorial representations, the spatial arrangement of the vertices and edges is of no concern; an edge may be drawn as a straight line segment or not. All that really matters is whether or not an edge exists joining v to w. Thus we are sometimes better off to deal instead with (triangular) *matrix representations* (Figure 7.3 or Figure 7.4) which give only this information. That way we are not misled into thinking that additional vertices exist at points where two edges appear to intersect in a particular pictorial representation (See Figure 7.1).

Commenting on the importance of graph theory is rarely necessary. Since applications are found mainly in the recently emerging sciences (programming, communication theory, electrical networks, switching circuits, biology, chemistry, and psychology, to name a few), present-day students have a distinct advantage over their predecessors. Ordinarily, they have no difficulty perceiving the relevance of graph theory, even at first exposure. Suffice it to say that whenever we encounter a network or diagram whose nodes (vertices) are joined by various arcs or lines (edges) it is quite likely that a graphical formalism is appropriate to the analysis. Beyond that observation, our main concern here is to illustrate the wide variety of instances in which covering problems arise in one way or another.

As one approach to examining the sources of covering problems, we concentrate on certain graphical indices that are important because they usually express some kind of optimality. Among these indices, we will discuss the covering, domination, independence, and chromatic numbers in turn. In the exercises, still more of these indices are introduced. Let us begin with the covering number as this leads most naturally to what we will eventually call a "covering problem." If $G = (V, E)$ is any graph, we say that the subset of vertices $\{v_1, v_2, \cdots, v_s\}$ constitutes a *covering set* if

$$\{v, w\} \in E \Rightarrow v_j \quad \text{or} \quad w = v_j$$

for some $1 \leq j \leq s$, i.e., if every edge of the graph has at least one of its feet in the subset. Then the *covering number* $\alpha = \alpha(G)$ is simply the size of a smallest covering set. The following discussion should provide a vivid illustration of the kind of applications to be expected here. Suppose that the edges

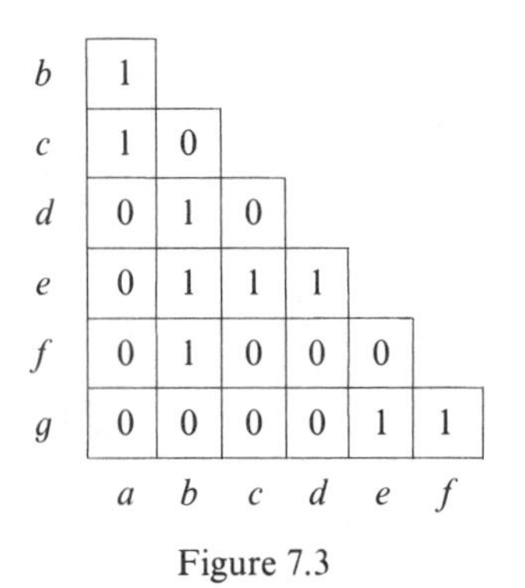

b	1					
c	1	0				
d	0	1	0			
e	0	1	1	1		
f	0	1	0	0	0	
g	0	0	0	0	1	1
	a	b	c	d	e	f

Figure 7.3

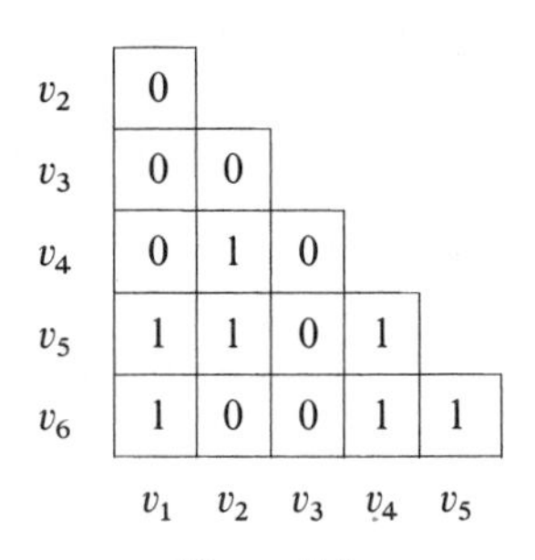

v_2	0				
v_3	0	0			
v_4	0	1	0		
v_5	1	1	0	1	
v_6	1	0	0	1	1
	v_1	v_2	v_3	v_4	v_5

Figure 7.4

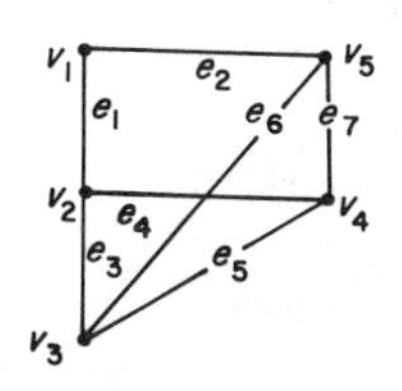

Figure 7.5

	e_1	e_2	e_3	e_4	e_5	e_6	e_7
v_1	1	1	0	0	0	0	0
v_2	1	0	1	1	0	0	0
v_3	0	0	1	0	1	1	0
v_4	0	0	0	1	1	0	1
v_5	0	1	0	0	0	1	1

Figure 7.6

of $G = (V, E)$ are streets (blocks) of a city map, and suppose further that it is desired that every street have a fire hydrant at one corner or another at the ends of the block. Then a minimum number of hydrants needed for this purpose is obtained by determining the covering number of the graph. At the same time, a covering set (of minimum size) will specify the precise location of hydrants for achieving a particular optimum arrangement.

If $G = (V, E)$ is a graph with $V = \{v_1, v_2, \cdots, v_n\}$ and we denote the edges as $E = \{e_1, e_2, \cdots, e_m\}$, then we may introduce the *incidence relation*

$$vRe \Leftrightarrow v \in e \quad \text{(as an unordered pair)}$$

(read v is "incident to" e)
for $v \in V$ and $e \in E$. For the graph of Figure 7.5, we have $n = 5$, $m = 7$, and we are thus led to the 5×7 relation matrix $R = (r_{ij})$ of Figure 7.6, i.e., we set entries r_{ij} to one if $v_i Re_j$, otherwise zero. We refer to the matrix so obtained as the (vertex-edge) *incidence matrix* of the underlying graph. In this context, we note that a subset of vertices (rows of the matrix) is a covering set if, among all its rows, a one is found in every column. Because none of the rows of Figure 7.6 has more than three ones and seven edges need to be covered, we see immediately that $\alpha = \alpha(G) > 2$ for the graph G pictured in Figure 7.5. Moreover, since $\{v_2, v_3, v_5\}$ is easily seen to be a covering set, we conclude that $\alpha(G) = 3$.

EXERCISES[1]

1. Consider the graphs $G = (V, E)$ represented by the following triangular matrices. In each case, provide a pictorial representation.

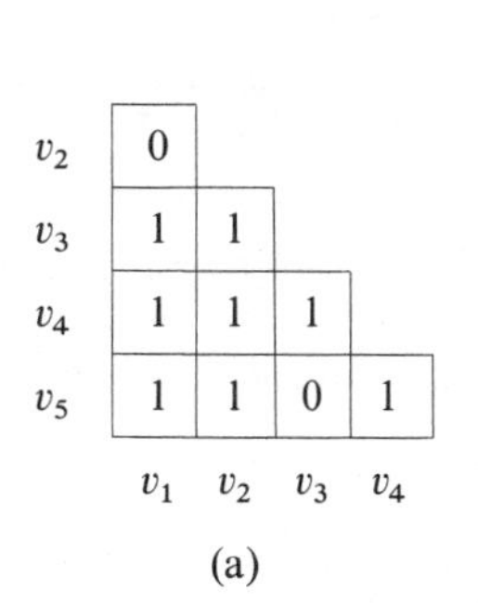

v_2	0			
v_3	1	1		
v_4	1	1	1	
v_5	1	1	0	1
	v_1	v_2	v_3	v_4

(a)

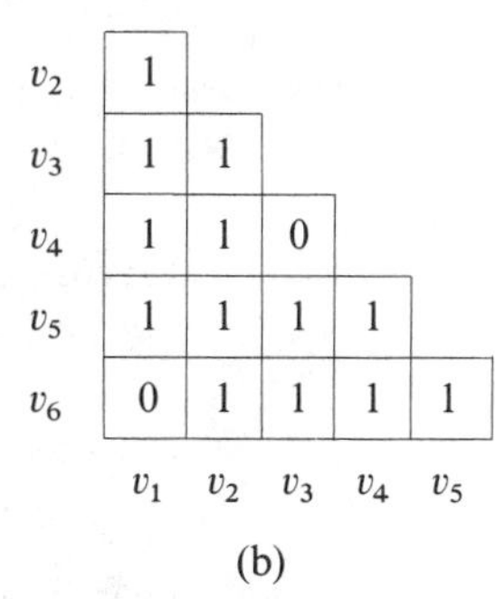

v_2	1				
v_3	1	1			
v_4	1	1	0		
v_5	1	1	1	1	
v_6	0	1	1	1	1
	v_1	v_2	v_3	v_4	v_5

(b)

[1] A dagger (†) marks the more challenging exercises.

2. Consider the graphs pictured below. In each case, determine the triangular matrix representation.

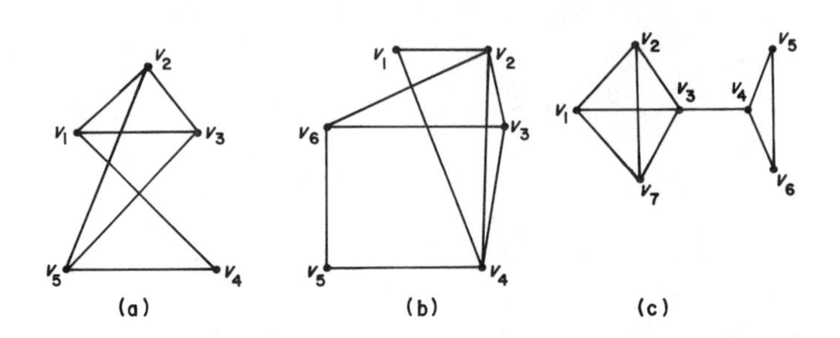

3. For each of the following graphs, label the edges $e_1, e_2, \cdots$, and determine the vertex-edge incidence matrix.
   (a) Figure 7.1.
   (b) Figure 7.2.
   (c) Exercise 1(a).
   (d) Exercise 1(b).
   (e) Exercise 2(a).
   (f) Exercise 2(b).
   (g) Exercise 2(c).

4. For the graphs of Exercise 3, use the vertex-edge incidence matrix to determine the covering number α and a corresponding covering set of minimum size.

5. Corresponding to each graph $G = (V, E)$ is a *complementary* graph $G' = (V, E')$ in which E' consists of precisely those edges which are not found in G. Provide a pictorial representation of the graph G' when G is the each of the following graphs.
   (a) Figure 7.1.
   (b) Figure 7.2.
   (c) Figure 7.5.
   (d) Exercise 1(a).
   (e) Exercise 1(b).
   (f) Exercise 2(a).
   (g) Exercise 2(b).
   (h) Exercise 2(c).

6. A graph $G = (V, E)$ is said to be *connected* if every pair of vertices is joined by a path (i.e., a sequence) of edges. Thus the graph of Figure 7.1 is connected but that of Figure 7.2 is not. Show by example that both G and G' (See Exercise 5) can be connected.

†7. In analogy to a set of vertices that cover all the edges, we may consider for connected graphs (See Exercise 6) a *covering set of edges* (they cover all the vertices) and a corresponding (minimal) *edge-covering number* α_e. For a graph with n vertices, prove that

$$n/2 \leq \alpha_e \leq n - 1.$$

Hint: The first inequality is straightforward. For the second, consult Ore [7].

8. For each of the following graphs, label the edges $e_1, e_2, \cdots$, and determine the *edge-vertex incidence matrix* (the transpose of the vertex-edge incidence matrix).
   (a) Figure 7.1.
   (b) Figure 7.5.
   (c) Exercise 1(a).
   (d) Exercise 1(b).
   (e) Exercise 2(a).
   (f) Exercise 2(b).
   (g) Exercise 2(c).

9. For the graphs of Exercise 8, use the edge-vertex incidence matrix to determine the edge covering number α_e (see Exercise 7) and a corresponding covering set of edges of minimum size.

3. The Diner's Problem

Suppose that we are going to order dinner in a restaurant from an a la carte menu. Certainly, we should be concerned with the nutritional aspect, and we want to make selections that add up to a well-balanced meal. To simplify the problem, suppose that the menu consists of the following items:

A. Spanish omelet,
B chicken enchilada,
C. Waldorf salad,
D. steak,
E. potato,
F. liver with onions.

Assume that these foods offer the nutritional benefits described in Figure 7.7, where the entry 1 or 0 indicates that the nutritional benefit in question is or is not provided, respectively.

Note the similarity to the covering problem of the previous section (see Figure 7.6). We must make a selection of rows in Figure 7.7 so that among all of those selected, there is a one in every column. Thus the diner may choose, for example,

A and B = Spanish omelet and chicken enchilada or
C and D and E = Waldorf salad and steak and potato,

either of these constituting a well-balanced meal. However, we do not permit the diner to overindulge. We do not allow a selection of rows (as A, B, C) for which a row may be deleted while still satisfying all the nutritional requirements. We are only interested in nonredundant coverings. Even so, many of these may exist, and we may wish to know all of them. For one thing, we could associate a cost with each entree A, B, C, D, E, F and then one might be interested in choosing a selection of minimum cost. These are the kind of problems we wish to treat in the following.

	Protein	Carbohydrates	Vitamins	Minerals
A	1	0	1	1
B	1	1	0	0
C	0	0	1	1
D	1	0	0	0
E	0	1	1	0
F	1	0	0	1

Figure 7.7

4. Sources of Covering Problems

The reader must already be of the opinion that covering problems arise in a wide variety of situations. We will reinforce this impression as we continue, but before proceeding further we should first formalize the whole class of problems under consideration. We are given two finite sets, a collection $A = \{a_1, a_2, \cdots, a_n\}$ of *cells* and another collection $B = \{b_1, b_2, \cdots, b_m\}$ whose elements will be called *points*. Note that we have deliberately chosen a neutral terminology here so as to allow for the various interpretations of the theory. In addition, each cell a has an associated *cost* $\#a$ which we take to be a nonnegative integer, for the sake of convenience. Finally, an *incidence relation* R from A to B is given and its associated *incidence matrix* $R = (r_{ij})$ consisting of zeros and ones, for which we assume that each column has at least one 1 entry. As usual, we take

$$r_{ij} = 1 \Leftrightarrow a_i R b_{ij},$$

and as a matter of terminology, we say that a_i *covers* b_j(b_j is *covered by* a_i) if $a_i R b_j$.

In this context, we extend the definition of the relation R to apply to subsets $C \subseteq A$ and $D \subseteq B$, writing CRD just in case it happens that for every point $b_j \in D$ a cell $a_i \in C$ exists with $a_i R b_j$. Through a slight abuse of notation, we permit the writing of $a_i RD$ or CRb_j when it is really the singleton sets $\{a_i\}$ or $\{b_j\}$ that are intended. If $C = \{a_\alpha\}$ is any collection of cells, we take the *cost* of the collection to be the sum

$$\#C = \sum_\alpha \#a_\alpha.$$

Other cost functions could also have been defined, but the one we have chosen is quite common and seems to be the most convenient. Given a *covering problem* (A, R, B) we will say that $C \subseteq A$ is a *minimal covering* if

(i) CRB (C is a *covering*)
(ii) $C'RB \Rightarrow \#C \leq \#C'$

i.e., if C is of minimum cost among all the coverings.

Our goal here is first to exhibit the variety of contexts in which such problems arise, and then in the next section, to develop the elements of a theory for handling such problems and to describe effective solutions techniques. We have certainly seen such problems before. In Section 2 (the discussion of the covering number $\alpha = \alpha(G)$ for a graph $G = (V, E)$), we encountered a covering problem (A, R, B) with $A = V$, $B = E$, and R the vertex-edge incidence relation. Here, we might think of the costs $\#v = 1$ for every vertex. Then the cost of covering is simply the number of vertices in the covering. It follows that a minimal covering is one employing the fewest number of vertices. We generally assume this cost situation (the cost of every cell is one) in the absence of other suggested weighting features.

In the diner's problem (A, R, B) we have A as the list of entrees, B the nutritional features, and R the relation which specifies the nutritional components of the various dishes. Here we might want to attach a numerical cost or price to each entree (that they may not be integers can be taken care of with a "scaling factor") so as to enable the diner to seek a most economical well-balanced meal. Once more, we have a covering problem in the sense outlined above.

As for additional examples, we will do well to return again to graph theory. If $G = (V, E)$ is a graph, then a subset of vertices $\{v_1, v_2, \cdots, v_s\}$ is said to constitute a *dominating set* if

$$v \in V \Rightarrow v = v_j \quad \text{or} \quad \{v, v_j\} \in E$$

for some $1 \leq j \leq s$, i.e., if every vertex is, or is adjacent to, a vertex of the subset. Again, the size of a smallest dominating set is called the *domination number* $\delta = \delta(G)$ of the graph G. If the vertices of G represent various trouble spots in a prison complex and the edges correspond to corridors, stairways, etc., then a smallest dominating set would provide an optimum location pattern for guards to be stationed and the domination number would give the minimum number of guards necessary for maintaining a watchful eye on these trouble spots. In any case, if we take $A = B = V$ and as R the relation

$$v_i R v_j \Leftrightarrow v_i = v_j \quad \text{or} \quad \{v_i, v_j\} \in E,$$

then the problem of determining a minimal dominating set is reduced once again to a covering problem.

If the graph of Figure 7.5 is reproduced as Figure 7.8, then as a covering problem for determining the domination number, we obtain the incidence matrix of Figure 7.9. Here we see immediately that $\{v_2, v_5\}$, for instance, is a dominating set (a covering for the derived covering problem). Since no coverings by a singleton set occur—as no row of the matrix consists entirely of ones—this must be a minimal covering, and accordingly $\delta = 2$ for this graph.

On a chessboard, as shown in Figure 7.10, we are able to place five queens in such a way that they dominate all 64 squares. Some experimentation is

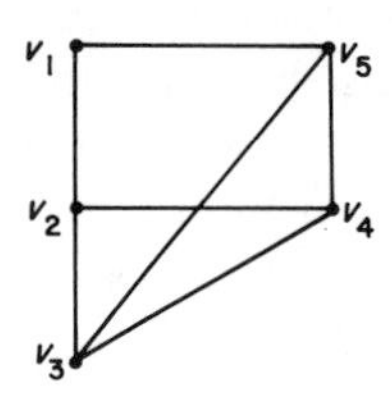

Figure 7.8

	v_1	v_2	v_3	v_4	v_5
v_1	1	1	0	0	1
v_2	1	1	1	1	0
v_3	0	1	1	1	1
v_4	0	1	1	1	1
v_5	1	0	1	1	1

Figure 7.9

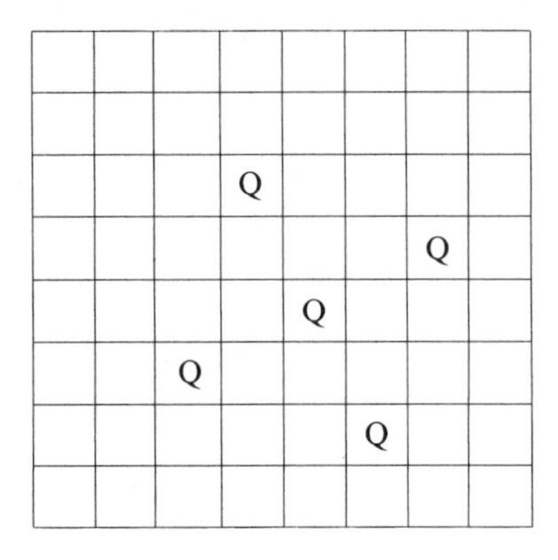

Figure 7.10

necessary to be convinced that there is no way to accomplish this feat with fewer queens, but then it is seen that $\delta = 5$ for the appropriate graph (on 64 vertices!).

Let us next introduce an important graphical index which seems at first not to be related to a covering problem at all. We let $G = (V, E)$ be any graph, and we refer to a subset of vertices $\{v_1, v_2, \cdots, v_s\}$ as an *independent set* provided that

$$\{v_i, v_j\} \in E \qquad \text{for all } 1 \leq i, j \leq s.$$

Then we may refer to *maximal* independent sets—those which are not strictly contained in another independent set. Note that a maximal independent set need not be an independent set of maximum size! In any case, the size of a largest independent set is called the *independence number* $\beta = \beta(G)$ for the graph G. Consider a communications channel for transmitting symbols from a given alphabet. Due to distortion in the channel and other anomalies, certain symbols may be confused or misinterpreted; that is, various pairs of symbols may be received as if they were the same. In such a situation it would be advisable not to use the entire alphabet, but to select instead a largest subalphabet having the property that no two of its letters could be confused. Such a selection would be a largest independent set (of size β) in a graph having edges that depict all the confusable pairs of symbols.

In as much as β is defined by a maximum condition and our covering problems instead involve minimality, any possible connection to covering problems is difficult to see at a first glance. The relationship is indeed a secondary one, but will follow rather easily from the second of two results we wish to establish connecting β to our earlier graphical indices, α and δ.

Theorem 1. *$\beta \geq \delta$ in any graph $G = (V, E)$.*

PROOF. We will show that an independent set is maximal if it is a dominating set. The inequality will then follow quite easily, since a largest independent set will certainly dominate, though it may not be a smallest set with this property.

If $\{v_1, v_2, \cdots, v_s\}$ is maximal as an independent set, then we cannot adjoin an additional vertex without losing the independence. Such a set must surely dominate because $v \neq v_j$ and $\{v, v_j\} \notin E$ for all $j = 1, 2, \cdots, s$ would contradict the maximality. Conversely, suppose $\{v_1, v_2, \cdots, v_s\}$ is a dominating independent set. For any $v \in V$ such that $v \neq v_j$ (all j) we must have $\{v, v_j\} \in E$ for some j, so that $\{v_1, v_2, \cdots, v_s\}$ cannot be independent. □

Theorem 2. *For any graph $G = (V, E)$ with n vertices,*

$$\alpha + \beta = n.$$

PROOF. This time we show that if $V = A \cup B$ with $A \cap B = \varnothing$, then A is a covering set iff B is an independent set. It will then follow immediately that $\alpha + \beta = n$.

With the assumption that $V = A \cup B$ and $A \cap B = \varnothing$, suppose that A is a covering set. Then consider any two elements $v, w \in B$. If it were true that $\{v, w\} \in E$, then we must have $v \in A$ or $w \in A$. However, either of these would contradict the disjointness of A and B. So we have $\{v, w\} \notin E$ for all $v, w \in B$, and B is independent. Conversely, if B is an independent set and $\{v, w\} \in E$, then we cannot have both vertices in B. It follows that one of them is in A, and A is a covering set. □

From the proof of this second theorem, we see that the maximal independent sets are precisely the set complements (with respect to V) of the nonredundant covering sets in any graph $G = (V, E)$. These are the covering sets for which the deletion of any vertex leads to a noncovering set. Thus in reviewing the analysis of Figures 7.5 and 7.6, we note that $\{v_2, v_3, v_5\}$ is nonredundant—the deletion of v_2 uncovers e_1, the deletion of v_3 uncovers e_5, and the deletion of v_5 uncovers e_2. Accordingly, the complementary set of vertices $V \sim \{v_2, v_3, v_5\} = \{v_1, v_4\}$ is a maximal independent set. In general, we can determine an independent set of maximum size (and hence β) if we obtain a minimal covering of the covering problem pertaining to covering sets. Moreover, we can find all the maximal independent sets if we are willing to obtain all the nonredundant coverings (see section 5) for this problem.

Before leaving the subject of independence, let us return to the chessboard and ask, "How many queens may we place on the board in such a position that none is threatened?" Apparently, we are asking for the independence number β for the graph of the queen's relation (again a graph with 64 vertices). According to Theorem 2.1, we have

$$\beta \geq \delta = 5.$$

Actually, $\beta = 8$ as shown by Figure 7.11 (and the fact that nine queens must yield two in the same row).

As one more important instance where covering problems arise in graph theory, we now introduce the chromatic number of a graph. We have tried to give vivid illustrations to help demonstrate the meaning of the domination, covering, and independence numbers. For the chromatic number, the illustration is in technicolor! Imagine that we have a box of crayons for the purpose of coloring the vertices of a graph $G = (V, E)$. We add the stipulation that adjacent vertices (those joined by an edge) must be given different colors. What is the fewest number of colors that will suffice? This number is the *chromatic number* $\kappa = \kappa(G)$ for the graph G. Phrased somewhat differently, let a *coloring* of G be a partition

$$V = \bigcup_{i=1}^{s} V_i, \qquad V_i \cap V_j = \varnothing \quad (i \neq j)$$

of the vertices into s blocks, subject to the condition

$$v, w \in V_i \Rightarrow \{v, w\} \notin E.$$

Then $\kappa = \kappa(G)$ is the fewest number of blocks (colors) to be found among all the colorings of G. In the field of cartography, where this problem first arose, a map of several countries is drawn in the plane, and we then consider

					Q		
Q							
				Q			
	Q						
							Q
		Q					
						Q	
			Q				

Figure 7.11

the associated graph $G = (V, E)$ obtained as follows. We take V to be the set of countries, and we put $\{v, w\} \in E$ just in case the countries v and w share a common boundary. Note that such a graph is *planar*: a pictorial representation will exist in which no edges intersect. A coloring of the graph provides a plan for coloring the countries of the map in such a way that countries sharing a common boundary are given different colors.

The chromatic number was the subject of perhaps the most celebrated unsolved problem in all of mathematics: the four color problem. The task was to determine the fewest number of colors that will serve in coloring every planar graph. Three colors were known not to be enough and five colors to always suffice, but the question remained, Are four colors always sufficient? The question, tantalizingly simple to state, survived the onslaught of the world's most capable mathematicians. Exhaustive computer analysis of all planar graphs having up to forty vertices revealed none requiring more than four colors. Finally, in 1977, Appel and Haken [1], [2] showed that four colors would always suffice, i.e., that $\kappa(G) \leq 4$ for every planar graph G.

We have yet to demonstrate that the coloring problems are covering problems. The trick is to note that the blocks V_i in a coloring must be independent sets (inasmuch as $v, w \in V_i \Rightarrow \{v, w\} \notin E$). This suggests that we set up a covering problem (A, R, B) in which the cells of A are all the maximal independent sets (crayons) for the graph $G = (V, E)$. If these are the subsets of vertices $U_1, U_2, \cdots, U_r$, then we take $B = V$ and set

$$U_i R v_j \Leftrightarrow v_j \in U_i.$$

Again taking the cost of each crayon equal to one, we have a covering problem (A, R, B) whose solution will yield as a minimal covering a subset $C = \{U_\alpha\}$ of the U_j, say $\{U_{\alpha_1}, U_{\alpha_2}, \cdots, U_{\alpha_k}\}$. Setting

$$\begin{aligned} V_1 &= U_{\alpha_1} \\ V_2 &= U_{\alpha_2} \sim V_1 \\ V_k &= U_{\alpha_k} \sim (V_1 \cup V_2 \cup \cdots \cup V_{k-1}) \end{aligned}$$

we obtain a coloring of G having the fewest number of colors. Note that in setting up this covering problem, we first need all the maximal independent sets of the graph. Fortunately, as already indicated, these are the complements of the nonredundant covering sets, to be obtained as the solution to yet another covering problem. We present an illustration of the entire process at the end of the next section.

EXERCISES

10. Let (A, R, B) be a covering problem. Then a covering $C \subseteq A$ is said to be *redundant* if a cell $a \in C$ exists such that $C \sim \{a\}$ is again a covering. Otherwise, C is called *nonredundant*. Prove that every minimal covering is nonredundant. The converse is not true. Give a counterexample showing this.

11. In a graph without isolated vertices, show that every covering set is also a dominating set.

12. In scheduling students for classes at a university, consider a graph whose vertices represent the classes and let there be an edge $\{v, w\}$ if a time conflict occurs, for example when a student must schedule both classes, v and w. Such classes should be arranged to meet at different times. Discuss this scheduling problem as a coloring problem.

13. Find an example showing that a maximal independent set need not be an independent set of maximum size.

14. Obtain the incidence matrix for the covering problem pertaining to dominating sets in case G is the graph of
    (a) Figure 7.1,
    (b) Figure 7.2,
    (c) Figure 7.5,
    (d) Exercise 1(a),
    (e) Exercise 1(b),
    (f) Exercise 2(a),
    (g) Exercise 2(b),
    (h) Exercise 2(c).

†15. (For chess players only) Show directly that $\beta \geq \delta$ for the graph of the knight's relation.

†16. Prove that in any graph with n vertices

$$n/\beta \leq \kappa \leq n - \beta + 1.$$

†17. In analogy to an independent set of vertices, we may consider for connected graphs (see Exercise 6) an *independent set of edges* (no two of them share a vertex) and a corresponding (maximal) *edge-independence number* β_e. For a graph with n vertices, prove that

$$\alpha_e + \beta_e = n.$$

(See Exercise 7.)

18. Characterize those graphs whose independence number is one.

19. Arguing from the definitions, obtain the domination number and a dominating set of smallest size for the graphs G listed in Exercise 14.

20. Arguing from the definitions, obtain the independence number and an independent set of maximum size for the graphs G listed in Exercise 14.

21. Arguing from the definitions, obtain the chromatic number and a coloring involving fewest colors for the graphs G listed in Exercise 14.

22. Given the map of Figure 7.12, determine the associated graph.

23. Repeat Exercise 22 for the countries of South America.

24. Let $C \subseteq A$ be a covering for the covering problem (A, R, B). Show that C is

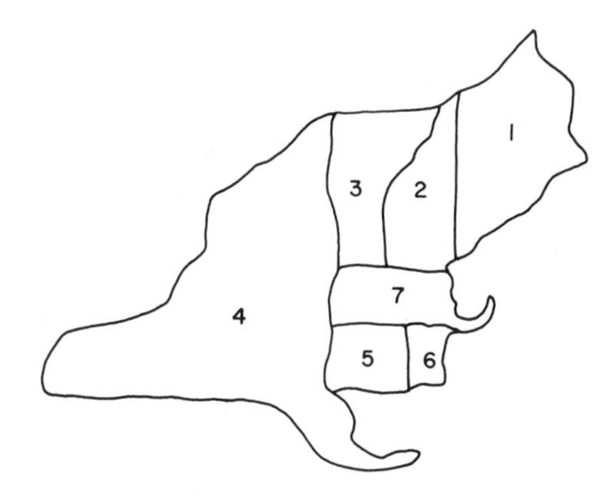

Figure 7.12

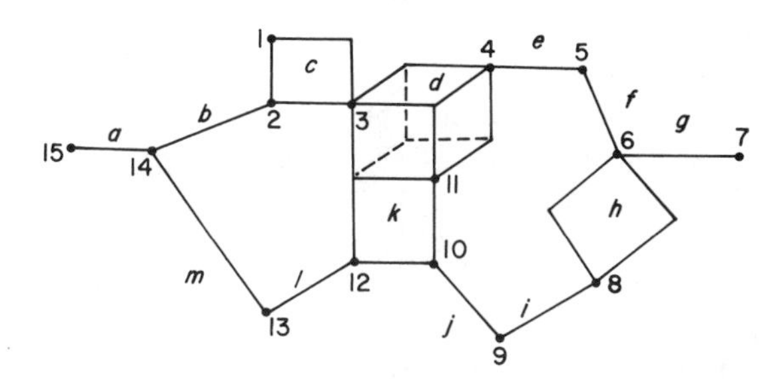

Figure 7.13

redundant (see Exercise 10) if a cell $a \in C$ exists such that

$$aRb \Rightarrow (C \sim \{a\})Rb,$$

i.e., such that every point covered by a is covered by cells other than a (in C).

25. In a cellular complex (see Figure 7.13), certain labeled points are incident to various cells (vertices, lines, faces, cubes, etc.) as given by the geometry. Such problems arise in switching theory, where the cost of vertices is assumed to be n, that of lines is $n - 1$, of faces $n - 2$, of cubes $n - 3$, etc., for some integer n. For the given complex, with $n = 6$, derive the appropriate covering matrix, listing the costs of the cells.

5. Methods of Solution

At the outset, let us distinguish three questions that may arise in connection with covering problems (A, R, B).

Problem A. Find all nonredundant coverings (see Exercise 10).
Problem B. Find any minimal covering.
Problem C. Determine the cost of a minimal covering.

From Exercise 10 we learn that all the minimal coverings stand among the nonredundant ones. Accordingly, a solution to Problem A yields a solution (in fact, all solutions) to Problem B. Moreover, a simple computation will determine the answer to Problem C if a solution to Problem B is given.

5.1 Reduction Techniques

In the 1950's, switching theorists came face to face with covering problems (see Exercise 25) in their attempts to minimize the cost of logical circuity used in implementing switching (Boolean or logical) functions. Their investigations led to an elementary theory for handling such questions, and in particular, to three now classical results which give rise to extremely powerful reduction techniques. In this section, we retrace the outlines of this theory and provide examples of their application to the three problems stated above. Needless to say, these techniques are then applicable to all classes of covering problems, regardless of their origin.

We begin with some basic definitions pertaining to a general covering problem (A, R, B). The cell a is said to be *essential* (to point b) in the covering problem (A, R, B) if a is the only cell of A which covers the point b. In the corresponding incidence matrix, column b will have but one 1 entry in the row corresponding to a. Our other definitions, similarly, are most easily understood in the matrix terminology. If $R = (r_{ij})$ is the incidence matrix with respect to $A = \{a_1, a_2, \cdots, a_n\}$ and $B = \{b_1, b_2, \cdots, b_m\}$, then b_j *dominates* b_k (written $b_j > b_k$) if

$$r_{ik} = 1 \Rightarrow r_{ij} = 1 \quad \text{for all } i,$$

i.e., if column j has ones in every row that column k has ones (and perhaps in others as well). Similarly, a_i *dominates* $a_k (a_i > a_k)$ provided that

$$r_{kj} = 1 \Rightarrow r_{ij} = 1 \quad \text{for all } j$$

and $\# a_i \leq \# a_k$.

Theorem 3. *Let a be essential in the covering problem (A, R, B). Then $C \cup \{a\}$ is a nonredundant covering of (A, R, B) if and only if C is a nonredundant covering of $(A \sim \{a\}, R, B \sim aR)$, where aR is the set of all points covered by a in the problem (A, R, B).*

Proof. Suppose C is a nonredundant covering for the problem $(A \sim \{a\}, R, B \sim aR)$. First we show that $C \cup \{a\}$ is a covering of (A, R, B). If $b \in B \sim aR$, then a cell exists in C which covers b; otherwise, i.e., if $b \in aR$, then b is covered by a. To see that this covering is nonredundant, we note that a could not possibly be removed—it is essential to some vertex $b \in B$. On the other hand, if we were able to remove some cell $c \in C$ and still have a covering of (A, R, B), then $C \sim \{c\}$ would be a covering of $(A \sim \{a\}, R, B \sim aR)$, contradicting the fact that C was nonredundant. We leave the converse as an exercise. □

Theorem 4. *Let $b > b'$ in the covering problem (A, R, B). Then C is a nonredundant covering of (A, R, B) if and only if it is a nonredundant covering of $(A, R, B \sim \{b\})$.*

PROOF. If C is a nonredundant covering of $(A, R, B \sim \{b\})$ and $b > b'$ in the problem (A, R, B), then as a covering of $(A, R, B \sim \{b\})$, the point b' is covered by some cell $c \in C$. Since c also covers b (because $b > b'$) we have that C is a covering of (A, R, B). If it were redundant, so that $C \sim \{a\}$ was also a covering for some $a \in C$, then it would be all the more a covering for the subset $B \sim \{b\}$, contradicting the nonredundancy of C here. Once again the converse is left as an exercise. □

Theorem 5. *Let $a' > a$ in the covering problem (A, R, B). If C is a minimal covering of $(A \sim \{a\}, R, B)$ then it is a minimal covering of (A, R, B).*

PROOF. First we see, under the stated assumption, that C is a covering of (A, R, B), clearly. For the minimality, suppose instead that a covering C' of (A, R, B) exists with $\#C' < \#C$. We then choose C'' as follows:

$$C'' = \begin{cases} C' & \text{if } a \notin C' \\ (C' \sim \{a\}) \cup \{a'\} & \text{if } a \in C'. \end{cases}$$

Then C'' is a covering for $(A \sim \{a\}, R, B)$—note the use of $a' > a$ here—with

$$\#C'' = \begin{cases} \#C' < \#C & \text{if } a \notin C' \\ \#C' - \#a + \#a' \le \#C' < \#C & \text{if } a \in C'. \end{cases}$$

contradicting the minimality of C as a covering for the problem $(A \sim \{a\}, R, B)$. □

The use of these three theorems as the basis for a matrix reduction technique should be clear. In looking for all the nonredundant coverings (Problem A) for the covering problem (A, R, B) with matrix $R = (r_{ij})$, we may *delete*

(i) *any essential row* (cell a) *and all columns in which that row has a one entry*, remembering to reinsert a after solving the reduced problem,

(ii) *any dominating column*, and if we are looking only for some minimal covering (Problem B), then we may also *delete*

(iii) *any dominated row*.

In the case of Problem B or equivalently Problem C, these three reduction rules can be applied iteratively, continually reducing the size of the matrix under consideration. We call them Rules (i), (ii), (iii), respectively, in the following.

Cost		b_1	b_2	b_3	b_4	b_5	b_6
2	a_1	0	0	0	1	1	1
3	a_2	0	0	0	1	0	1
4	a_3	1	1	0	0	1	0
1	a_4	1	0	0	0	0	0
3	a_5	0	0	1	0	0	1
2	a_6	0	0	1	0	0	0

Figure 7.14

Cost		b_3	b_4	b_6
2	a_1	0	1	1
3	a_2	0	1	1
1	a_4	0	0	0
3	a_5	1	0	1
2	a_6	1	0	0

Figure 7.15

Cost		b_3	b_4
2	a_1	0	1
3	a_2	0	1
1	a_4	0	0
3	a_5	1	0
2	a_6	1	0

Figure 7.16

	b_1	b_2	b_3	b_4
a_1	1	0	0	1
a_2	0	0	1	1
a_3	0	1	1	0
a_4	1	1	0	0

Figure 7.17

In Figure 7.14 we are given a covering problem in which Rule (i) is applicable; row a_3 is essential to column b_2. After the reduction (the deletion of row a_3 and columns b_1, b_2, b_5), we obtain the matrix of Figure 7.15. Now we note that column b_6 dominates column b_4; according to Rule (ii) we may delete b_6, yielding Figure 7.16. Here, row a_4 is dominated (trivially, since it has no one entries) by any row, and may be deleted by Rule (iii). We note that such trivial applications of Rule (iii) may be invoked even in the case of Problem A. Finally, we observe that $a_1 > a_2$ and $a_6 > a_5$, leading us to the minimal covering $\{a_1, a_3, a_6\}$ of cost 8. Note the reinsertion of the essential cell a_3.

It often happens that we are led by these three reduction rules to a matrix in which none of the rules applies, one having no essential rows and no domination of rows or columns. (See Figure 7.17.) Such a covering matrix is said to be *cyclic*. The very existence of such matrices shows that the reduction techniques thus far presented are insufficient for solving the general covering problem. In what follows, we describe two distinct approaches for obtaining solutions even in these cases.

5.2 The Branching Method

The method to be described here is applicable to a general covering problem (A, R, B), though one usually exhausts the reduction techniques of the previous section before resorting to branching. That is because the amount of computation required may grow exponentially with the size of the problem, an unfortunate but unavoidable feature. The method depends on the following theorem whose proof is so similar to those of the previous section that it can safely be omitted.

	b_1	b_2	b_3	b_4	b_5	b_6	b_7
a_1	1	0	0	1	1	1	0
a_2	0	0	1	1	0	0	0
a_3	0	1	0	0	1	0	1
a_4	1	1	0	0	0	0	1
a_5	1	0	0	1	0	0	1
a_6	1	0	0	0	1	0	1
a_7	0	1	1	0	0	1	0
a_8	0	1	0	0	1	1	0

Figure 7.18

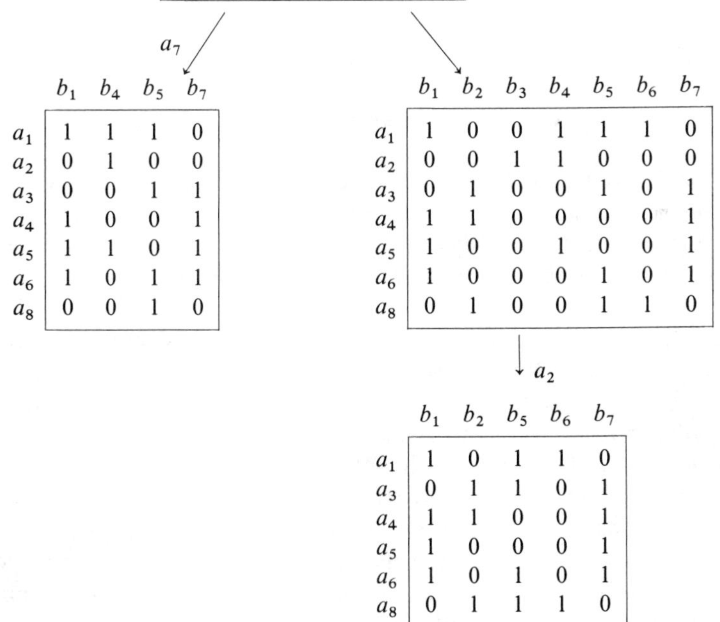

	b_1	b_2	b_3	b_4	b_5	b_6	b_7
a_1	1	0	0	1	1	1	0
a_2	0	0	1	1	0	0	0
a_3	0	1	0	0	1	0	1
a_4	1	1	0	0	0	0	1
a_5	1	0	0	1	0	0	1
a_6	1	0	0	0	1	0	1
a_7	0	1	1	0	0	1	0
a_8	0	1	0	0	1	1	0

a_7

	b_1	b_4	b_5	b_7
a_1	1	1	1	0
a_2	0	1	0	0
a_3	0	0	1	1
a_4	1	0	0	1
a_5	1	1	0	1
a_6	1	0	1	1
a_8	0	0	1	0

	b_1	b_2	b_3	b_4	b_5	b_6	b_7
a_1	1	0	0	1	1	1	0
a_2	0	0	1	1	0	0	0
a_3	0	1	0	0	1	0	1
a_4	1	1	0	0	0	0	1
a_5	1	0	0	1	0	0	1
a_6	1	0	0	0	1	0	1
a_8	0	1	0	0	1	1	0

a_2

	b_1	b_2	b_5	b_6	b_7
a_1	1	0	1	1	0
a_3	0	1	1	0	1
a_4	1	1	0	0	1
a_5	1	0	0	0	1
a_6	1	0	1	0	1
a_8	0	1	1	1	0

Figure 7.19

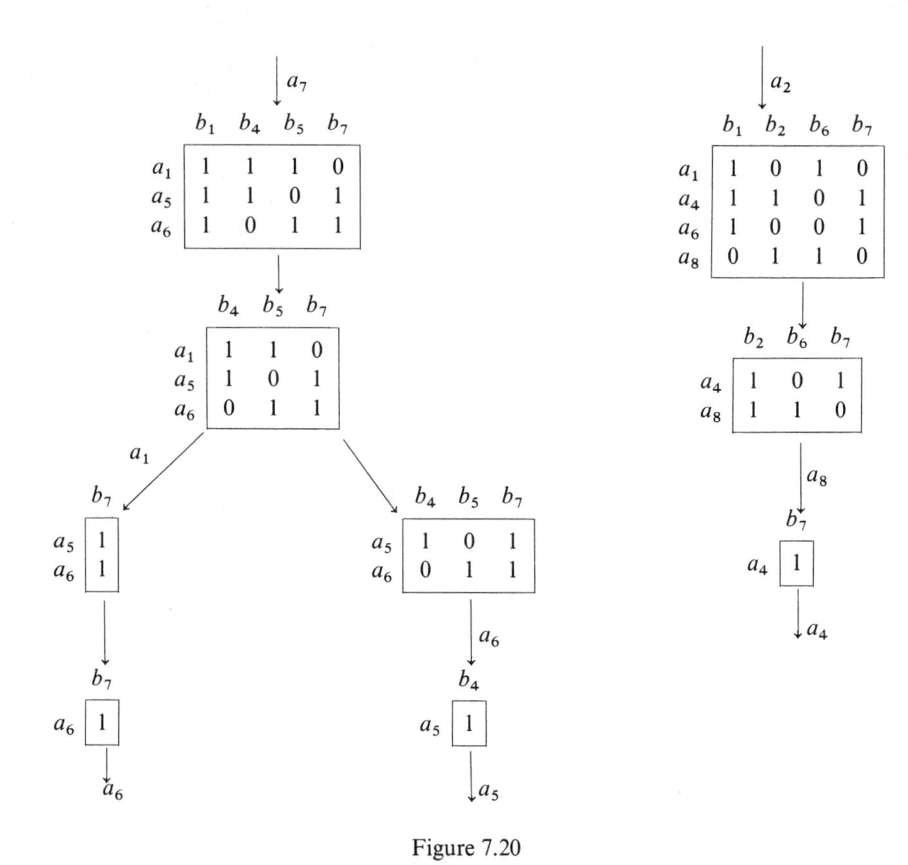

Figure 7.20

Theorem 6. *Let (A, R, B) be a covering problem with inessential cell $a \in A$. Then the nonredundant coverings take the form:*
$C \cup \{a\}$ with C a nonredundant covering of $(A \sim \{a\}, R, B \sim aR)$; or
C with C a nonredundant covering of $(A \sim \{a\}, R, B)$.

Note that if R is a cyclic matrix, then every cell is inessential, but the method is particularly effective if a column exists with exactly two ones. Then in branching into two covering problems as indicated in the theorem, the second problem will immediately exhibit an essential cell, provided that a is taken as one of the two rows in which these ones appeared. For example, consider the covering problem of Figure 7.18. For simplicity, we assume that each cell has cost one. Column b_3 has exactly two ones, in rows a_2 and a_7. Suppose we take $a = a_7$ in Theorem 6. The branching is then into two subsidiary problems, one supposing that a_7 is chosen, the other supposing that it is not. In the second alternative, we arrive at a problem in which a_2 is essential (to b_3) with the resulting reduction via Rule (i). The results are summarized in Figure 7.19. The original problem has been reduced to that of solving two smaller problems.

Now assume that we are only interested in finding some minimal covering. Then we may use Rule (iii), in the first of these problems to delete rows a_2, a_3, a_4, a_8. In the second problem, we may delete column b_5 (dominating column b_6) and then rows a_3 and a_5, giving the matrices at the top of Figure 7.20. Then at the left, column b_1 may be deleted; at the right we may delete b_1 followed by a_1 and a_6 as indicated. Continuing at the right, we now find that first a_8, then a_4 are essential. At the left we must once again apply Theorem 6 to branch according to whether a_1, say, is chosen or not. All in all, *a tree is developed, and we are assured that one of the paths will be labeled by the cells of a minimal covering.* In this case, we obtain $\{a_1, a_6, a_7\}$, $\{a_5, a_6, a_7\}$, $\{a_2, a_4, a_8\}$, each of which is minimal.

5.3 The Algebric Method

Just as in the preceeding section, the method to be presented here is perfectly general and is especially appropriate when we are interested in Problem A—to find all the nonredundant coverings. First, a brief digression is in order. Consider simple English-language statements that are either true or false. Furthermore, suppose we are allowed to form compound statements by connecting two statements with the conjunctions "and" or "or." We assume that such compound statements obey the everyday rules of truth or falsity as follows.

x	y	x and y
false	false	false
false	true	false
true	false	false
true	true	true

x	y	x or y
false	false	false
false	true	true
true	false	true
true	true	true

We further propose that the words "and" and "or" be replaced by the operation symbols $\cdot$ and $+$, respectively. Then an appropriate choice of "variables" for the simple statements a, b, $\cdots$, x, y, etc., will allow the compound statements to be represented by corresponding "polynomials."

We intend to use these ideas in writing a polynomial expression which will accurately describe the conditions under which one obtains a covering for the problem (A, R, B). First we assign a variable (a_i seems the most suitable here) to signify the selection of a given cell of A. For column j of the incidence matrix, we define a polynomial

$$f_j(a_1, a_2, \cdots, a_n) = \sum_{r_{ij}=1} a_i$$

indicating precisely the cells which can cover the point b_j. Thus in the case of Figure 7.17 we have

$$f_1 = a_1 + a_4$$
$$f_2 = a_3 + a_4$$
$$f_3 = a_2 + a_3$$
$$f_4 = a_1 + a_2$$

where the linguistic interpretation is clear: we can cover b_1 with a_1 *or* a_4, b_2 with a_3 *or* a_4, etc. In order to form a polynomial for symbolizing the conditions under which we select cells constituting a covering, we simply form the product of the f_i, i.e., we set

$$f(a_1, a_2, \cdots, a_n) = \prod_{j=1}^{m} f_j(a_1, a_2, \cdots, a_n).$$

For the covering problem of Figure 7.17 we obtain

$$f = (a_1 + a_4)\cdot(a_3 + a_4)\cdot(a_2 + a_3)\cdot(a_1 + a_2),$$

indicating that we must select a_1 *or* a_4, *and* a_3 *or* a_4, *and* a_2 *or* a_3, *and* a_1 *or* a_2. Note that if

$$a_1 = a_3 = \text{true}, \qquad a_2 = a_4 = \text{false},$$

i.e., if we select a_1 and a_3 but not a_2 or a_4, then

$$\begin{aligned} f &= (\text{true} + \text{false})\cdot(\text{true} + \text{false})\cdot(\text{false} + \text{true})\cdot(\text{true} + \text{false}) \\ &= \text{true}\cdot\text{true}\cdot\text{true}\cdot\text{true} \\ &= \text{true}, \end{aligned}$$

indicating that we thereby obtain a covering. On the other hand, assignment

$$a_1 = a_2 = \text{true}, \qquad a_3 = a_4 = \text{false}$$

results in the value

$$\begin{aligned} f &= (\text{true} + \text{false})\cdot(\text{false} + \text{false})\cdot(\text{true} + \text{false})\cdot(\text{true} + \text{true}) \\ &= \text{true}\cdot\text{false}\cdot\text{true}\cdot\text{true} \\ &= \text{false}, \end{aligned}$$

showing that we do not have a covering in this case.

We claim that all the nonredundant coverings of the covering problem (A, R, B) can be obtained by "multiplying out" the polynomial $f(a_1, a_2, \cdots, a_n)$ as just defined. In multiplying out to eventually express f as a sum of products, we of course impose a *distributive law*:

$$x\cdot(y + z) = x\cdot y + x\cdot z.$$

(Note that we have already tacitly assumed the use of *associative* and *commutative* laws:

$$(x + y) + z = x + (y + z), \qquad (x \cdot y) \cdot z = x \cdot (y \cdot z)$$

$$x + y = y + x, \qquad x \cdot y = y \cdot x$$

in this connection.) Furthermore, we employ the following *idempotent* and *absorption* laws:

$$x + x = x, \qquad x \cdot x = x$$

$$x + (x \cdot y) = x, \qquad x \cdot (x + y) = x$$

to shorten the length of our polynomial expressions, thereby removing redundance as well. If we think of the economics of the situation, say in the diner's problem (Section 3), and we suppose that apple pie *or* (apple pie *and* berry pie) meets a certain set of nutritional requirements, then an economical solution would not include both ($x + x : y = x$). The other laws have similar interpretations, both here and in the more general context of the covering problem (A, R, B).

In the case of the diner's problem, we transform the polynomial f from a product of sums to a sum of products as follows:

$$\begin{aligned} f &= (A + B + D + F)(B + E)(A + C + E)(A + C + F) \\ &= (B + AE + DE + EF)(A + C + E)(A + C + F) \\ &= (AB + BC + BE + AE + DE + EF)(A + C + F) \\ &= AB + AE + BC + EF + CDE. \end{aligned}$$

Let us give a detailed description of our first multiplication, just to be sure that the mechanics of the process are clearly understood. Using the distributive law, we obtain

$$\begin{aligned} &(A + B + D + F)(B + E) \\ &= (A + B + D + F)B + (A + B + D + F)E \\ &= AB + BB + DB + FB + AE + BE + DE + FE \\ &= AB + B + BD + BF + AE + BE + DE + EF \\ &= B + AE + DE + EF. \end{aligned}$$

Note particularly the use of the idempotent law ($BB = B$) and the subsequent absorption of the products AB, BD, BF, and BE by B. The last equality for f which is to be read "*A and B, or A and E, or,* $\cdots$," gives a list of all the well-balanced meals (nonredundant coverings). The diner will choose from the meals

A and B = Spanish omelet and chicken enchilada
A and E = Spanish omelet and potato
B and C = chicken enchilada and Waldorf salad
E and F = potato and liver with onions
C and D and E = Waldorf salad and steak and potato.

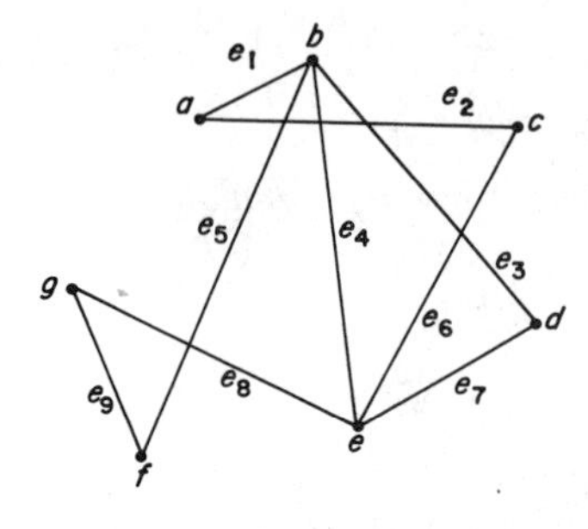

Figure 7.21

	e_1	e_2	e_3	e_4	e_5	e_6	e_7	e_8	e_9
a	1	1	0	0	0	0	0	0	0
b	1	0	1	1	1	0	0	0	0
c	0	1	0	0	0	1	0	0	0
d	0	0	1	0	0	0	1	0	0
e	0	0	0	1	0	1	1	1	0
f	0	0	0	0	1	0	0	0	1
g	0	0	0	0	0	0	0	1	1

Figure 7.22

Note once more that in each case, the selected meals include a 1 from every column—they are indeed coverings.

As indicated in the foregoing example, to multiply two (nonredundant) sums of products polynomials to obtain another is a straightforward procedure. By *nonredundancy* here, we mean one in which the use of idempotent and absorption laws can no longer be applied to shorten the polynomial expression. Our claim for the resulting sum of products representation of f is the following statement, which we offer without proof.

Theorem 7. *Let (A, R, B) be a covering problem with product of sums polynomial $f(a_1, a_2, \cdots, a_n)$. Then the nonredundant sum of products representation of f contains as its summands all the nonredundant coverings of the problem (A, R, B).*

As a final illustration, we consider the graph of Figure 7.1 and we first seek to determine all of the nonredundant covering sets. The graph is reproduced as Figure 7.21 with the incidence matrix for the appropriate covering problem given in Figure 7.22. Conversion of the polynomial $f(a, b, c, d, e, f, g)$ from a product of sums to a sum of products is summarized in the calculation.

$$\begin{aligned}
f &= (a+b)(a+c)(b+d)(b+e)(b+f)(c+e)(d+e)(e+g)(f+g) \\
&= (a+bc)(b+d)\cdots \\
&= (ab+bc+ad)(b+e)\cdots \\
&= (ab+bc+ade)(b+f)\cdots \\
&= (ab+bc+adef)(c+e)\cdots \\
&= (bc+abe+adef)(d+e)\cdots \\
&= (bcd+adef+bce+abe)(e+g)\cdots \\
&= (adef+bce+abe+bcdg)(f+g) \\
&= adef+bcef+abef+bceg+abeg+bcdg.
\end{aligned}$$

	a	b	c	d	e	f	g
A	0	0	1	1	0	0	1
B	0	0	1	1	0	1	0
C	1	0	0	1	0	0	1
D	1	0	0	1	0	1	0
E	1	0	0	0	1	1	0
F	0	1	1	0	0	0	1

Figure 7.23

At this stage, we can state that the covering number $\alpha = 4$. Furthermore, according to our earlier discussion surrounding the proof of Theorem 2, we may immediately claim that the complementary sets

$$A = \{c, d, g\}, \qquad D = \{a, d, f\}$$
$$B = \{c, d, f\}, \qquad E = \{a, e, f\}$$
$$C = \{a, d, g\}, \qquad F = \{b, c, g\}$$

are the maximal independent sets of the graph.

If we now wish to determine the minimal colorings of the graph, then as discussed earlier, we may consider the covering problem given by the incidence matrix of Figure 7.23.
The same algebraic method can then be used to compute

$$\begin{aligned}
f &= (C + D + E)F(A + B + F)(A + B + C + D)E(B + D + E)(A + C + F) \\
&= (CF + DF + EF)(A + B + F) \cdots \\
&= (CF + DF + EF)(A + B + C + D) \cdots \\
&= (CF + DF + AEF + BEF)E \cdots \\
&= (CEF + DEF + AEF + BEF)(B + D + E) \cdots \\
&= (CEF + DEF + AEF + BEF)(A + C + F) \\
&= CEF + DEF + AEF + BEF
\end{aligned}$$

We conclude that $k = 3$ and we may use CEF, for example, to color the vertices with three colors

$$C: a, d, g \qquad E: e, f \qquad F: b, c$$

according to ideas presented earlier. Note in conclusion, however, that cells E, F were essential, and by our reduction techniques, this information could have been used to achieve a simplification of the problem at the outset.

The algebraic method of solution presented here leads naturally to a discussion of the so-called "free distributive lattices." The elements of these finite lattices $L(n)$ are formal polynomial expressions in symbols $x_1, x_2, \cdots, x_n$, such as those we have considered here. The elements of $L(3)$ are as listed below:

$$\begin{array}{ccc}
0 & x_1 + x_2 & x_1x_2 + x_3 \\
1 & x_1 + x_3 & x_1x_3 + x_2 \\
x_1 & x_2 + x_3 & x_2x_3 + x_1 \\
x_2 & x_1x_2 & x_1x_2 + x_1x_3 \\
x_3 & x_1x_3 & x_1x_2 + x_2x_3 \\
x_1x_2x_3 & x_2x_3 & x_1x_3 + x_2x_3 \\
x_1 + x_2 + x_3 & & x_1x_3 + x_2x_3 + x_1x_2.
\end{array}$$

More generally, we have

$$|L(3)| = 20$$
$$|L(4)| = 168$$
$$|L(5)| = 7581$$
$$|L(6)| = 7{,}828{,}354$$

whereas a general formula for the number of elements in $L(n)$ is not known. The reader who would like to learn more about these lattices or about lattices in general may wish to refer to the book by Birkhoff. See also Prather [8, Section 4.3].

Exercises

26. In each of the following, use the three reduction rules either to find a minimal covering or to reduce the given incidence matrix to a cyclic matrix.

(a)

cost		b_1	b_2	b_3	b_4	b_5	b_6	b_7	b_8
4	a_1	1	1	0	1	1	0	0	0
7	a_2	0	0	1	0	0	0	1	0
4	a_3	0	1	0	0	1	1	0	0
3	a_4	0	0	0	0	0	0	0	1
4	a_5	1	0	0	1	0	1	1	0
6	a_6	0	1	0	0	0	0	0	1
2	a_7	0	1	0	0	0	0	0	0

(b)

cost		b_1	b_2	b_3	b_4	b_5	b_6	b_7
4	a_1	1	0	1	0	1	0	0
2	a_2	0	0	1	0	0	0	1
4	a_3	1	1	0	0	1	1	0
7	a_4	0	0	1	1	1	0	0
3	a_5	0	1	0	1	0	1	0
3	a_6	1	0	0	1	1	0	0
5	a_7	0	1	1	0	0	1	1

(c)

cost		b_1	b_2	b_3	b_4	b_5	b_6	b_7	b_8	b_9	b_{10}	b_{11}
4	a_1	1	0	0	0	0	0	0	0	0	0	0
8	a_2	0	0	1	0	0	0	1	0	1	1	0
2	a_3	0	1	0	0	0	1	0	0	0	0	0
8	a_4	0	1	1	0	0	1	0	0	1	0	0
6	a_5	0	0	0	1	1	0	0	0	0	0	1
5	a_6	0	1	1	0	0	0	0	1	0	0	0
4	a_7	0	0	0	0	1	0	0	1	0	0	1
2	a_8	0	0	0	0	1	0	1	1	0	1	0
4	a_9	0	0	1	1	0	0	0	0	0	1	1

(d) The incidence matrix resulting from Exercise 10(b).

†27. Prove the converse for Theorem 3.

†28. Prove the converse for Theorem 4.

†29. Prove Theorem 5.

†30. Prove Theorem 6.

31. Use the branching method to find all the nonredundent dominating sets for the following graphs.
    (a) Figure 7.1.
    (b) Figure 7.2.
    (c) Figure 7.3.
    (d) Exercise 1(a).
    (e) Exercise 1(b).
    (f) Exercise 2(a).
    (g) Exercise 2(b).
    (h) Exercise 2(c).

32. Repeat using the algebraic method.

33. Use the branching method to find all the nonredundant covering sets for the following graphs.
    (a) Figure 7.1.
    (b) Figure 7.2.
    (c) Figure 7.5.
    (d) Exercise 1(a).
    (e) Exercise 1(b).
    (f) Exercise 2(a).
    (g) Exercise 2(b).
    (h) Exercise 2(c).

34. Repeat using the algebraic method.

35. Use the branching method to find all the nonredundant covering sets *of edges* (see Exercise 2.7) for the following graphs.
    (a) Figure 7.1.
    (b) Figure 7.2.
    (c) Figure 7.5.
    (d) Exercise 1(a).

(e) Exercise 1(b).
(f) Exercise 2(a).
(g) Exercise 2(b).
(h) Exercise 2(c).

36. Repeat using the algebraic method.

37. Use the results of Exercise 33 and the branching method to find all minimal colorings for the following graphs.
(a) Figure 7.1.
(b) Figure 7.2.
(c) Figure 7.5.
(d) Exercise 1(a).
(e) Exercise 1(b).
(f) Exercise 2(a).
(g) Exercise 2(b).
(h) Exercise 2(c).

38. Repeat using the results of Exercise 34 and the algebraic method.

39. Use the branching method to find all the well-balanced meals for the diner's problem (Section 3).

40. Let $G = (V, E)$ be any graph. A collection $\{v_1, v_2, \cdots, v_s\}$ of vertices is called a *clique* if $\{v_i, v_j\} \in E$ for all $1 \leq i, j \leq s$, $i \neq j$. Discuss methods for finding the following.
(a) All maximal cliques.
(b) Some clique of largest size.
(c) The size of a largest clique (i.e., the clique number $\beta' = \beta'(G)$).
Hint: See Exercise 5.

41. For the incidence matrix resulting from Exercise 25, find a minimal covering using the branching method.

†42. Develop a computer program for implementing the branching method.

†43. Repeat for the algebraic method.

†44. Develop a computer program for implementing the reduction techniques—Rules (i), (ii), (iii).

45. Using the branching method, find a minimal coloring of the following maps.
(a) Exercise 22.
(b) Exercise 23.

46. Repeat using the algebraic method.

References

[1] K. Appel and W. Haken, "Every planar map is four colorable. Part 1: Discharging," *Illinois J. Math.* vol. 21, pp. 429–490, 1977.
[2] ——, "Every planar map is four colorable. Part II: Reducibility," *Illinois J. Math.*, vol. 21, pp. 491–567, 1977.

[3] C. Berge, *The Theory of Graphs*. New York: Wiley, 1962.
[4] G. Birkhoff, *Lattice Theory*, 3rd. ed. Providence, RI: AMS Colloquium Publications, American Mathematical Society, 1973.
[5] S. Even, *Algorithmic Combinatories*. New York: MacMillan, 1973.
[6] D. Givone, *Introduction to Switching Theory*. New York: McGraw-Hill, 1970.
[7] O. Ore, *Theory of Graphs*. Providence, RI: AMS Colloquium Publications, American Mathematical Society, 1962.
[8] R. Prather, *Discrete Mathematical Structures for Computer Science*. Boston: Houghton Mifflin, 1976.

Notes for the Instructor

Objectives. This chapter presents a survey of the occurrence of covering problems in various fields, particularly those arising from graph-theoretic models. Two distinct solution methodologies are treated, a branching method and an algebraic method. This chapter is designed to supplement courses in discrete mathematical modeling, applications of graph theory, discrete mathematical structures, and the like.

Prerequisites. Familiarity with standard set-theoretic notation and with elementary algebraic manipulation.

Time. This chapter can be treated in two or three class periods depending on the depth of coverage.

CHAPTER 8

A Pulse Process Model of Athletic Financing

E. L. Perry*

1. Introduction

This module illustrates the process of mathematical modeling using pulse processes as a tool. We assume that students have received some instruction in the cyclical nature of math modeling such as the discussion given in Roberts [1, ch. 1]. Other prerequisites include a knowledge of matrix algebra (simple multiplication and addition of matrices) and, in the last section only, the ability to compute the eigenvalues of an $n \times n$ matrix. Students without a background in eigenvalues may simply omit this section. Any interested reader that does not have this background should first consult any standard text in linear algebra or Roberts [1].

Our model is of a hypothetical university and should not be taken as a true model of any one school. It is given only to illustrate the general modeling process, although we do feel that modifications of the model could actually be used in the early stages of the decisionmaking process in any university to decide the general directions and changes that should be made in athletic financing. Since we are modeling a ficticious university, no data are available to actually test the predictions. In actual practice, the predictions should be tested against past data. We will be content with illustrating the process by two trips through the modeling cycle. Although our first model, a signed digraph model, will yield some information, it will be inadequate to answer the questions that we have in mind. The second model, a pulse process model, will give us some answers to our questions but we shall make no attempt to test their validity.

The reader should do as many of the exercises as possible as some new

* TRW Inc., Webster, TX 77598.

material is introduced in the problems. A dagger (†) preceding the exercise number means the exercise is slightly more difficult than the others. Although some computer problems are included, these are clearly marked and may be omitted by students without a computer background. The student is also urged to modify this model to fit his or her own school.

2. The Problem

Many colleges today are facing problems with athletic financing. In order to compete on a national level, the schools are forced to lay out huge sums on scholarships, dormitory facilities, stadium and field houses, and coaches' salaries while the only major sources of income are the football gate receipts and donations. Other sports such as baseball, basketball, tennis, and track may produce some revenue, but in general they do not pay their own way. We hope to construct a model which will show the relationships between the variables associated with financing an athletic program at a typical private university of about 8000 students which plays in a major athletic conference and has a minimal women's athletic program in basketball, track, and tennis along with the usual nonincome producing men's sports program in track, baseball, basketball, tennis, and golf.

We also want our model to be able to make forecasts about the effects of making changes in the levels of certain variables. For example, suppose that the amount budgeted for men's nonincome sports is cut by 10%, and the amount budgeted for women's nonincome sports is increased by 10%. What effect will this have on the levels of the variables over the next ten years? In general, we hope to be able to solve the "forecasting problem." If a "pulse" (positive or negative) is introduced in one or more of the variables, what will the effect be on other variables in the future?

3. The First Model

The first problem encountered is the identification of the variables. Through interviews with coaches, athletic directors, administrative officials, and students one can produce a huge list of measurable (and some not so easily measurable) quantities that could be called variables, such as amount of donations, athletic revenue, quality of play, expenditures to run the program, recruiting expenses, size of staff, total number of scholarships, average home attendance, budgets for women's track, basketball, and tennis, and budgets for men's track, basketball, tennis, and golf. We want our list of variables to be a manageable size, so we have decided to consolidate the amount budgeted for women's track, women's basketball, and women's tennis into one variable called "the amount budgeted for women's athletics." Similarly, we have one variable called "the amount budgeted for men's nonincome sports" which includes the budgets for track, baseball, basketball, tennis, and golf. If one so desired, another model could be constructed in which some of these variables were broken into parts. For example, it could be argued that the amount budgeted for basketball should be included as a separate variable. We choose not to do this, because at the school under consideration, basketball does not produce enough revenue to pay its own way. The major athletic revenue is derived from football and we choose to concentrate on the income producing aspect by including six variables which deal with the football program. The complete list of ten major variables that we will consider is as follows.

v_1 Amount of annual donations and gifts given to the athletic program.
v_2 Amount budgeted for women's athletics.
v_3 Athletic revenue, including gate receipts from all sports at the university (the most income comes from football).
v_4 Quality of play in football. Although this variable is difficult to measure, it could be done by assigning a number corresponding to the number of teams which finished below the university team in final conference standings.
v_5 Amount budgeted for male nonincome sports.
v_6 Expenditures to run program including salaries.
v_7 Recruiting expenditures for football.
v_8 Size of coaching staff for football.
v_9 Total scholarships given for football.
v_{10} Average home attendance at football games. This variable is included separately because it is thought to be a good measure of the influence of the football program on the alumni and university.

In general, to produce a signed digraph model or a pulse process model, one gathers a group of "experts" in the field and either uses a questionnaire or interviews to identify variables. If this list is too large, one then consolidates (those which behave similarly) or eliminates (those which have little effect) some of these to reduce the list to a manageable size. More formal methods can be used to produce the list of variables, but it is not our purpose in this chapter to go into these aspects of building the model. The interested reader should consult Roberts [2].

In our model, we will look at the status of the system at discrete times $t = 0, 1, 2, 3, \cdots$ which in this case may be thought of as years. Although our unit of time could have been hours, days, months, and so on, it seems appropriate to take increments of one year because most of the changes we want to study would not appear in a shorter period. Also, in a university, most budgets, schedules, etc., are made on a yearly basis. Furthermore, we will not need to be able to actually put a numerical value on the variables at any one time since we will only be concerned with increases or decreases which occur in the values of the variables rather than the values themselves. To know that such a value could be found if one so desires is sufficient.

We assume that a pulse (an increase or decrease) in one of these variables at time t leads directly to an increase or decrease (another pulse) in some of the other variables at time period $t + 1$. These relationships are allowed to take two forms.

1) An increase (decrease) at v_i leads directly to an increase (decrease) at v_j within one time period. For example, in our list of variables, it seems reasonable that an increase in v_1 at time t would lead to an increase in v_7 at time $t + 1$, while a decrease in v_1 would lead to a decrease in v_7 within one year. This type of relationship is recorded in our model as follows. We let $v_1, v_2, \cdots, v_{10}$ be represented by dots (or vertices), and then a solid line with an arrow, called a *positive arc*, is drawn from v_i to v_j. For example, in our model a positive arc will occur from v_1 to v_7.

2) An increase (decrease) at v_i leads directly to a decrease (increase) at v_j within one time period. For example, in our list of variables, it seems reasonable that an increase at v_6 at time t would lead directly to a decrease at v_7 at time $t + 1$, while a decrease at v_6 at time t would lead to an increase at v_7 at time $t + 1$. Relationships of this type are recorded in our model by drawing a dotted line called a *negative arc* from v_i to v_j. For example, a negative arc will occur from v_6 to v_7 in our model.

A pulse at v_i at time t may cause no direct change at v_j at time $t + 1$. In this case we do not put any line from v_i to v_j.

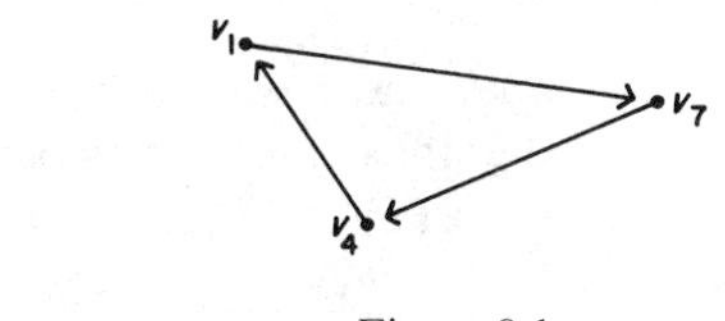

Figure 8.1

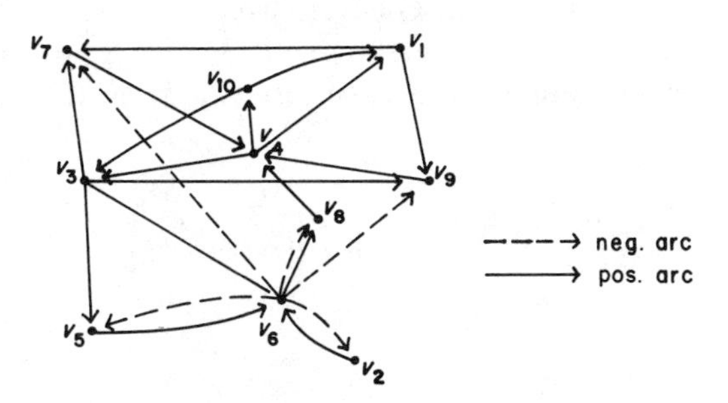

Figure 8.2. v_1: Amount of Donations; v_2: Amount Budgeted for Women's Athletics; v_3: Athletic Revenue; v_4: Quality of Play in Football; v_5: Amount Budgeted for Male Nonincome Sports; v_6: Expenditures To Run Program; v_7: Recruiting Expenditures for Football; v_8: Size of Coaching Staff for Football; v_9: Total Scholarships Given for Football; v_{10}: Average Home Attendance at Football Games.

We also must understand what we mean when we say that a pulse at v_i leads *directly* to a certain type of pulse at v_j. Let us illustrate by an example. An increase at v_7 taking place at time t might eventually cause an increase at v_1. However, this is not a "direct" effect because the increase at v_7 causes an increase at v_4 which, in turn, causes an increase at v_1. The relationship between v_1, v_7, and v_4 in our model is summarized in Figure 8.1. Note that no line of any kind exists from v_7 to v_1 because we do not feel that a pulse at v_7 will directly affect v_1 within one time period. Of course, all judgements of this type are subjective and hence open to debate. Figure 8.2 shows the relationships as the author views them. You can no doubt argue that changes should be made in this diagram, and in fact you will be asked to do this in some of the exercises. A model such as those shown in Figures 8.1 and 8.2, in which exist a set of points (or vertices) $v_1, v_2, \cdots, v_n$ and a set of lines (or arcs) $e_1, e_2, \cdots, e_m$, each connecting a pair of points, is called a *directed graph* or simply a *digraph*. In the particular case when each line may be thought of as positive or negative, we have a signed digraph. Let us associate a "+" sign with positive arcs and a "−" sign with negative arcs in our model so that the resulting structure, shown in Figure 8.2 is a signed digraph.

Another way to record the relationships between the variables would be to use a matrix $M = (a_{ij})$ where

$$a_{ij} = \begin{cases} 1, & \text{if there is a solid line from } v_i \text{ to } v_j. \\ -1, & \text{if there is a dotted line from } v_i \text{ to } v_j. \\ 0, & \text{if there is no line from } v_i \text{ to } v_7. \end{cases}$$

This matrix is called the *signed adjacency matrix* for the signed digraph. Some examples of simple signed digraphs and their associated signed adjacency matrices are shown in Figure 8.3. (The adjacency matrix of an unsigned digraph is the same as the signed adjacency matrix if we assume there are no dotted (negative) lines.) Note especially the second example in Figure 8.3, where an arc (positive in this case) occurs from v_1 to itself. Such an arc is called a *loop*. A loop in a signed digraph indicates that a pulse at that vertex leads directly to a new pulse at the same vertex in the next time period.

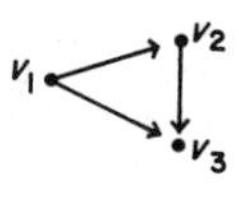

$$M = \begin{pmatrix} 0 & 1 & 1 \\ 0 & 0 & -1 \\ 0 & 0 & 0 \end{pmatrix}$$

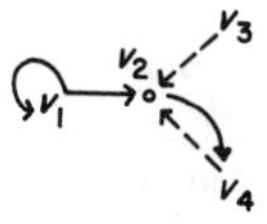

$$M = \begin{pmatrix} 1 & 1 & 0 & 0 \\ 0 & 0 & 0 & 1 \\ 0 & -1 & 0 & 0 \\ 0 & -1 & 0 & 0 \end{pmatrix}$$

Figure 8.3

EXERCISES

1. For each of the signed digraphs below, find the signed adjacency matrix.

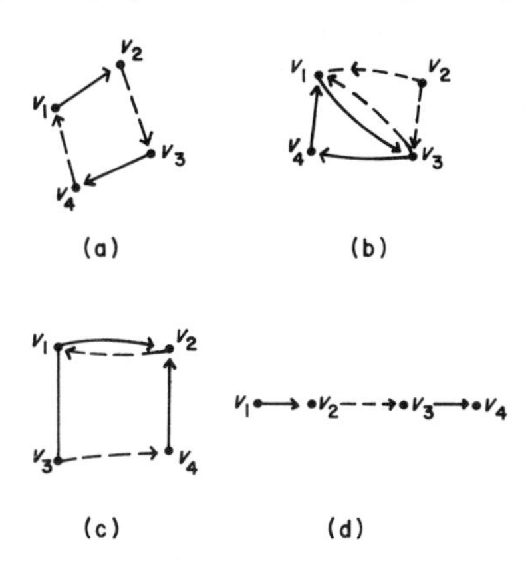

2. The *transpose* of a matrix M is obtained by interchanging the rows and columns. For a signed digraph, the transpose of the signed adjacency matrix is called the *cross-impact matrix*. Find the cross-impact matrix for each signed digraph in Exercise 1.

3. Argue that a loop should occur at v_4 in the signed digraph in Figure 8.2.

4. Do you feel the positive arc from v_9 to v_4 in Figure 8.2 is justified?

5. Should a positive arc exist from v_8 to v_{10} in Figure 8.2? Explain.

6. Should a negative arc exist from v_4 to v_7 in Figure 8.2? Explain.

7. Let D denote a digraph (not signed) in which the vertices $v_1, v_2, \cdots, v_n$ represent children in a certain classroom and an arc (positive) is drawn from v_i to v_j if and only if v_i chooses v_j as a friend. If M denotes the adjacency matrix for D and M' denotes the transpose of M, explain the significance of the i, jth entry in M'.

†8. Let D, M, M' be the same as in Exercise 7. If N denotes the matrix product $M \cdot M'$, explain the significance of the i, jth entry in N.

9. Let D, M, M', N be the same as in Exercise 8. If P denotes the matrix product $M' \cdot M$, will P and N be the same? Explain the significance of the i, jth entry of P.

†10. If M denotes the adjacency matrix for a digraph (unsigned), and $M^2 = M \cdot M$, what is the significance of the i, jth entry of M^2? What if M is the adjacency matrix for a signed digraph?

4. Analysis of the First Model

Figure 8.2 gives the first model that we will analyze. The signed adjacency matrix for Figure 8.2 is given in Figure 8.4.

Let (v_i, v_j) denote the arc from v_i to v_j. A sequence of vertices and arcs

	v_1	v_2	v_3	v_4	v_5	v_6	v_7	v_8	v_9	v_{10}
v_1	0	0	0	0	0	0	1	0	1	0
v_2	0	0	0	0	0	1	0	0	0	0
v_3	0	0	0	0	1	1	1	0	1	0
v_4	1	0	1	0	0	0	0	0	0	1
v_5	0	0	0	0	0	1	0	0	0	0
v_6	0	−1	0	0	−1	0	−1	−1	−1	0
v_7	0	0	0	1	0	0	0	0	0	0
v_8	0	0	0	1	0	1	0	0	0	0
v_9	0	0	0	1	0	0	0	0	0	0
v_{10}	1	0	1	0	0	0	0	0	0	0

Figure 8.4

such as v_1, (v_1, v_9), v_9, (v_9, v_4), v_4, (v_4, v_{10}), v_{10} is called a *path*. The idea is to follow the arcs from v_1 to v_{10}.

More generally, if D is a signed (or unsigned) digraph with vertices v_1, $v_2, \cdots, v_n$, a sequence $u_1, (u_1, u_2), u_2, \cdots, u_p, (u_p, u_{p+1}), u_{p+1}$, where each u_i is one of $v_1, v_2, \cdots, v_n$ and each (u_i, u_{i+1}) is an arc of D, is called a *path* in D. The length of the path is P and the path is *simple* if $u_1, u_2, \cdots, u_p$ are all distinct. The path is *closed* if $u_{p+1} = u_1$. A simple closed path is sometimes called a *cycle*. When there is no danger of confusion, we simply write the vertices that appear in a path or a cycle and supress the arcs. For example, in Figure 8.2, v_1, v_9, v_4, v_{10} would be a simple path from v_1 to v_{10}, while $v_1, v_9, v_4, v_{10}, v_1$ would be a cycle beginning and ending at v_1. Also, $v_3, v_6, v_5, v_6, v_8, v_4, v_3$ would be a closed path beginning and ending at v_3, but it is not simple because v_6 is repeated. The cycles play an important part in the analysis of a signed digraph model in that they give "feedback" resulting from pulses. For example, in Figure 8.2, we observe that an increase at v_1 at $t = 0$ will come back to v_1 as another increase at $t = 3$ through the cycle v_1, v_9, v_4, v_1. Observe that the increase at v_1 at $t = 0$ causes v_9 to increase at $t = 1$ which in turn causes v_4 to increase at $t = 2$. Then the pulse is transmitted as an increase at v_1 at $t = 3$. Also, the increase in v_1 at $t = 0$ would return to v_1 as further increases at $t = 4$ and $t = 5$ respectively through the cycles $v_1, v_9, v_4, v_{10}, v_1$ and $v_1, v_9, v_4, v_3, v_{10}, v_1$. Can you identify other cycles which begin and end at v_1?

Associated with any cycle or path is a sign obtained by multiplying the signs of all arcs in the cycle or path (count a positive arc as $+1$ and a negative arc as -1). Positive cycles are sometimes called *augmenting cycles* because any increase at a vertex on a positive cycle comes back through the cycle to the vertex as a further increase (why?). For example, in Figure 8.2, v_1, v_9, v_4, v_1 and v_1, v_9, v_{10}, v_1 are both augmenting cycles. Similarly, a negative cycle is sometimes called an *inhibiting cycle* because an increase at any vertex on the cycle will return through the cycle as a decrease. In Figure 8.2, we see that v_6, v_5, v_6 and v_6, v_7, v_4, v_3, v_6 are inhibiting cycles. As a matter of fact, a little reflection will show that *any cycle involving* v_6

Time Period	v_1	v_2	v_3	v_4	v_5	v_6	v_7	v_8	v_9	v_{10}
0	+	0	0	0	0	0	0	0	0	0
1	0	0	0	0	0	0	+	0	+	0
2	0	0	0	+ +	0	0	0	0	0	0
3	+	0	+	0	0	0	0	0	0	+
4	+	0	+	0	+	+	+ +	0	+ +	0
5	0	−	0	+ +	− +	+ +	− + +	−	− + +	0

Figure 8.5

will be an inhibiting cycle. This follows because (1) the only arcs in our model leaving v_6 are negative and any cycle involving v_6 must use exactly one of these, and (2) no other negative arcs exist. Since the cycles through v_6 are of lengths 2 and 4, we can conclude that any increase in expenditures to run the athletic program will be followed in two- and four-year periods by decreases in expenditures. However, these decreases would later return to v_6 as increases through the inhibiting cycles. Thus our model predicts that an increase at v_6 will lead to an "oscillating" effect with periodic increases and decreases at v_6. Note however that this does not take into account the magnitude of the increases and decreases. Will the increases grow larger each time? Will they grow smaller? Our model leaves these questions unanswered.

Let us now take an increase at v_1, the amount of donations to the program, and trace the pulses through a five-year period (Figure 8.5). Here a + (or −) under v_i at time t indicates an increase (or decrease) in v_i at t. If two + signs occur under v_i at time t, this means two lines terminating at v_i contributed increases to v_i at time t. Similarly, a + and − sign indicate two lines terminating at v_i, one contributing an increase, the other a decrease. A zero indicates no change. In general, the two corresponding to time t is obtained by looking at row $t - 1$. The nonzero values in row $t - 1$ indicate increases or decreases which then lead to increases or decreases in row t. Two plus signs are treated as a single increase, an entry consisting of some plus signs is treated as a single increase, and an entry consisting of some plus signs together with some minus signs is ignored because we cannot tell if it is an increase or a decrease or if the changes cancel each other. The results after five years can be obtained by looking at the columns. If the column headed by v_i contains all plus signs, it is certain that v_i will increase. If the column contains all negative signs, it is certain that v_i will decrease. The model where a column contains some pluses together with some negatives will not tell us whether an increase or a decrease takes place in that variable, for the magnitude of the changes must be considered.

Since we hear a great deal of talk today about increasing the budget for women's athletics, let us observe what would happen if v_2 were increased at $t = 0$ by some outside donation. It appears from Figure 8.6 that v_1, v_3,

Time	v_1	v_2	v_3	v_4	v_5	v_6	v_7	v_8	v_9	v_{10}
0	0	+	0	0	0	0	0	0	0	0
1	0	0	0	0	0	+	0	0	0	0
2	0	−	0	0	−	0	−	−	−	0
3	0	0	0	− − −	0	− − −	0	0	0	0
4	−	+	−	0	+	0	+	+	+	−
5	−	0	−	+ + +	−	+ − + +	− −	0	− −	0

Figure 8.6

Time	v_1	v_2	v_3	v_4	v_5	v_6	v_7	v_8	v_9	v_{10}
0	0	+	0	0	−	0	0	0	0	0
1	0	0	0	0	0	− +	0	0	0	0

Figure 8.7

and v_{10} would all decrease during the five-year period and the other vertices including v_2 would be relatively unchanged. Thus our model predicts that it would be undesirable to put a huge amount of money into women's athletics without otherwise changing the structure of the program.

Our next example (Figure 8.7) shows that this model cannot always give meaningful information. At $t = 0$, we increase v_2 and decrease v_5. Since the only change at $t = 1$ is in v_6 and we cannot call it an increase or decrease, no further information is obtained. We might point out that in Figure 8.7 an increase in v_2, the women's athletic budget, coupled with a decrease in v_5, the men's nonincome sports budget, is exactly what many groups are now proposing as an answer for the problem of financing women's sports programs in college. To handle situations where the key vertices contain both plus and minus signs in the same entry, one needs to know the magnitudes of the increases and decreases. A model that takes this into account is obtained in the next section.

Exercises

11. Identify all cycles that pass through vertex v_3 in Figure 8.2. How many are augmenting cycles and how many are inhibiting cycles?

12. Identify all cycles that pass through vertex v_4 in Figure 8.2. How many are positive cycles and how many are negative cycles?

13. Suppose our short-range goal is to increase the average home attendance at football game in two years. List three ways that our model says we could accomplish this. Which of the three do you feel would be the easiest to actually do? Why?

14. Use a chart similar to that shown in Figure 8.5 to trace pulses through the signed digraph model of Figure 8.2 for five years beginning with an increase in v_5 at $t = 0$.

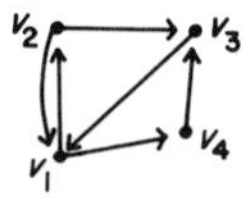

Fig. 8.8

15. Use a chart similar to that shown in Figure 8.5 to trace pulses through the signed digraph model of Figure 8.2 for five years beginning at $t = 0$ with a simultaneous decrease at v_1 and an increase at v_2. Summarize (in words) the results of your chart. Would this be a good strategy to adopt in order to increase v_2?

16. Consider the digraph in Figure 8.8. Let M be its adjacency matrix. Compute M^2 and M^3. What does the 1 in the first row, first column of M^2 tell you? What does the 1 in the 2, 2 position of M^2 tell you? Why is there a 0 in the 3, 3 position of M^2? What is the significance of the numeral in the i, i positive of M^3?

†17. Show that if D is a digraph (not signed) and M is its adjacency matrix, then the i, i entry in M^K gives the number of closed paths of length K there are in D beginning and ending at v_i.

18. Let M and D be as in Exercise 17. What would the i, i entry in $M + M^2 + M^3 + \cdots + M^P$ tell you?

†19. Suppose D is a signed digraph and M is its signed adjacency matrix. What information does the i, i entry in M^K give?

†20. Suppose D is a signed digraph and M is its signed adjacency matrix. What would the i, i entry in $M + M^2 + \cdots + M^P$ tell you?

21. Make a signed digraph model similar to that shown in Figure 8.2 except add the following variables.

    v_{11} Amount budgeted for basketball.
    v_{12} Quality of play in basketball.
    v_{13} Total number of scholarships given for basketball.
    v_{14} Recruiting expenses for basketball.

    You will need to determine relationships between these variables as well as relationships between these variables and v_1 through v_{10}.

5. The Second Model

Since our signed digraph model proved to be inadequate to answer some of our questions, we try to construct a better model. Because the inability of our model to measure magnitudes of pulses seems to cause problems in the signed digraph model, we will assume that for each i and j such that an arc (v_i, v_j) exists in our model, there is a "weight" a_{ij} which can be associated with the arc. Further, when the value of v_i is increased K units at time t, v_j will increase (or decrease) Ka_{ij} units at time $t + 1$. In other words, the pulse in v_i at time t is related in a linear fashion to the pulse in v_j at time $t + 1$. A digraph which has a weight associated with each arc is called a *weighted digraph*. Note that a signed digraph can be considered as a weighted digraph in which the weights are chosen from $+1$ and -1. Figure 8.9 gives an illustration of a weighted digraph.

Associated with each weighted digraph is a matrix (a_{ij}), where

$$a_{ij} = \begin{cases} \text{weight of arc } (v_i, v_j), \text{ when the arc is in the digraph.} \\ 0, \text{ if the arc } (v_i, v_j) \text{ is not in the digraph.} \end{cases}$$

This matrix is called the (*weighted*) *adjacency matrix* for the digraph. The weighted adjacency matrix for Figure 8.9 is given in Figure 8.10.

Suppose, in Figure 8.9, a pulse of 1 unit is introduced at $t = 0$ in v_1. At $t = 1$, v_2 will be increased by (0.3) (1) = 0.3 units while v_5 will be decreased (note the negative sign) by (1.1) (1) = 1.1 units. The pulse of 0.3 at v_2 at $t = 1$ will lead to an increase of (2.1) (0.3) = 0.63 units at v_4 and also to an increase of (1.2) (0.3) = 0.36 in v_3 at $t = 2$. Also the pulse of -1.1 units at v_5 at $t = 1$ will lead to a pulse (decrease) of $(-1.1)(1.2) = -1.32$ units in v_4 at $t = 2$. Thus v_4 is influenced by two pulses at $t = 2$ and the total

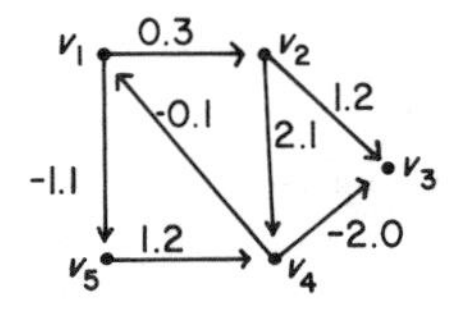

Figure 8.9

$$\begin{array}{c} \\ v_1 \\ v_2 \\ v_3 \\ v_4 \\ v_5 \end{array} \begin{array}{c} \begin{array}{ccccc} v_1 & v_2 & v_3 & v_4 & v_5 \end{array} \\ \begin{bmatrix} 0 & 0.3 & 0 & 0 & -1.1 \\ 0 & 0 & 1.2 & 2.1 & 0 \\ 0 & 0 & 0 & 0 & 0 \\ -0.1 & 0 & -2.0 & 0 & 0 \\ 0 & 0 & 0 & 1.2 & 0 \end{bmatrix} \end{array} = A$$

Figure 8.10

Time	v_1	v_2	v_3	v_4	v_5
$t = 0$	1	0	0	0	0
$t = 1$	1	0.3	0	0	-1.1
$t = 2$	1	0.3	0.36	-0.69	-1.1
$t = 3$	1.07	0.3	1.74	-0.69	-1.1
$t = 4$	1.07	0.32	1.74	-0.69	-1.18
$t = 5$	1.07	0.32	1.76	-0.74	-1.18

Figure 8.11

effect is $-1.32 + 0.63 = -0.69$ units. Assuming the value of each variable was 0 to begin and v_1 was increased 1 unit at $t = 0$, Figure 8.11 shows the value of the variables at times $t = 0$ to $t = 5$. (All values rounded to nearest hundredth.)

Since our purpose is to study only the changes in the values of the variables from their initial values and not the values themselves, we may assume each variable has a starting value of 0. Note that

$$\begin{aligned}(\text{New value of } v_i \text{ at time } t+1) &= (\text{old value of } v_i \text{ at time } t)\\ &\quad + (\text{pulse in } v_i \text{ at time } t+1).\end{aligned} \tag{1}$$

Also,

$$(\text{Pulse in } v_i \text{ at time } t+1) = \sum_j a_{ji}\,(\text{Pulse in } v_j \text{ at time } t), \tag{2}$$

where the sum is taken over all j such that (v_j, v_i) is an arc of our model. Since we assumed that $a_{ji} = 0$ when no arc exists from v_j to v_i, the sum can be taken over all j. Let us agree to let $P(t) = (p_1(t), p_2(t), \cdots, p_n(t))$ be a vector which gives the pulses at time t. Thus $p_i(t)$ gives the pulse at vertex i at time t. Also $V(t) = (v_1(t), v_2(t), \cdots, v_n(t))$ will be a vector which gives the values of the vertices at time t. So, $v_i(t)$ will be a symbol for the value of vertex v_i at time t. By convention, $V(0)$ will denote the values after the initial pulses have been added. With this notation, equation (1) can be rewritten,

$$V(t+1) = V(t) + P(t+1). \tag{1$'$}$$

Also, (2) becomes

$$P(t+1) = \left(\sum_{j=1}^{n} a_{j1}p_j(t), \sum_{j=1}^{n} a_{j2}p_j(t), \cdots, \sum_{j=1}^{n} a_{jn}p_j(t)\right). \tag{2$'$}$$

However, note that $\Sigma_{j=1}^{n} a_{ji}p_j(t)$ is simply the sum of the products of the elements of $P(t)$ times the corresponding elements in the ith column of the weighted adjacency matrix. Thus (8.2′) can now be written as a product of a vector and a matrix:

$$P(t+1) = P(t)A, \tag{2$''$}$$

where A is the weighted adjacency matrix. Note that in order to use equations (1′) and (2″) one must know four things:

(1) the vector $V(0)$ showing initial values after the outside pulses have been added;
(2) the vector of "outside initial pulses," $P(0)$;
(3) the weighted adjacency matrix A;
(4) the meaning of "one unit" change in the variables.

We take care of (4) first. In each of the variables v_1–v_{10} listed in Figure 8.2, let one unit be 10% of the starting value of the variable. Thus a pulse of $+2$ in v_1 would indicate an increase in the value of v_1 by 20% of its starting

	Amount of Annual Donations and Gifts	Amount Budgeted for Women's Athletics	Athletic Revenue	Quality of Play	Amount Budgeted for Male Non-Income Sports	Expenditures to Run Program and Salaries	Recruiting Expenditures	Size of Coaching Staff	Total Scholarships	Average Home Attendence
1. Amount of Annual Donations and Gifts	0	0	0	0	0	0	0.4	0	0.1	0
2. Amount Budgeted for Women's Athletics	0	0	0	0	0	0.1	0	0	0	0
3. Athletic Revenue	0	0	0	0	1.0	0.4	0.7	0	0.2	0
4. Quality of Play	0.31	0	0.50	0	0	0	0	0	0	1.1
5. Amount Budgeted for Male Nonincome Sports	0	0	0	0	0	0.2	0	0	0	0
6. Expenditures to Run Program and Salaries	0	−0.5	0	0	−1.0	0	−0.8	−0.6	−0.07	0
7. Recruiting Expenditures	0	0	0	0.5	0	0	0	0	0	0
8. Size of Coaching Staff	0	0	0	0.2	0	0.4	0	0	0	0
9. Total Scholarships	0	0	0	0.75	0	0	0	0	0	0
10. Average Home Attendence	0.5	0	0.8	0	0	0	0	0	0	0

Figure 8.12

value, while a decrease, -3, in v_1 would indicate a decrease in v_1 by 30% of its starting value. With this understanding, we can usually take $V(0) = P(0)$, which actually assumes a starting value of 0 for each variable and then adds the initial pulse $P(0)$. Then the values $V(t)$ give us the change in each variable at time t as a percentage change of their initial values.

To illustrate, let A be as given in Figure 8.10, $V(0) = (1, 0, 0, 0, 0)$ and $P(0) = (1, 0, 0, 0, 0)$. Note that the first element of $P(0)$ being 1 while the other elements are all 0 indicates that initially v_1 is increased by 1 unit while no other value is pulsed. Also, the starting values of the vertices are assumed to be 0 but when v_1 is increased by 1 unit at $t = 0$ we get $V(0) = (1, 0, 0, 0, 0)$. Thereafter, (1′) and (2″) can be used to compute the pulses and values.

$$P(1) = P(0)A = (0, .3, 0, 0, -1.1)$$
$$V(1) = V(0) + P(1) = (1, .3, 0, 0, -1.1)$$
$$P(2) = P(1)A = (0, 0, .36, -.69, 0)$$
$$V(2) = V(1) + P(2) = (1, .3, .36, -.69, -1.1).$$

Note that $V(1)$ and $V(2)$ agree with $t = 1$ and $t = 2$ in Figure 8.11.

Thus to modify the signed digraph model of the preceding section to fit this model we need to determine the weights a_{ij}. This is a subjective thing much as was the determination of the signs of the arcs in our earlier model. In general, two methods can be combined to produce the weights.

1) Analysis of past data will often provide a hint as to the weights. By observing where outside increases and decreases in values were made, the mathematician can get a hint as to the weights. This is complicated by the fact that outside pulses do *not* usually occur one at a time. Changes in values of the vertices are usually multiple in any one time period.

2) Consultation with "experts" such as coaches, athletic directors, and university officials will often give insights into what the weights should be. Roberts [1] gives information on combining the judgements of a group to reach a common opinion. We shall not go into these methods here.

Young [3] used a combination of the above techniques to conclude that the weights in Figure 8.12 would be reasonable for a typical private school such as the one we are modeling. We choose to give the weights in the form of the weighted adjacency matrix.

As an example, take A to be as given in Figure 8.12, $V(0) = (1, 0, 0, 0, 0, 0, 0, 0, 0, 0)$, and $P(0) = V(0)$ and use (1′) and (2″) to get two values and pulses:

$$P(1) = P(0)A = (0, 0, 0, 0, 0, 0, .4, 0, .1, 0)$$
$$V(1) = V(0) + P(1) = (1, 0, 0, 0, 0, 0, .4, 0, .1, 0)$$
$$P(2) = P(1)A = (0, 0, 0, .095, 0, 0, 0, 0, 0, 0)$$
$$V(2) = V(1) + P(2) = (1, 0, 0, .095, 0, 0, .4, 0, .1, 0).$$

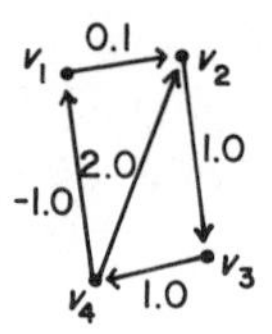

Figure 8.13

Thus our model predicts that an increase of 10% (one unit) in the yearly donations to the athletic department will result in two years in increases of 95% in the quality of play, a 4% increase in recruiting expenditures and a 1% increase in scholarships, provided that no other outside pulses are added to the system during this period.

EXERCISES

22. In the weighted digraph in Figure 8.13 assume $V(0) = (1, 0, 0, 0) = P(0)$ and compute $V(1)$, $V(2)$, $V(3)$, $V(4)$ and $P(1)$, $P(2)$, $P(3)$, $P(4)$.

23. In a weighted digraph, one can speak of augmenting and inhibiting cycles as was done in signed digraphs. Simply consider the product of the signs of the weights (not the magnitudes of the weights). However, in a weighted digraph, we can also classify cycles as *expanding* or *damped.* An expanding cycle is one in which the product of the weights on the cycle is greater than one in absolute value. A cycle is damped if the product of the weights on the cycle is less than one in absolute value. Classify the following cycles as expanding or damped. Also in each one, let $V(0) = P(0) = (1, 0)$ and compute $V(1)$, $V(2)$, $V(3)$, $V(4)$, $V(5)$ and $P(1)$, $P(2)$, $P(3)$, $P(4)$, $P(5)$.

What can one say about a pulse as it passes through a damped cycle? Expanding cycle?

24. Give an example of a cycle that is neither expanding *n* or damped (refer to Exercise 23 for definitions of these terms).

25. In the weighted digraph model given in Figure 8.12, determine what the model predicts would happen over a five-year period if an increase of 10% is made in recruiting expenditures at $t = 0$ and no other outside pulses are introduced during this period.

26. In Figure 8.12, find what the model predicts will happen over a five-year period if the amount budgeted for women's athletics is increased by 10% and at the same time the amount budgeted for men's nonincome sports is decreased by 10%.

†27. Let D be a weighted digraph with weighted adjacency matrix A. Show that $P(t) = P(0)\, A^t$ holds for $t = 1, 2, 3, \cdots$.

†28. Let D be a weighted digraph with n vertices and weighted adjacency matrix A. Assume further that $V(0) = P(0)$. Show that $V(t) = P(0)\,[I + A + \cdots + A^t]$, $t = 1, 2, 3, \cdots$ where I the $n \times n$ identity matrix.

29. Redraw Figure 8.2, putting weights on all the arcs as given in Figure 8.12. Consider all cycles that pass through v_1. Classify them as augmenting or inhibiting and then as expanding, damped, or neither. (See Exercise 23.)

30. (Computer Problem) Write a computer program which use equations (1′) and (2″) to compute $V(t)$ and $P(t)$ for our model of athletic financing.

31. (Computer Problem) Write a computer program which uses the results of Exercises 27 and 28 to compute $V(t)$ and $P(t)$ for our model of athletic financing. Compare this program with the one done in Exercise 30. Which is more efficient?

6. Analysis of the Second Model

The weighted digraph model given in Figure 8.12 allows us to obtain an answer to the forecasting problem described at the end of Section 2. Using a high-speed computer and equations (1′) and (2″), we can forecast values of the vertices at future times corresponding to any initial outside pulses. Consider again the situation where the budget for women's sports is increased by 10%, while the budget for men's nonincome sports is decreased by 10% at $t = 0$. Figure 8.14 shows a forecast that was obtained using the basic program given in the Appendix, the weighted adjacency matrix as shown in Figure 8.12, and $V(0) = P(0) = (0, 1, 0, 0, -1, 0, 0, 0, 0, 0)$. To conserve space, not all of the values of t are reprinted here even though the computer printed the values for all t from 1 through 25.

Notice that the value of v_2 stays slightly above 1 while the value of v_5 stays slightly above -1, indicating that over the next 25-year period we could expect the budget for women's athletics to remain up about 10% while the budget for men's nonincome sports would remain down but it would be down only around 9% from the eighth year through the twenty-fifth year. Another interesting observation is that the model predicts that v_1, donations to the program, would actually increase by a small fraction from the fourth year to the twenty-fifth year under this arrangement. This information can most readily be seen in Figure 8.15 where we have graphed v_1, v_2, and v_5 against t for the 25-year period.

Although we have not graphed it, we can also observe that v_3, athletic revenue, would increase about 1% during the eighth through twenty-fifth year. All of these predictions assume no further "external" changes in the system.

Our model also allows us to compare strategies. Consider, for example,

t Equals		1							
0	1	0	0	−1	−0.1	0	0	0	0
t Equals		2							
0	1.05	0	0	−0.9	−0.1	0.08	0.06	0.007	0
t Equals		3							
0	1.05	0	0.05725	−0.9	−0.051	0.08	0.06	0.007	0
t Equals		4							
0.0177475	1.0255	0.028625	0.05725	−0.949	−0.051	0.0408	0.0306		
0.00357	0.062975								
t Equals		5							
0.049235	1.0255	0.079005	0.0291975	−0.920375	−0.06356	0.0679365			
0.0306	0.0110697	0.062975							
t Equals		8							
0.0543382	1.013532	0.0872988	0.065211	−0.9230478	−0.0485839				
0.0689967	0.0162379	0.0149782	0.0983732						
t Equals		12							
0.0781182	1.012243	0.1253839	0.0631131	−0.8892081	−0.0421386				
0.1014929	0.0146919	0.0243476	0.1085612						
t Equals		16							
0.0878662	1.014813	0.1409511	0.0653931	−0.8529121	−0.0313255				
0.1351812	0.0177758	0.0328804	0.100426						
t Equals		20							
0.0864403	1.016232	0.1386209	0.0772206	−0.8269404	−0.0171617				
0.1594244	0.0194783	0.0391507	0.0870253						
t Equals		24							
0.0792178	1.01416	0.1270403	0.095309	−0.8208116	−0.0046836				
0.165878	0.0169924	0.0415595	0.0802615						
t Equals		25							
0.0696765	1.002342	0.1118637	0.1175071	−0.8682762	−0.0051332				
0.1243622	0.0028101	0.0336577	0.1048399						

Figure 8.14

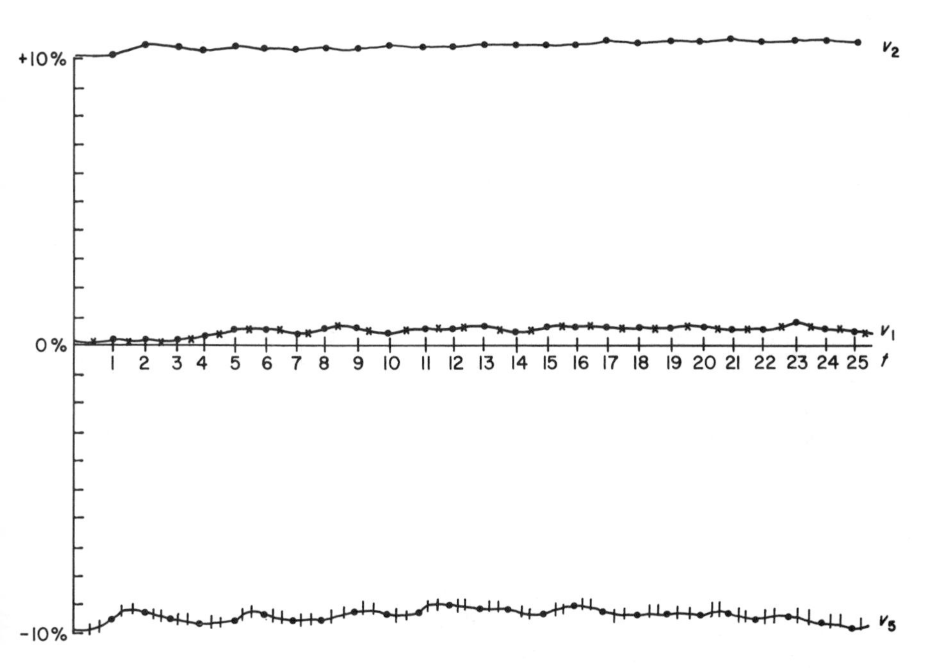

Figure 8.15. Values of v_1, v_2, v_5 under Pulse Process in Which v_2 Is Increased 10 Percent and v_5 Is Decreased 10%.

what would happen if we simply increase v_2 by 10% without making other changes. (A pulse process in which one vertex is increased one unit while the others are held fixed is sometimes called a *simple pulse process.*) Figure 8.16 shows that the graphs of v_1, v_2, v_3 under this change. Note that v_2 stays just under a 10% increase most of the time while v_1 and v_3 both drop off at $t = 3$ and are down about 1% from $t = 4$ through $t = 25$. Thus it seems that it might actually be better for revenue to the program to decrease v_5 as we increase v_2 rather than seeking to increase v_2 alone. However, a small change (like 1%) may not be meaningful in view of the many "guesses" that have been made as to the various weights.

Of course, some question remains about the validity of the model. Some of the assumptions made originally may or may not be valid. For example, is it valid to assume that a pulse of K units leads in a linear fashion to pulses at adjacent vertices? Perhaps it would be more realistic to associate a function, $F_{ij}(t)$, of the time t with each arc and assume that a pulse of K units at vertex i leads through the arc (v_i, v_j) to a pulse of $KF_{ij}(\mathrm{t})$ units at v_j in the next time period. Another option would be to associate a function, $F_{ij}(v_i(t))$, dependent on the level of v_i, with the arc (v_i, v_j) and assume that a pulse of K units at v_i at time t leads through the arc (v_i, v_j) to a pulse of $KF_{ij}(v_i(t))$ units at v_j at time $t + 1$. This would certainly seem justified for certain arcs such as (v_9, v_4) in view of the limit of 30 scholarships imposed by the NCAA on member colleges. We might, for example, take

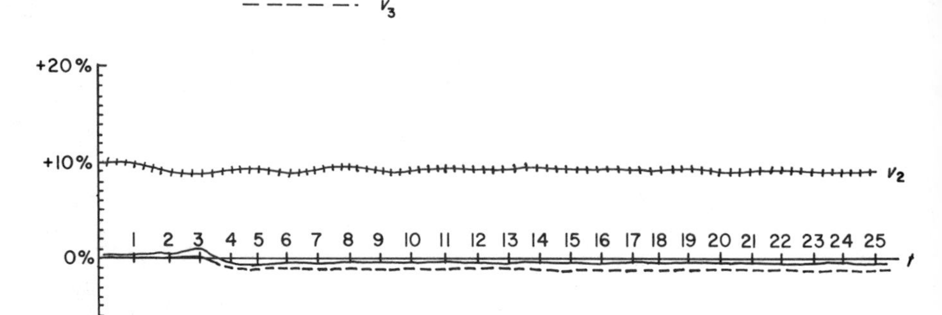

Figure 8.16. Values of v_1, v_2, v_3 under Simple Pulse Process in Which v_2 Is Increased 10% at $t = 0$. v_2 in General Stays Just under +10%; v_1 and v_5 Both Drop Off at $t = 3$; v_3 Is Down Approximately 1% from $t = 4$ on; v_1 Is Down about 0.7% from $t = 4$ on.

$$F_{94}(v_9) = \begin{cases} 0, & \text{when } v_9(t) \geq 30 \\ 0.75, & \text{when } v_9(t) < 30. \end{cases}$$

Another change that one might wish to make in the model is the time period. It may be the case that some changes are slower to take effect than others. For example, an increase at v_4 is very likely to affect v_1 within one year while it may take more than one year for an increase at v_7 to affect v_4. One could associate with each arc (v_i, v_j) both a function $F_{ij}(v_i(t))$ and a number t_{ij} and assume that a pulse of K units at time t leads to a pulse of $KF_{ij}(v_i(t))$ units in v_j at time $t + t_{ij}$. In such a model, it might be desirable to make the basic time period shorter. These are all ways in which one might make the model more complicated but, at the same time, remove some of the simplifying assumptions which were made. Is this a desirable thing to do? In some cases the answer may be no because the true test of any model is the validity of its predications. Before making the model more complicated (and hence more expensive to build and use), one should take some of the predictions of the present model and test them against past data. Since our model is of a hypothetical university, we do not have any past data with which to work. However, if this model were adapted to a real university, the testing step would now be in order.

Exercises

32. Use the computer printout given in Figure 8.14 and graph the values of v_3, v_4, and v_{10}. What predictions does the model yield? Do they seem reasonable?

33. We suggested in the module that it might be quite reasonable to attach a function

to the arc (v_9, v_4) rather than just a weight. Are there other such arcs? If so, tell which ones and explain your reasoning.

34. Identify some arcs (v_i, v_j) in the pulse process model where it may take longer than one year for a change in v_i to affect v_j.

35. One shortcoming of the model we have produced is that it does not allow for outside pulses at any time except $t = 0$. Modify the model to allow an outside pulse at any time at any vertex. *Hint:* Use a vector $P^0(t) = (p_1^0(t), p_2^0(t), \cdots, p_n^0(t))$ in which $p_i^0(t)$ represents the outside pulse at v_i at time t. What changes should be made in equations (1′) and (2″)?

36. (Computer Problem) Write a computer program to compute $V(t)$ and $P(t)$ in the model described in Exercise 35.

37. What would be the effect over the next five-year period of the simple pulse in which v_8 is increased by 1 unit?

38. Describe the effect over the next five-year period of the simple pulse in which v_7 is increased by 1 unit.

39. Write analogs of equations (1) and (2) for the model in which two numbers a_{ij} and t_{ij} are associated with each arc (v_i, v_j) and it is assumed that a pulse of K units at v_i at time t leads through the arc to a pulse of Ka_{ij} units at v_j at time $t + t_{ij}$.

†40. Form a new model in which the weights a_{ij} are replaced with functions F_{ij}, each of which depends on the level of vertex i. List all the functions (some of which might be constant) and justify your reasoning.

41. A pulse process model is said to be *pulse stable under all simple pulse processes* if there exists some constant K such that $|p_i(t)| \leq K$ holds for $t = 1, 2, 3, \cdots$ and for all i under any simple pulse process. Which of the following are pulse stable under all simple pulse processes?

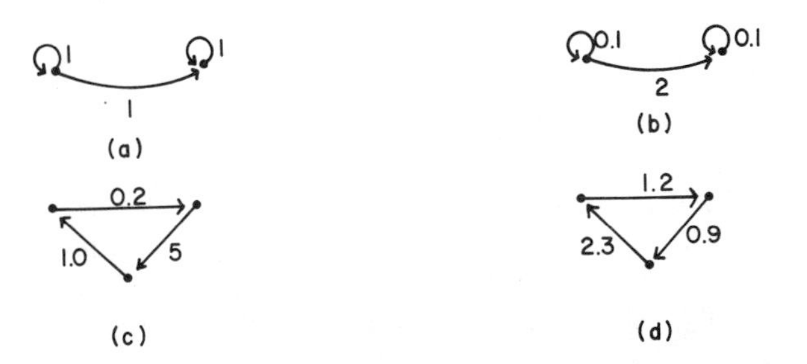

42. Suppose a weighted digraph D consists entirely of a cycle. Under what conditions will it be pulse stable under all simple pulse processes? Justify your answer.

†43. Determine for which values of a, b, c, d the weighted digraph is pulse stable under all simple pulse processes.

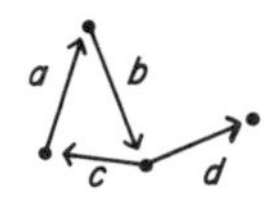

7. Some More Difficult Questions

Recall that a simple pulse process is one in which $P(0)$ has one entry 1 and all others zero. A weighted digraph D is said to be *pulse stable under all simple pulse processes* if a constant K exists such that for each vertex v_i and each value of t, $|p_i(t)| \leq K$ holds under all simple pulse processes. Similarly, a weighted digraph D is said to be *value stable under all simple pulse processes* if a constant K exists such that for each vertex v_i and for each value of t, $|v_i(t)| \leq K$ holds under all simple pulse processes. Intuitively speaking, a weighted digraph is pulse stable under all simple pulse processes if under simple pulse processes, the pulses do not keep growing without bound. The weighted digraph is value stable under all simple pulse processes if, under simple pulse processes, the values do not keep growing without bound. For example, consider the weighted digraph in Figure 8.17.

If a single vertex v_i is pulsed by one unit at $t = 0$ (a simple pulse process), then at $t = 3$, v_i will again be pulsed, but this time by 0.2. Also, at $t = 6$, the value of v_i is increased by 0.04. In general at $t = 3p$, v_i will be pulsed by $(0.2)^p$. Since $(0.2)^p \leq 1$ for each value of p, the system is pulse stable under all simple pulse processes. Assuming that $v_i = 0$ for $i = 1, 2, 3$, we have that

$$v_i(t) = 1 + (0.02) + (0.02)^2 + \cdots + (0.02)^p \tag{3}$$

when $t = 3p$. Note that

$$(0.02)v_i(t) = (0.02) + (0.02)^2 + \cdots + (0.02)^{p+2}. \tag{4}$$

Subtracting (8.4) from (8.3), one obtains

$$0.08v_i(t) = 1 - (0.02)^{p+1}. \tag{5}$$

Thus

$$v_i(t) = \frac{1}{0.08} - \frac{(0.02)^{p+1}}{0.08} < 12.5. \tag{6}$$

Equation (6) shows that for $t = 3p$ we have $v_i(t)$ bounded by 12.5. It follows that $v_i(t) < 12.5$ for other values of t as well. (Why?) Similar reasoning will show that $v_j(t) < 12.5$ for all j and all t. Thus Figure 8.17 gives an example of a weighted digraph that is both pulse and value stable under all simple pulse processes.

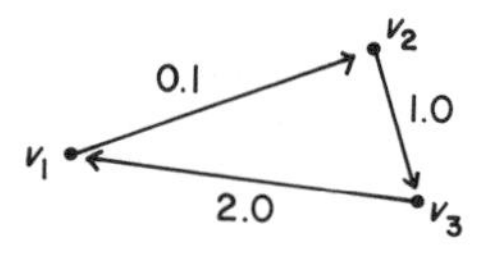

Figure 8.17

If all the weights of Figure 8.17 were changed to 1.0 then the system would be pulse stable but not value stable under all simple pulse processes.

Note that a system which is value stable under all simple pulse processes is necessarily pulse stable as well because if K is a bound for the values,

$$\begin{aligned} |p_i(t)| &= |v_i(t) - v_i(t-1)| \\ &\leq |v_i(t)| + |v_i(t-1)| \\ &\leq K + K = 2K. \end{aligned}$$

Thus $2K$ is a bound for the pulses.

It turns out that the eigenvalues of the weighted adjacency matrix can give valuable clues as to whether a given system is pulse and/or value stable. We state, without proof, three theorems which relate eigenvalues to stability. Proofs may be found in Roberts, [1] or Roberts and Brown [4].

Theorem 1 (Roberts and Brown). *If a weighted digraph is pulse stable under all simple pulse processes, then every eigenvalue of D has magnitude at most one.*

Theorem 2 (Roberts and Brown). *Suppose D is a weighted digraph with all nonzero eigenvalues distinct. If every eigenvalue has magnitude at most one, than D is pulse stable under all simple pulse processes.*

Theorem 3 (Roberts and Brown). *Suppose D is a weighted digraph. Then D is value stable under all simple pulse processes if and only if D is pulse stable under all simple pulse processes and one is not an eigenvalue of D.*

For example, note the adjacency matrix for Figure 8.17 shown in Figure 8.18. The eigenvalues are computed as follows.

$$\begin{aligned} \det(A - \lambda I) &= \det\begin{pmatrix} -\lambda & 0.1 & 0 \\ 0 & -\lambda & 1.0 \\ 2.0 & 0 & -\lambda \end{pmatrix} \\ &= -\lambda \det\begin{pmatrix} -\lambda & 1.0 \\ 0 & -\lambda \end{pmatrix} - 0.1 \det\begin{pmatrix} 0 & 1.0 \\ 2.0 & -\lambda \end{pmatrix} \\ &= -\lambda^3 + 0.1(2) \\ &= -\lambda^3 + 0.2. \end{aligned}$$

$$A = \begin{pmatrix} 0 & 0.1 & 0 \\ 0 & 0 & 1.0 \\ 2.0 & 0 & 0 \end{pmatrix}$$

Figure 8.18

The eigenvalues are the cube roots of 0.2. Since these are all distinct and have magnitude less than one, by Theorem 2, the system is pulse stable under all simple pulse processes and by Theorem 3, it is value stable under all simple pulse processes since one is not an eigenvalue.

Pulse and value stability are assumed to be necessary characteristics in a real life system. It seems reasonable that there should be some limits to growth in most real world situations. Using a computer program, we can compute the eigenvalues for the matrix of Figure 8.12. They are (correct to four decimals) 0, 0, 0, 0, $0.1454 - 0.5163i$, $0.1454 + 0.5163i$, -0.7746, 0.8604, $-0.1883 + 0.9748i$, $-0.1883 - 0.9748i$. Here the nonzero enginvalues are distinct, and they all have magnitude less then one so (by Theorem 2) pulse stability and (by Theorem 3) value stability exist under all simple pulse processes.

Since pulse and value stability are desirable in most real world situations, it would be nice if we knew what to do when a model is found in which there is pulse and/or value instability. Suppose, for example, in our model of athletic financing, it turned out that we did not have value stability. Administrative officials of the university might be interested in pinpointing changes that could be made to introduce value stability to the system. In order to give this information with our present knowledge, the mathematician must be able to tell which values in the weighted adjacency matrix could be changed to make all nonzero eigenvalues distinct and less than one in magnitude. More generally, what is needed here is a geometric type of theorem which tells us from the structure of the digraph how to change the weights to achieve stability. Suppose we keep all weights fixed except one, say a_{ij}, and allow this one to vary. What happens to the eigenvalues? More generally, what happens to the pulse and value stability as this weight varies? These questions, which are unanswered by mathematicians at the present time, are sometimes called *sensitivity analysis* problems. How sensitive is the system to changes in the weights? In the exercises are some questions dealing with sensitivity analysis. For a recent reference on the subject, see Roberts [3].

Exercises

44. In each of the following, decide if the weighted digraph is pulse stable or value stable under all simple pulse processes. Justify your answer using Theorems 1–3.

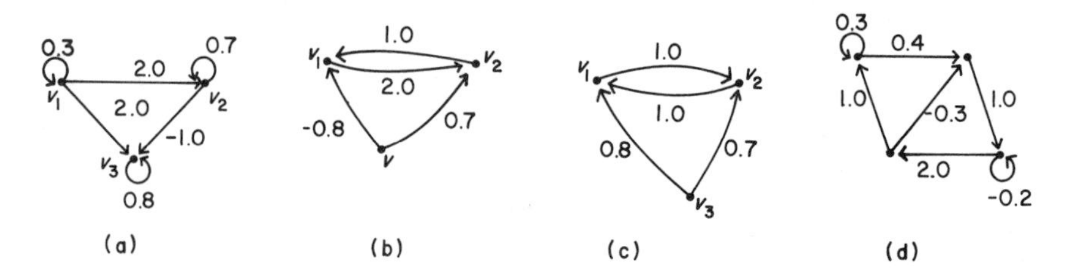

45. For what values of x will the following digraph be pulse stable under all simple pulse processes?

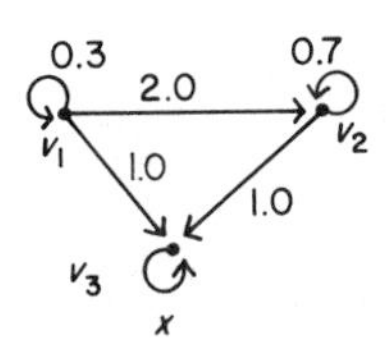

46. For what values of y will the following digraph be pulse stable under all simple pulse processes?

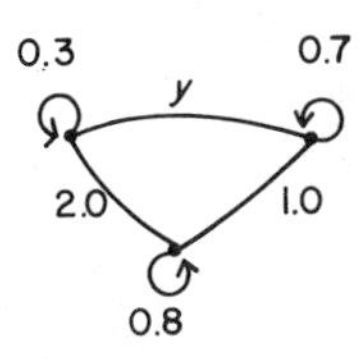

47. Consider the following weighted digraph.

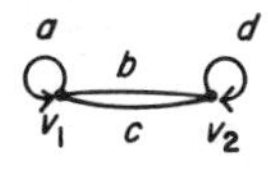

(a) Suppose $a = b = d = 1$. For what values of c will the system be pulse stable under all simple pulse processes?
(b) Suppose $a = c = d = 1$. For what values of b will the system be pulse stable under all simple pulse processes?
(c) Suppose $a = 0.5$, $b = 0$, $c = 1$. For what values of d will the system be pulse stable under all simple pulse processes?
†(d) If $(a + d)^2 - 4(ad - bc) = 0$, what can be said about pulse and/or value stability of the system?

48. Suppose the eigenvalues of a weighted digraph are as given below. What can be said about pulse and value stability under simple pulse processes? Justify your answer.
(a) $0, 1, -1, i, -i$.
(b) $0, 0, 1, -1, 2$.
(c) $0, 0.6i, -0.6i$.
(d) $0, 0, -0.9i, 0.9i$.

†49. What can be said about the structure of weighted digraph D in which all the eigenvalues are 0?

References

[1] F. S. Roberts, *Discrete Mathematical Models*. Englewood Cliffs, NJ: Prentice-Hall. 1976.
[2] ——, *Weighted Digraph Models for Energy Use and Air Pollution in Transportation*

Systems. The Rand Corporation # R-1578-NSF. Santa Monica, CA, Dec. 1974. (Abridged version appeared in *Environment and Planning*, vol. 7, pp. 703–724, 1975.)

[3] ——, "Structures and stability in weighted digraph models," *Annals New York Academy Science*, vol. 32, pp. 64–77, 1979.

[4] F. S. Roberts and T. A. Brown, "Signed digraphs and the energy crisis," *American Math. Monthly*, vol. 82, pp. 577–594, 1975.

[5] Dean, Young, "A weighted digraph model for private collegiate athletic finances," unpublished paper, Baylor University, Waco, TX, 1975.

Appendix

Basic Program for Computing Values

```
10   Mat Input A(10, 10)
20   Mat Input P(1, 10)
30   Mat Input V(1, 10)
40   FOR I = 1 to 25
50   MAT Q = P * A
60   MAT W = V + Q
70   PRINT "T EQUALS", I
80   MAT PRINT W;
90   MAT P = Q
100  MAT V = W
110  NEXT I
120  END
```

Notes for the Instructor

Objectives. This chapter is intended for use in a finite mathematics or a modeling course at the sophomore level with or without a computer prerequisites.

Prerequisites. A knowledge of matrix algebra is necessary. In order to complete the last section, the student needs to know how to compute the eigenvalues of an $n \times n$ matrix. (However, this section can be omitted for students without this knowledge.) Although the computer programming problems are included in the exercises, these may be omitted if the chapter is used in a noncomputer-oriented course.

Time. It can be completed in about three hours of class time or it could be done in a one hour class period if students are mature enough to do some work on their own.

CHAPTER 9

Traffic Equilibria on a Roadway Network

Elmor L. Peterson*

PART I

1. The Scenario

Blairsville and Bolivar are two small residential towns that house most of the workers at the nearby Consolidated Coal Mines. As indicated by the road map in Figure 9.1, the surrounding area has only three connecting roads because of its extremely mountainous, uninhabitable nature. Travelers between the two towns and the coal mines generally use Valley Road and Mountainside Road. Although Highbridge Road provides a shortcut from Bolivar to both Blairsville and the Consolidated Coal Mines, only vehicles lighter than 2500 pounds can use its relatively weak bridge.

The area residents are generally satisfied with their roadway system, except for the severe traffic congestion during the morning and evening rush hours. The traffic during those two hours is due almost exclusively to workers traveling between their homes and the mines. Since the mines operate only during the day, such traffic is predominately toward them in the morning and away from them in the evening.

Bolivar residents have noted that the travel times between their homes and the mines during the rush hours are much less via Highbridge Road and Valley Road than via Mountainside Road. They feel certain that the excessive congestion on Mountainside Road could be alleviated by strengthening the weak bridge on Highbridge Road, so that vehicles heavier than 2500 pounds would no longer be confined to using Mountainside Road.

* Department of Mathematics and Graduate Program in Operations Research, North Carolina State University, Raleigh, NC 27650.

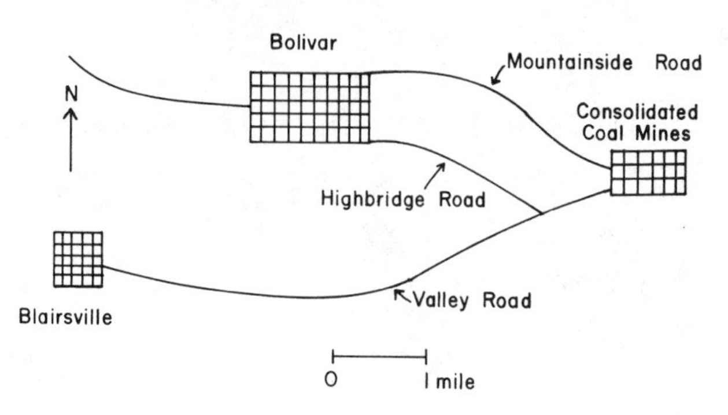

Figure 9.1. Area Road Map.

However, strengthening that bridge would be rather costly, so the Bolivar residents first want an estimate of the resulting benefits—in particular, the resulting travel times between their homes and the mines during the rush hours. The problem of predicting those travel times has been given to consulting engineer, Joe Doakes.

2. A Preliminary Analysis

Joe begins with a decision to concentrate on predicting the morning rush-hour travel times. He hopes to justify this concentration eventually by showing that the evening rush-hour travel times would be the same as their morning counterparts. If this conjecture is not eventually established, he will simply have to perform an additional analysis and calculation of the evening rush-hour travel times.

Joe realizes that to predict any natural phenomenon, an underlying mechanism must first be uncovered and then thoroughly understood. In an attempt to uncover the relevant mechanism for predicting travel times, he chooses the physical sciences' time-honored way of empirical observation.

Since traffic cannot be studied in a laboratory under completely controlled conditions, Joe must limit his observations for the most part to the existing traffic situation, supplementing them with a poll to determine driver route-selection processes. He expects to induce from the existing traffic situation how the travel times are related to the traffic flows. If he can also induce from the driver route-selection processes how the traffic flows are determined, he should then be able to deduce the desired travel times.

3. Empirical Observations

After observing several buildups of the morning rush-hour traffic from the vantage point of a helicopter, Joe notes that the average travel time for traffic moving on a given road segment in the general direction of the mines depends only on the corresponding traffic flow (measured in vehicles per minute traversing the segment midpoint). For Highbridge Road and the segment of Valley Road between Blairsville and the Highbridge Road intersection, the average travel times also include the average times needed to negotiate the intersection via a control light, times that are small relative to the times required to traverse the road segments.

Joe's helicopter observations confirm his intuitive feeling that such travel times should increase (or at least should not decrease) with each increase in the corresponding traffic flow (because of the increase in congestion caused by an increase in traffic flow). Joe now has little doubt that he can construct from further measurements a set of curves that relate the various travel times to the corresponding traffic flow.

Joe's helicopter observations also seem to confirm his intuitive feeling that the peak traffic flows during the morning rush hour are in "steady-state equilibrium"; in particular, the total flow into the intersection of Valley Road and Highbridge Road equals the total flow out of that intersection. Joe is now confident that analyses similar to those used to predict steady-state equilibrium flows in electrical networks will enable him to predict peak traffic flows during the morning rush hour. Such flows would of course provide the peak travel times, which presumably would be the only travel times of any real interest to the Bolivar residents.

Joe's driver poll confirms his suspicion that from among the routes legally available to a given driver, the driver selects, by trial and error, the route or routes that minimize his travel time. To maintain minimality, drivers occasionally try alternative routes, especially when they have information that indicates a different route could result in a shorter travel time.

Since collecting precise numerical data can be a time-consuming and costly undertaking, Joe now decides to pinpoint the data required for his predictions. Needless to say, such data requirements are actually determined by the mathematical model to be used.

4. Model Construction

In any modeling attempt, the relevant entities and their interrelations must be clearly identified. In particular, Joe must take into account the individual drivers, with special attention focused on their origin and vehicle weight. He must also consider the road segments over which they travel, as well as the routes they might possibly choose.

In any modeling attempt, the danger exists of including too much detail, which makes the resulting mathematical developments unnecessarily difficult (if not impossible) without much (if any) compensating increase in understanding or numerical accuracy. In particular, Joe should obviously create only three driver classifications: 1) drivers going from Blairsville to the mines; 2) drivers going from Bolivar to the mines with vehicles weighing

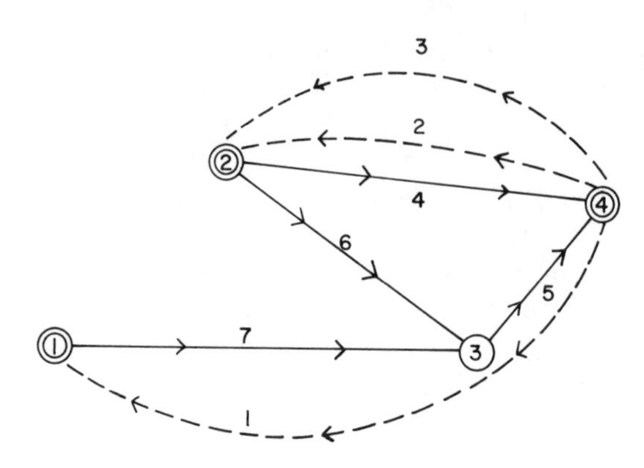

Figure 9.2. Graphical Representation.

less than 2500 pounds; 3) drivers going from Bolivar to the mines with vehicles weighing at least 2500 pounds.

This "aggregation" of thousands of drivers into "commodities" 1, 2, and 3, along with an "abstraction" of only the essential features of the area road map, is conveniently represented in Figure 9.2. Blairsville, Bolivar, the intersection of Highbridge and Valley Roads, and the Consolidated Coal Mines are represented by "nodes" 1, 2, 3, and 4, respectively. The double circles around nodes 1, 2, and 4 serve as reminders that these three nodes represent origins or destinations of traffic. Commodities 1, 2, and 3 are represented by "links" 1, 2, and 3, respectively, each of which is "directed" by an arrow from the corresponding commodity destination (node) back to its origin (node). Needless to say, the relevant road segments are represented by links 4, 5, 6, and 7, each of which is directed by an arrow that indicates the previously observed direction of the morning rush hour traffic flow. The dotted lines along links 1, 2, and 3 serve as reminders that these three "return links" represent commodities and are purely mathematical artifices—in contrast with "real links" 4, 5, 6, and 7, which actually correspond to real road segments. However, it shall prove mathematically helpful to imagine that all traffic due to a given commodity flows from its destination back to its origin over its return link, even though in reality it is absorbed at its destination.

Note that all details having to do with the town and coal mine street maps have been omitted, as have all nontopological details having to do with the road segments. Although some of those details must yet be added to make a workable model, most will never be incorporated unless preliminary predictions via the model indicate that further accuracy is desired.

Now, with each of the three commodities is associated the family of all possible "paths" over which that particular commodity might conceivably flow (over real links from its origin to its destination). In particular, the family of all possible paths over which commodity 1 might conceivably flow consists of the path over links 7 and 5, and the path over links 7, 6, and 4.

(Note, however, that links 7, 5, 4, 6, and 5 do not constitute such a path; the reason is that a path, by definition, cannot cross itself. This formal rejection of such possible "routes" is realistic because commodity 1—in fact, any commodity—would obviously be foolhardy to follow one of them.) Likewise, the family of all possible paths over which commodity 2 might conceivably flow, more concisely termed the "possible path family" for commodity 2, clearly consists of the path over links 6 and 5, and the path over link 4. Finally, the possible path family for commodity 3 is obviously identical to the possible path family for commodity 2 (because commodities 2 and 3 share a common origin and destination).

To eliminate any ambiguity that might occur when two or more commodities share the same possible path family, each possible path is extended into a possible "circuit" by appending to it the given commodity's return link. For example, the possible path over links 6 and 5 is extended into the possible circuit around links 6, 5, and 2 when it is to be associated with commodity 2, but is extended into the possible circuit around links 6, 5, and 3 when it is to be associated with commodity 3. Naturally, the family of all circuits obtained by appending a given commodity's return link to each of its possible paths is termed the commodity's "possible circuit family."

There are numerous reasons why a given commodity might eliminate from consideration some of its possible paths (circuits). In particular, Joe's helicopter observations indicate that commodity 1 has eliminated its possible path over links 7, 6, and 4, probably because the drivers comprising commodity 1 are convinced that its use could only result in increased travel time compared with their other possible path over links 7 and 5. Although commodity 2 has not eliminated either of its possible paths, commodity 3 has eliminated its possible path over links 6 and 5 because of the vehicle weight restriction that originally motivated the partitioning of Bolivar drivers into these two commodities. In all cases, the remaining subfamily of a given commodity's possible path (circuit) family is termed its "feasible path (circuit) family."

Clearly, a more concise description is needed for paths and circuits. The most convenient way to represent a path (circuit) is to first associate each link k with the kth component of the vectors in seven-dimensional Euclidean space E_7. Then a given path (circuit) can be represented by the vector whose kth component is either 1 or 0, depending, respectively, on whether link k is or is not part of the given path (circuit). For example, the possible path over links 6 and 5 is represented by the vector $(0, 0, 0, 0, 1, 1, 0)$ in E_7. Moreover, its extension to the corresponding possible circuit around links 6, 5, and 2 for commodity 2 is represented by the vector $(0, 1, 0, 0, 1, 1, 0)$ in E_7, while its extension to the corresponding possible circuit around links 6, 5, and 3 for commodity 3 is represented by the vector $(0, 0, 1, 0, 1, 1, 0)$ in E_7.

Now, the feasible "circuit vectors" for commodities 1, 2, and 3 can be conveniently enumerated as follows:

$$\delta^1 \triangleq (1, 0, 0, 0, 1, 0, 1)$$
$$\begin{cases} \delta^2 \triangleq (0, 1, 0, 0, 1, 1, 0) \\ \delta^3 \triangleq (0, 1, 0, 1, 0, 0, 0) \end{cases}$$
$$\delta^4 \triangleq (0, 0, 1, 1, 0, 0, 0),$$

where the symbol $\triangleq$ means equality by definition.

In particular then, the feasible circuit family for commodity 1 is (represented by) the set $\{\delta^1\}$, while the feasible circuit family for commodity 2 is the set $\{\delta^2, \delta^3\}$, and the feasible circuit family for commodity 3 is the set $\{\delta^4\}$.

Although the location of 1 among the first three components of a given feasible circuit vector δ^j indicates the commodity i for which it is a feasible circuit vector, we also use the index sets

$$[1] \triangleq \{1\}$$
$$[2] \triangleq \{2, 3\}$$
$$[3] \triangleq \{4\}$$

to directly relate the enumeration of the commodity feasible circuit families to the enumeration of the individual feasible circuit vectors.

In summary, a total of four feasible circuits exists over which traffic can flow; and given a feasible circuit j (namely, an integer in the circuit index set $\{1, 2, 3, 4\}$), a unique commodity i exists (in the commodity index set $\{1, 2, 3\}$) such that $j \in [i]$, which means that commodity i (and only commodity i) flows over the feasible circuit j. Moreover, the corresponding feasible circuit vector δ^j has components

$$\delta_k^j = \begin{cases} 1, & \text{when } k = \text{that } i \text{ for which } j \in [i] \\ 0, & \text{when } k \neq \text{that } i \text{ for which } j \in [i] \text{ but } 1 \leq k \leq 3 \\ 1, & \text{when } 4 \leq k \leq 7 \text{ and real link } k \text{ is part of circuit } j \\ 0, & \text{when } 4 \leq k \leq 7 \text{ and real link } k \text{ is not part of circuit } j. \end{cases}$$

Now, a "possible circuit flow" is just a vector

$$z \in E_4$$

whose jth component z_j is simply the input (traffic) flow on circuit j of that commodity i for which $j \in [i]$. [For example, the possible circuit flow $z = (3, 5, 4, 2)$ means that three flow units are assigned (by commodity 1) to circuit 1 (the only feasible circuit for commodity 1); five flow units are assigned (by commodity 2) to circuit 2, while four flow units are assigned (also by commodity 2) to circuit 3 (the only other feasible circuit for commodity 2); two flow units are assigned (by commodity 3) to circuit 4 (the only feasible circuit for commodity 3).] Since each commodity flows only in the given link directions, it is obvious that each possible circuit flow z must also satisfy the (component-wise) vector inequality

$$z \geq 0. \tag{1}$$

For a steady-state flow (e.g., flow at the peak of the morning rush hour), each such z generates a "possible total flow"

$$x \triangleq \sum_{j=1}^{4} z_j \delta^j \in E_7, \tag{2}$$

whose kth component x_k is clearly the resulting total flow of all commodities on link k for $k = 1, 2, \cdots, 7$. [For example, the possible circuit flow $z = (3, 5, 4, 2)$ generates, via this defining equation (2), the possible total flow $x = (3, 9, 2, 6, 8, 5, 3)$, whose components are clearly the resulting total flows on the corresponding links.]

Given that

d_i is the total (nonnegative) input flow of commodity i

and that

b_k is the total capacity of real link k (which may be $+\infty$),

a "feasible circuit flow" is just a possible circuit flow z that generates a "feasible total flow" x, namely a possible total flow x such that

$$x_k = d_k, \qquad \text{for } k = 1, 2, 3 \tag{3}$$

while

$$0 \leq x_k < b_k, \qquad \text{for } k = 4, 5, 6, 7. \tag{4}$$

[For example, given the total input flows $d_1 = 3$, $d_2 = 9$, and $d_3 = 2$, and given the total real link capacities $b_4 = 7$, $b_5 = 10$, $b_6 = 9$, and $b_7 = 8$, the possible circuit flow $z = (3, 5, 4, 2)$ and the resulting possible total flow $x = (3, 9, 2, 6, 8, 5, 3)$ are clearly feasible.] Even though conditions (1) and (2) clearly imply that the inequality $0 \leq x_k$ in condition (4) is redundant, that inequality is explicitly included to indicate the interval over which the travel time functions c_k of x_k are to be defined.

Although a given feasible circuit flow z generates a unique feasible total flow x, it is obvious that a given feasible total flow x can generally be generated by more than a unique feasible circuit flow z; in fact, the set of all such feasible circuit flows z is clearly identical to the set of all solutions z to (the linear) conditions (1) and (2).

Since Joe still has some doubt about whether the travel time y_k over a given link k can always be given with reasonable accuracy by a function c_k of *only* the corresponding link total flow x_k, he decides to ask Bill Ivory, Professor of Transportation Science at nearby Tower University. After receiving a fairly detailed physical description of the network links and their intersections, Bill tells Joe that although such functions c_k do not really exist, the travel time y_k over such a link k can always be related with reasonable accuracy to only the corresponding link total flow x_k by a curve with the general shape shown in Figure 9.3.

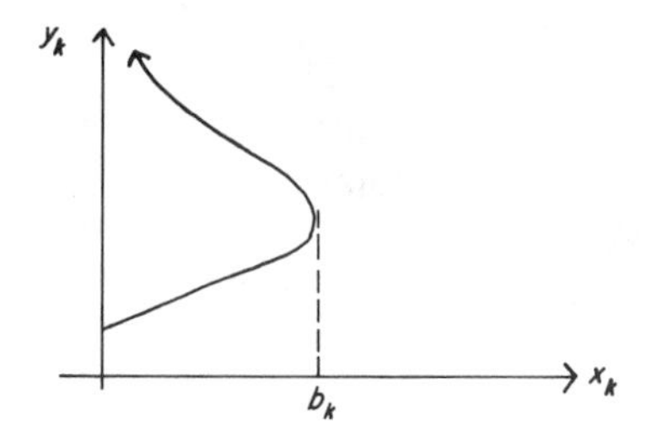

Figure 9.3. Travel Time Related to Traffic Flow.

Evidently, Joe's helicopter observations have detected only the increasing lower parts of such curves, which can of course be described by functions c_k of only x_k. However, when the total flow x_k on a given link k reaches the link's capacity b_k, a "traffic jam" can occur. During a traffic jam, the travel time y_k continues increasing to infinity while the total flow x_k actually decreases back to zero, as indicated by the upper part of the corresponding curve. Consequently, Joe can now feel confident that travel-time functions c_k do exist and can be used in his model as long as traffic jams are not being observed or predicted (i.e., as long as each $x_k < b_k$, as required in condition (4)).

Bill also tells Joe that the information gained in his driver poll indicates that the network being studied will either be in or tend toward a state of "Wardrop equilibrium," that is, a state in which the travel time is the same on each path used by a given commodity and is not greater than the travel time on each path feasible for but unused by the given commodity. In such a state, each individual driver has clearly achieved a minimal travel time, so there is no incentive for the circuit flows and hence the total flows to change.

Finally, Bill tells Joe that the model constructed so far is a special case of the relatively sophisticated model recently constructed by Hall and Peterson [12], whose mathematization of the preceding Wardrop statement in the context of the present problem would be as follows. By definition, the only feasible circuit flows z that place the present roadway network in a state of Wardrop-equilibrium are those z that generate a feasible total flow x for which both (fictitious) "return link travel times"

$$y_k \in E_1, \qquad \text{for } k = 1, 2, 3 \tag{5a}$$

and "real link travel times"

$$y_k = c_k(x_k), \qquad \text{for } k = 4, 5, 6, 7 \tag{5b}$$

exist such that the resulting "travel-time vector"

$$y = (y_1, y_2, y_3, y_4, y_5, y_6, y_7)$$

satisfies the following "inner product" conditions

$$\langle \delta^j, y \rangle \begin{cases} = 0, & \text{if } z_j > 0, \\ \geq 0, & \text{if } z_j = 0, \end{cases} \qquad \text{for } j = 1, 2, 3, 4. \tag{6}$$

To verify that the preceding definition does indeed mathematize the Wardrop statement, note from the defining formulas for the feasible circuit vectors δ^j that (6) simply asserts the following traffic situation. For each of the feasible circuits actually used by a given commodity i (namely, each circuit $j \in [i]$ for which $z_j > 0$), the travel time is the same, in fact, just $-y_i$ (the negative of the corresponding return link travel time y_i), which in turn does not exceed the travel time for each of the feasible circuits not used by the given commodity i (namely, each circuit $j \in [i]$ for which $z_j = 0$).

Each vector z that satisfies (1)–(6) is termed a "Wardrop-equilibrium circuit flow," and each such flow z generates a "Wardrop-equilibrium total flow" x via equation (2). Although a given Wardrop-equilibrium circuit flow z generates a unique Wardrop-equilibrium total flow x, it is obvious that a given Wardrop-equilibrium total flow x can generally be generated by more than a unique Wardrop-equilibrium circuit flow z; in fact, the set of all such Wardrop-equilibrium circuit flows z is clearly identical to the set of all solutions z to both (the linear) conditions (1)–(2) and the "complementary slackness" conditions

$$\text{either } \langle \delta^j, y \rangle = 0 \text{ or } z_j = 0, \qquad \text{for } j = 1, 2, 3, 4. \tag{7}$$

5. Data Requirements

An examination of conditions (1)–(7) shows that the preceding model requires input data consisting of the feasible circuit vectors δ^j, the commodity input flows d_i, the real-link capacities b_k, and the travel-time functions c_k.

The feasible circuit vectors δ^j have already been determined by inspection, but the remaining input data must be determined empirically by measuring the flows and resulting travel times on each link, preferably at various times during a buildup of the morning rush hour traffic.

The commodity input flows d_i can be determined by measuring for each commodity i the number of vehicles per minute entering the network at the peak of the morning rush hour. The real-link capacities b_k can, of course, be inferred from the travel-time functions c_k.

The travel-time functions c_k can be approximated through "interpolation" and "extrapolation," by fitting a finite number of observed points on their graphs with formulae that exhibit the observed monotonicity. The total flows at which the resulting formulas become infinite then approximate the real-link capacities b_k.

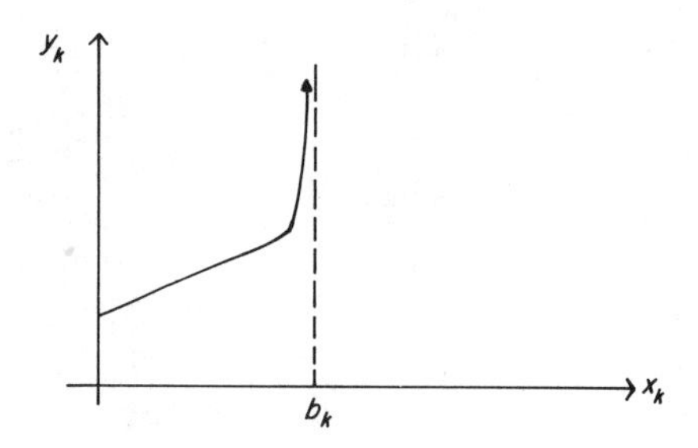

Figure 9.4. Travel Time as an Approximate Function of Traffic Flow.

Needless to say, the original curve relating the travel time y_k to the link total flow x_k (as illustrated in Figure 9.3) is replaced via this approximation by the curve shown in Figure 9.4. Of course, the lower part of the original curve is accurately approximated by this curve, except near $x_k = b_k$ (when traffic jams begin to form).

6. Data Collection

After checking with the traffic bureaus of Blairsville, Bolivar, and the county in which they are located, Joe finds that the required input data are not available. Although no reason has existed previously for collecting much of it, the remainder is cautiously withheld because the bureaucrats in charge are concerned about Joe's motives. Consequently, Joe and his aides must collect all of the required input data by making their own measurements.

To do so, they further monitor various buildups of the morning rush hour traffic. To supplement their helicopter observations, they measure some of the flows with the aid of roadside counting devices, and they measure some of the travel times via test drives. In fact, the ensuing cost of hardware usage and personnel time is actually larger than the total remaining cost for the whole project.

The resulting commodity input flows, expressed in hundreds of vehicles per minute, are

$$d_1 = 10, \qquad d_2 = 6, \qquad d_3 = 10.$$

The resulting real-link capacities, also expressed in hundreds of vehicles per minute, are

$$b_4 = 25, \qquad b_5 = 20, \qquad b_6 = 10, \qquad b_7 = 15.$$

The resulting travel-time functions have the formula

$$c_k(x_k) = t_k(1 - x_k/b_k)^{-1}, \qquad \text{for } k = 4, 5, 6, 7,$$

where the "zero-flow travel times," expressed in minutes, are

$$t_4 = 15, \qquad t_5 = 1, \qquad t_6 = 2, \qquad t_7 = 5.$$

Needless to say, *all flows are to be expressed in hundreds of vehicles per minute, and all travel times are to be expressed in minutes.*

Although Bill Ivory recommended using the preceding travel-time formula with the exponent -1 replaced by an additional parameter $-p_k < 0$, Joe found that the other two parameters b_k and t_k provide enough flexibility for obtaining reasonably accurate fits to his empirically determined points on the travel-time curves.

7. Model Validation

Joe's data include observed circuit flows at the peak of the morning rush hour, namely

$$z_1^* = 10, \qquad z_2^* = 6, \qquad z_3^* = 0, \qquad z_4^* = 10.$$

As a test for his model, Joe decides to see whether these flows actually place the present roadway network in a state of Wardrop-equilibrium. To do so, he need of course only check to see if conditions (1)–(6) are satisfied.

The observed vector z^* obviously satisfies condition (1) and clearly generates, via condition (2), a possible total flow

$$x_1^* = 10, \qquad x_2^* = 6, \qquad x_3^* = 10$$

$$x_4^* = 10, \qquad x_5^* = 16, \qquad x_6^* = 6, \qquad x_7^* = 10.$$

This total flow x^* is very close to the observed total flow at the peak of the morning rush hour—further evidence that such flows are essentially in steady-state equilibrium.

Using the data of the preceding section, Joe notes that this total flow x^* satisfies conditions (3) and (4); so z^* is a feasible circuit flow, and x^* is a feasible total flow. Condition (3) is satisfied with complete precision because the input flows d_k were actually computed from the observed circuit flows z_j^* via this condition (rather than having been measured by independent observations). Moreover, the relatively large difference between x_k^* and b_k for $k = 4, 5, 6, 7$ indicates that traffic jams are not imminent.

Using the data and formulas of the preceding section, Joe notes that the calculated total flow x^* provides via condition (5b) the real link travel times

$$y_4^* = 25, \qquad y_5^* = 5, \qquad y_6^* = 5, \qquad y_7^* = 15.$$

These travel times are very close to the observed travel times at the peak of the morning rush hour because the input parameters b_k were actually computed from the observed total flows and travel times via condition (5b).

The question of whether the present roadway network is in a state of Wardrop-equilibrium now hinges on whether return link travel times y_1^*, y_2^*, and y_3^* exist such that conditions (5a) and (6) are also satisfied. Of course, (5a) is automatically satisfied by any set of numbers y_1^*, y_2^*, and y_3^*, so the question actually hinges on whether a solution vector $(y_1^*, y_2^*, y_3^*, 25, 5, 5, 15)$ to condition (6) exists.

Using the known values for δ^j, z^*, and y^*, Joe notes that condition (6) simply requires that

$$\begin{aligned} y_1 \qquad\quad + 20 &= 0 \\ y_2 \quad + 10 &= 0 \\ y_2 \quad + 25 &\geq 0 \\ y_3 + 25 &= 0. \end{aligned}$$

Clearly, this linear system has a unique solution

$$y_1^* = -20, \qquad y_2^* = -10, \qquad y_3^* = -25,$$

so the present roadway network is indeed in a state of Wardrop-equilibrium.

The Wardrop-equilibrium travel time for commodity 1 is $20(= -y_1^*)$, while the Wardrop-equilibrium travel time for commodity 2 is $10(= -y_2^*)$, and the Wardrop-equilibrium travel time for commodity 3 is $25(= -y_3^*)$. Note that commodities 1 and 3 use all of their feasible circuits (actually only one each) while commodity 2 uses only one of its two feasible circuits. The reason is that the strict inequality induced into the preceding linear system by the solution y^* shows that any flow (i.e., any vehicles) assigned to circuit 3 (by commodity 2) would experience a travel time of 25, which is of course much larger than the Wardrop-equilibrium travel time of 10 for commodity 2.

The strictly increasing nature of the functions c_k guarantees the uniqueness of both the Wardrop-equilibrium total flow x^* and the Wardrop-equilibrium travel-time vector y^*, a fact that is easy to establish only after subsequent theoretical developments. This uniqueness in turn implies via the analysis in the final paragraph of Section 4 that the Wardrop-equilibrium circuit flow z^* is unique if and only if it is the only solution z to the resulting conditions (1), (2) and (7), namely, the conditions

$$z_1 \geq 0 \qquad z_2 \geq 0 \qquad z_3 \geq 0 \qquad z_4 \geq 0$$

$$\left.\begin{aligned} 10 &= z_1 \\ 6 &= z_2 + z_3 \\ 10 &= z_4 \\ 10 &= z_3 + z_4 \\ 16 &= z_1 + z_2 \\ 6 &= z_2 \\ 10 &= z_1 \end{aligned}\right\}$$

$$z_3 = 0.$$

Inspection shows that these conditions do indeed have a unique solution z^*, a situation that is generally not the case for more complicated networks.

Joe is now convinced that this network model involving conditions (1)–

(7) provides a valid mechanism for predicting the flows and travel times at the peak of the morning rush hour. Consequently, he decides to use it to predict the peak flows and travel times that would result if the relatively weak bridge on Highbridge Road were strengthened sufficiently to handle vehicles heavier than 2500 pounds, that is, if the circuit vector

$$\delta^5 \triangleq (0, 0, 1, 0, 1, 1, 0)$$

were also made feasible for commodity 3.

8. Problem Solution

Since the preceding analysis shows that the present peak travel time of commodity 3 is 25 while the present peak travel time over links 5 and 6 is only 10, it is clear that commodity 3 would like to reassign some of its input flow d_3 (i.e. some of its vehicles) from path 4 to path 5. Due to the monotonicity of the travel-time functions c_k, the peak travel time on path 4 decreases and the peak travel time on path 5 increases during such a reassignment. Needless to say, a Wardrop-equilibrium is reestablished when the reassignment is sufficiently large to make these two travel times equal.

It may now be beneficial for the reader to help Joe predict the new peak flows and travel times, simply by decreasing z_4 from z_4^* with a compensating increase in z_5 from 0, until the travel time $c_4(x_4)$ equals the travel time $c_5(x_5) + c_6(x_6)$. In doing so, the reader should realize that conditions (1)–(6) are again being solved, this time with an extra circuit flow variable z_5 and an extra circuit vector δ^5. Using the (yet to be established) fact that both the new Wardrop-equilibrium total flow x^{**} and the new Wardrop-equilibrium travel-time vector y^{**} are unique, the reader might also benefit from studying, via the appropriate conditions (1), (2), and (7), the uniqueness of the new Wardrop-equilibrium circuit flow z^{**}. In any event, a complete solution is provided in the Appendix at the end of this section.

Even though such a solution (x^{**}, y^{**}, z^{**}) is exact through seven digits (determined by computer calculation), considerable error could still result from using y^{**} to predict the peak travel times. The largest source of such error is likely to be (implicit) extrapolation via the formulae for $c_5(x_5)$ and $c_6(x_6)$ as x_5 and x_6 increase beyond any values previously observed. Relatively little error should result from (implicit) interpolation via the formula for $c_4(x_4)$ as x_4 decreases through values previously observed (during buildups of the morning rush-hour traffic).

However, after a conversation with Bill Ivory, Joe decides against the large additional expense that would be required to "simulate" larger flows x_5 and x_6 to obtain (possibly) more accurate formulae for $c_5(x_5)$ and $c_6(x_6)$. The reason is that, with relatively little additional expense, Joe can eventually test the sensitivity of y^{**} to variations in the travel time functions c_k. If y^{**} turns out to be relatively insensitive to such variations, Joe can feel more confident about his predictions and not bother to estimate the c_k with additional accuracy.

In the meantime, Joe communicates his tentative predictions to all Bolivar residents, and on that basis alone they decide to strengthen the weak bridge on Highbridge Road. Although the decrease in peak travel time across link 4 (for heavy vehicles) is much smaller than the increase in peak travel time across links 5 and 6 (for light vehicles), the relatively wealthy and influential drivers of heavy vehicles prevail.

Appendix

The feasible circuit vectors for commodities 1, 2, and 3 are

$$\delta^1 \triangleq (1, 0, 0, 0, 1, 0, 1)$$

$$\begin{cases} \delta^2 \triangleq (0, 1, 0, 0, 1, 1, 0) \\ \delta^3 \triangleq (0, 1, 0, 1, 0, 0, 0) \end{cases}$$

$$\begin{cases} \delta^4 \triangleq (0, 0, 1, 1, 0, 0, 0) \\ \delta^5 \triangleq (0, 0, 1, 0, 1, 1, 0). \end{cases}$$

The corresponding index sets are

$$[1] \triangleq \{1\}$$

$$[2] \triangleq \{2, 3\}$$

$$[3] \triangleq \{4, 5\}.$$

The commodity input flows are

$$d_1 = 10, \qquad d_2 = 6, \qquad d_3 = 10,$$

and the real-link capacities are

$$b_4 = 25, \qquad b_5 = 20, \qquad b_6 = 10, \qquad b_7 = 15.$$

The zero-flow travel times are

$$t_4 = 15, \qquad t_5 = 1, \qquad t_6 = 2, \qquad t_7 = 5,$$

and the travel-time functions have the formula

$$c_k(x_k) = t_k(1 - x_k/b_k)^{-1}, \qquad \text{for } k = 4, 5, 6, 7.$$

The appropriate conditions for characterizing Wardrop-equilibria are

$$z \geq 0 \tag{A1}$$

$$x \triangleq \sum_{j=1}^{5} z_j \delta^j \tag{A2}$$

$$x_k = d_k, \qquad \text{for } k = 1, 2, 3 \tag{A3}$$

$$0 \leq x_k < b_k, \qquad \text{for } k = 4, 5, 6, 7 \tag{A4}$$

$$y_k \in E_1, \qquad \text{for } k = 1, 2, 3 \tag{A5a}$$

$$y_k = c_k(x_k), \qquad \text{for } k = 4, 5, 6, 7 \tag{A5b}$$

$$\langle \delta^j, y \rangle \begin{cases} = 0, & \text{if } z_j > 0, \\ \geq 0, & \text{if } z_j = 0, \end{cases} \qquad \text{for } j = 1, 2, 3, 4, 5. \tag{A6}$$

A solution to these conditions—obtained via a computer algorithm [17] that implements the theory to be developed in Part II—is

$$z_1^{**} = 10, \quad z_2^{**} = 6, \quad z_3^{**} = 0, \quad z_4^{**} = 7.83133, \quad z_5^{**} = 2.16867,$$

$$x_1^{**} = 10, \quad x_2^{**} = 6, \quad x_3^{**} = 10,$$

$$x_4^{**} = 7.83133, \quad x_5^{**} = 18.16867, \quad x_6^{**} = 8.16867, \quad x_7^{**} = 10,$$

$$y_1^{**} = -25.92105, \quad y_2^{**} = -21.84211, \quad y_3^{**} = -21.84211,$$

$$y_4^{**} = 21.84211, \quad y_5^{**} = 10.92105, \quad y_6^{**} = 10.92105, \quad y_7^{**} = 15.$$

A comparison of (x^*, y^*, z^*) and (x^{**}, y^{**}, z^{**}) shows that commodities 1 and 2 do not reassign any of their own flows. However, commodity 3 reassigns 2.16867 units of its own flow from circuit 4 to circuit 5. In doing so, it decreases its own peak travel time by only 3.15789 while increasing the peak travel times of commodities 1 and 2 by 5.92105 and 11.84211, respectively.

Although x^{**} and y^{**} are uniquely determined, z^{**} is not uniquely determined, a fact that the reader can establish by calculating all possible z^{**} (via the analysis in the final paragraph of Section 4). This lack of uniqueness is clearly due to the fact that commodities 2 and 3 now have identical feasible path families. In essence, commodities 2 and 3 now constitute a single (aggregated) commodity, say commodity 2, with a feasible circuit family $\{\delta^2, \delta^3\}$ and a *total* input flow $d_2 = 6 + 10 = 16$. Viewing the problem this way reduces the dimension of z from 5 to 3 and results in a uniquely determined $z^{**\prime}$, which the reader should be able to compute from his knowledge of z^{**}.

PART II

9. Another Scenario

In addition to the many Bolivar residents with light vehicles, all Blairsville residents are upset by Bolivar's decision to strengthen the bridge on Highbridge Road. The reason is that Joe has predicted a sizeable increase in their peak travel time, an increase that is due entirely to the additional congestion on Valley Road (link 5). The reaction of Blairsville residents includes the hiring of Joe Doakes to predict the peak travel times that would result if an additional road were constructed between Blairsville and Bolivar to alleviate the congestion on Valley Road.

Confronted with the possibility of having to analyze increasingly complicated networks, Joe decides to delve more deeply into the work of Hall and Peterson [12], with special emphasis on the theory that leads to computer algorithms capable of solving problems not easily solvable by hand calculation.

10. A General Model

A general roadway network can be conveniently represented by a directed graph (consisting of nodes and directed links) on which there is multicommodity flow. Each commodity (i.e., each traffic type) is associated with a (fictitious) "return link" over which only that particular commodity flows. Such a return flow takes place only in the prescribed return link direction, which is of course from the commodity's given destination (node) back to its given origin (node). Each "real link" represents a collection of unidirectional lanes over which any of the given commodities might flow from a certain intersection (node) to an adjacent intersection (node). Such a (multicommodity) flow can take place only in the prescribed real link direction, which is dictated of course by local traffic regulations. Thus two-way streets are represented by at least two real links.

In the most general model to be studied here, we consider a directed graph with a total of n links. In particular,

$$\text{the return links are numbered } 1, 2, \cdots, r,$$

and

$$\text{the real links are numbered } r + 1, r + 2, \cdots, n.$$

Hence a total of r commodities and a total of $(n - r)$ real links exist over which the r commodities can flow. For convenience, we number the r commodities so that

$$\text{commodity } i \text{ has return link } i \text{ for } i = 1, 2, \cdots, r.$$

With each commodity i is associated the family of all possible "paths" over which that particular commodity might conceivably flow (over real links from its given origin to its given destination). Since a path, by definition, cannot cross itself, each *possible path family* is finite. Needless to say, *we assume that the network is sufficiently "connected" to guarantee that the possible path family for each commodity i is not empty.*

It is obvious that two or more different commodities exhibit the same possible path family if and only if they have the same origin and destination. To eliminate this ambiguity, each possible path is extended into a possible "circuit" by appending to each such path the given commodity's return link. Naturally, the family of all circuits obtained by appending a given commodity's return link to the end of each of its possible paths is termed the commodity's *possible circuit family*.

Numerous reasons can be found why a given commodity usually eliminates from consideration some of the paths (circuits) in its possible path (circuit) family. For example, a complete scanning of them all might be unrealistic for even moderate-size networks (as found in moderate-size urban or metropolitan areas), in which event only those that show the most promise of being reasonably short can actually be considered. Moreover, local traffic regulations may not permit a given commodity (i.e., a given traffic type, such as heavy vehicles) on certain real links, in which event only those of its possible paths that do not include such real links can actually be considered. Finally, other factors (such as personal safety, driving pleasure, etc.) may cause a given commodity to eliminate some of its possible paths from further consideration. In any event, *we assume that each commodity i selects at least one of its possible paths for further consideration*, and we term the resulting (not necessarily proper) subfamily of its possible path family its *feasible path family*. Needless to say, we also term the corresponding subfamily of its possible circuit family its *feasible circuit family*.

For our purposes, the most convenient way to represent a path (circuit) is to first associate each network link k with the kth component of the vectors in n-dimensional Euclidean space E_n. Then, a given path (circuit) can be represented by the vector whose kth component is either 1 or 0, depending, respectively, on whether link k is or is not part of the given path (circuit).

In the most general model to be studied here, *we suppose that commodity i has a nonempty feasible circuit (path) family that is enumerated by the integer index set*

$$[i] \triangleq \{m_i, m_i + 1, \cdots, n_i\}, \qquad \text{for } i = 1, 2, \cdots, r,$$

where

$$1 = m_1 \leq n_1, \qquad n_1 + 1 = m_2 \leq n_2, \cdots, \qquad n_{r-1} + 1 = m_r \leq n_r \triangleq m.$$

Thus a total of m feasible circuits exists over which traffic can flow, and given a feasible circuit j (namely, an integer in the circuit index set $\{1, 2, \cdots, m\}$), a unique commodity i exists (in the commodity index set $\{1, 2, \cdots, r\}$) such that $j \in [i]$, which means that commodity i (and only commodity i) flows over the feasible circuit j. Moreover, *the vector δ^j representing a given feasible circuit j* has components

$$\delta_k^j = \begin{cases} 1, & \text{when } k = \text{that } i \text{ for which } j \in [i], \\ 0, & \text{when } k \neq \text{that } i \text{ for which } j \in [i] \text{ but } 1 \leq k \leq r, \\ 1, & \text{when } r + 1 \leq k \leq n \text{ and real link } k \text{ is part of circuit } j, \\ 0, & \text{when } r + 1 \leq k \leq n \text{ and real link } k \text{ is not part of circuit } j. \end{cases}$$

Now, a *possible circuit flow* is just a vector

$$z \in E_m$$

whose jth component z_j is simply the input flow on circuit j of that commodity i for which $j \in [i]$. Since each commodity can flow only in the given link directions, it is obvious that each possible circuit flow z must also satisfy the (vector) inequality

$$z \geq 0. \tag{1}$$

Of course, each such z generates a *possible total flow*

$$x \triangleq \sum_{j=1}^{m} z_j \delta^j \in E_n, \tag{2}$$

whose kth component x_k is clearly the resulting total flow of all commodities on link k for $k = 1, 2, \cdots, n$.

Given that

d_i is the total (positive) input flow of commodity i

and that

b_k is the total (positive) capacity of real link k (which may be $+\infty$),

a *feasible circuit flow* is just a possible circuit flow z that generates a *feasible total flow* x, namely, a possible total flow x such that

$$x_k = d_k, \qquad \text{for } k = 1, 2, \cdots, r, \tag{3}$$

while

$$0 \leq x_k < b_k, \qquad \text{for } k = r + 1, r + 2, \cdots, n. \tag{4}$$

[Even though conditions (1) and (2) clearly imply that the inequality $0 \leq x_k$ in condition (4) is redundant, it is explicitly included to indicate the interval over which travel-time functions c_k of x_k are to be defined.] Although a given feasible circuit flow z generates a unique feasible total flow x, it is obvious that a given feasible total flow x can generally be generated by more than a unique feasible circuit flow z; in fact, the set of all such feasible circuit flows z is clearly identical to the set of all solutions z to (the linear) conditions (1)–(2).

It is worth noting that the preceding definition of feasible flows (as well as all definitions and theorems to follow) does not invoke "nodal conservation laws," though it is easy to see that such laws are in fact implicitly satisfied.

On each real link k *we suppose that the link's node-to-node travel time* y_k

is a positive strictly increasing function c_k (of only the total flow x_k) that is continuous and approaches $+\infty$ at b_k—hypotheses that are not very restrictive in the context of roadway networks.

We further assume that the (real link) capacities b_k are sufficiently large to handle the commodity input flows d_k; that is, *we assume that there exists at least one feasible circuit flow z*. Actually, for nontrivial roadway networks of practical interest, there are usually infinitely many feasible circuit flows z.

By definition, the only feasible circuit flows z that place a given roadway network in a state of *Wardrop-equilibrium* are those z that generate a feasible total flow x for which both (fictitious) *return link travel times*

$$y_k \in E_1, \qquad \text{for } k = 1, 2, \cdots, r \tag{5a}$$

and *real link travel times*

$$y_k = c_k(x_k), \qquad \text{for } k = r + 1, r + 2, \cdots, n \tag{5b}$$

exist such that the resulting travel-time vector

$$y = (y_1, y_2, \cdots, y_n),$$

satisfies the inner product conditions

$$\langle \delta^j, y \rangle \begin{cases} = 0, & \text{if } z_j > 0, \\ \geq 0, & \text{if } z_j = 0, \end{cases} \qquad \text{for } j = 1, 2, \cdots, m. \tag{6}$$

To properly interpret the preceding definition, note from the defining formula for the feasible circuit vectors δ^j that condition (6) simply asserts the following traffic situation: for each of the feasible circuits actually used by a given commodity i (namely, each circuit $j \in [i]$ for which $z_j > 0$) the total origin-to-destination travel time is the same, in fact, just $-y_i$ (the negative of the corresponding "return link travel time" y_i), which in turn does not exceed the total origin-to-destination travel time for each of the feasible circuits not used by the given commodity i (namely, each circuit $j \in [i]$ for which $z_j = 0$).

Each vector z that satisfies conditions (1–6) is termed a *Wardrop-equilibrium circuit flow*, and each such flow z generates a *Wardrop-equilibrium total flow x* via equation (2). Although a given Wardrop-equilibrium circuit flow z generates a unique Wardrop-equilibrium total flow x it is obvious that a given Wardrop-equilibrium total flow x can generally be generated by more than a unique Wardrop-equilibrium circuit flow z; in fact, the set of all such Wardrop-equilibrium circuit flows z is clearly identical to the set of all solutions z to both (the linear) conditions (1–2) and the *complementary slackness conditions*

$$\text{either } \langle \delta^j, y \rangle = 0 \text{ or } z_j = 0, \qquad \text{for } j = 1, 2, \cdots, m. \tag{7}$$

The properties of Wardrop-equilibria are most easily inferred from the properties of "quasi-equilibria"—a slightly more general concept that is defined in terms of (point-to-set) functions γ_k with function-values

$$\gamma_k(x_k) \triangleq E_1, \qquad \text{for } x_k = d_k, \qquad \text{for } k = 1, 2, \cdots, r$$

$$\gamma_k(x_k) \triangleq \begin{cases} \{y_k \le c_k(0)\}, & \text{for } x_k = 0, \\ \{c_k(x_k)\}, & \text{for } x_k \in (0, b_k), \end{cases} \qquad \text{for } k = r+1, r+2, \cdots, n.$$

(and $\gamma_k(x_k) = \varnothing$ otherwise). Note that the condition $y_k \in \gamma_k(x_k)$ for $k = 1, 2, \cdots, r$ provides a convenient characterization of conditions (3) and (5a). Note also that the condition $y_k \in \gamma_k(x_k)$ for $k = r+1, r+2, \cdots, n$ provides a convenient characterization of conditions (4) and (5b) when $x_k > 0$. Although this new condition is more general than conditions (4) and (5b) when $x_k = 0$ (a generalization that is a prerequisite for subsequent mathematical developments), the mathematical difference between them turns out to be inconsequential in the determination of Wardrop-equilibria from quasi-equilibria.

By definition, the only feasible circuit flows z that place a given roadway network in a state of *quasi-equilibrium* are those z that generate a feasible total flow x for which both (fictitious) *return link quasi-travel times*

$$y_k \in \gamma_k(x_k), \qquad \text{for } k = 1, 2, \cdots, r \tag{5a$'$}$$

and *real link quasi-travel times*

$$y_k \in \gamma_k(x_k), \qquad \text{for } k = r+1, r+2, \cdots, n \tag{5b$'$}$$

exist such that the resulting *quasi-travel time vector*

$$y = (y_1, y_2, \cdots, y_n)$$

satisfies the inner product conditions

$$\langle \delta^j, y \rangle \begin{cases} = 0, & \text{if } z_j > 0, \\ \ge 0, & \text{if } z_j = 0, \end{cases} \qquad \text{for } j = 1, 2, \cdots, m. \tag{6$'$}$$

To properly interpret the preceding definition, note from the defining formula for the feasible circuit vectors δ^j that condition (6′) simply asserts the following traffic situation: for each of the feasible circuits actually used by a given commodity i (namely, each circuit $j \in [i]$ for which $z_j > 0$) the total origin-to-destination quasi-travel time is the same, in fact just $-y_i$ (the negative of the corresponding "return link quasi-travel time" y_i), which in turn does not exceed the total origin-to-destination quasi-travel time for each of the feasible circuits not used by the given commodity i (namely, each circuit $j \in [i]$ for which $z_j = 0$).

Each vector z that satisfies conditions (1–4, 5′, 6′) is termed a *quasi-equilibrium circuit flow*, and each such flow z generates a *quasi-equilibrium total flow* x via equation (2). Although a given quasi-equilibrium circuit flow z generates a unique quasi-equilibrium total flow x, it is obvious that a given quasi-equilibrium total flow x along with the corresponding quasi-travel time vector y can generally be generated by more than a unique quasi-equilibrium circuit flow z; in fact, the set of all such quasi-equilibrium circuit flows z is clearly identical to the set of all solutions z to both (the linear) conditions (1)–(2) and the complementary slackness conditions

$$\text{either } \langle \delta^j, y \rangle = 0 \text{ or } z_j = 0, \qquad \text{for } j = 1, 2, \cdots, m. \tag{7'}$$

Obviously, from the definition of γ_k, *each Wardrop-equilibrium flow* (x, z) *is a quasi-equilibrium flow*. Moreover, each quasi-equilibrium flow (x, z) clearly becomes a Wardrop-equilibrium flow simply by increasing to $c_k(0)$ each $y_k < c_k(0)$ for which $x_k = 0$. Thus, each *quasi-equilibrium flow* (x, z) is *actually a Wardrop-equilibrium flow*. Consequently, *we shall henceforth be concerned (without any loss of generality) with the properties of quasi-equilibria.*

11. The Key Characterization

First, we incorporate the feasibility conditions (1)–(2) into a set

$$X \triangleq \left\{ x = \sum_1^m z_j \delta^j \middle| z_j \geq 0 \qquad \text{for } j = 1, 2, \cdots, m \right\}, \tag{8}$$

which is obviously a "convex polyhedral cone generated by" the feasible circuit vectors δ^j. Then, we introduce its "dual cone"

$$Y \triangleq \{ y \in E_n | 0 \leq \langle x, y \rangle \qquad \text{for each } x \in X \}, \tag{9a}$$

which clearly has the representation formula

$$Y = \{ y \in E_n | 0 \leq \langle \delta^j, y \rangle \qquad \text{for } j = 1, 2, \cdots, m \}. \tag{9b}$$

This representation formula (9b) shows that Y is also a convex polyhedral cone (whose generators can be computed in any given situation with the aid of a linear-algebraic algorithm due to Uzawa [27]).

By definition, the *network-equilibrium conditions* (for the general quasi-equilibrium problem) are

(i) $x \in X$ and $y \in Y$
(ii) $0 = \langle x, y \rangle$
(iii) $y_k \in \gamma_k(x_k)$ for $k = 1, 2, \cdots, n$.

Each vector (x, y) that satisfies these conditions is termed a *network-equilibrium vector*.

Theorem 1 is the key to almost everything that follows.

Theorem 1. *Each quasi-equilibrium total flow x along with the corresponding quasi-travel time vector y produces a network-equilibrium vector (x, y). Conversely, each network-equilibrium vector (x, y) produces a quasi-equilibrium total flow x along with a corresponding quasi-travel time vector y.*

PROOF. First, note from the defining equation (8) for X that conditions (1, 2, 3, 4, 5′) are equivalent to the first part of condition (i) and all of condition (iii). Consequently, we can complete our proof by showing that condition (6′) is equivalent to the second part of condition (i) and condition (ii).

Toward that end, simply note from the defining equation (8) for X that

$$\langle x, y\rangle \equiv \sum_{1}^{m} z_j\langle \delta^j, y\rangle \qquad \text{for } x \in X. \tag{10}$$

An immediate consequence of this identity (10) is that condition (6′) implies the second part of condition (i) and condition (ii). On the other hand, an immediate consequence of the representation formula (9b) for Y and this identity (10) is that condition (6′) is also implied by the second part of condition (i) and condition (ii). □

Of course, a consequence of (the key) Theorem 1 is that *we can henceforth concentrate* (*without any loss of generality*) *on the properties of network-equilibrium vectors* (x, y). There are at least two reasons why such a concentration pays extremely high dividends: 1) the network-equilibrium conditions (i)–(iii) are mathematically simpler than the quasi-equilibrium conditions (1, 2, 3, 4, 5′, 6′), in that the latter also explicitly involve equilibrium circuit flows z; 2) the network-equilibrium conditions (i)–(iii) turn out to be essentially the "extremality conditions" for (generalized) "geometric programming" and hence provide a mechanism through which the powerful theory of geometric programming can be applied to the study of traffic equilibria.

The next section is devoted to actually constructing those geometric programming "dual problems" whose corresponding extremality conditions coincide with the network-equilibrium conditions (i)–(iii).

12. Dual Variational Principles

For simplicity, *we suppose that the improper integral* $\int_0^{b_k} c_k(s)\,ds$ *diverges for* $k = r + 1, r + 2, \cdots, n$, a hypothesis that is not very restrictive in view of our previous discussions (in Sections 4 and 5) concerning the general travel-time functions c_k.

Imitating the theory of "monotone networks" [16], [1], [23], we "integrate" each (point-to-set) function γ_k simply by computing the (Riemann) integral of any (point-to-point) function $\mathscr{c}_k$ whose domain coincides with the domain of γ_k and whose function values $\mathscr{c}_k(x_k) \in \gamma_k(x_k)$. Of course, for $k = r + 1, r + 2, \cdots, n$, the given travel-time function c_k is a legitimate choice for $\mathscr{c}_k$. In any event, our previous hypotheses concerning c_k guarantee that such integrals exist and do not depend on the particular function-value choice $\mathscr{c}_k(x_k) \in \gamma_k(x_k)$. Consequently, we denote such integrals by the symbol $\int^x \gamma_k(s)\,ds$ even though γ_k is not a real-valued (point-to-point) function.

With a judicious choice of the constant of integration, the integral $\int^{x_k} \gamma_k(s)\,ds$ clearly produces a function g_k whose domain

$$C_k \triangleq \begin{cases} \{d_k\}, & \text{for } k = 1, 2, \cdots, r \\ [0, b_k), & \text{for } k = r + 1, r + 2, \cdots, n \end{cases} \tag{11a}$$

and whose function-values are

$$g_k(x_k) \triangleq \begin{cases} 0, & \text{for } k = 1, 2, \cdots, r \\ \displaystyle\int_0^{x_k} \gamma_k(s)\, ds, & \text{for } k = r+1, r+2, \cdots, n. \end{cases} \tag{11b}$$

Each function g_k is "strictly convex" by virtue of both the singleton nature of C_k for $k = 1, 2, \cdots, r$, and the (hypothesized) strictly increasing nature of c_k for $k = r+1, r+2, \cdots, n$.

For future use, the reader should derive from the specific travel-time formula

$$c_k(x_k) = t_k(1 - x_k/b_k)^{-1}, \qquad \text{for } k = r+1, r+2, \cdots, n$$

introduced in Section 10 the correspondingly specific formula

$$g_k(x_k) = -b_k t_k \log(1 - x_k/b_k), \qquad \text{for } k = r+1, r+2, \cdots, n.$$

For our purposes, the differential calculus of the functions g_k should be expressed in terms of the (point-to-set) "subderivative" functions ∂g_k, whose function-values are

$$\partial g_k(x_k) \triangleq \{y_k | g_k(x_k) + y_k(x_k' - x_k) \le g_k(x_k') \qquad \text{for each } x_k' \in C_k\}.$$

To gain insight into this concept, note that the defining inequality for $\partial g_k(x_k)$ simply requires that the line with "affine" representation $y_k' = g_k(x_k) + y_k(x_k' - x_k)$ not be "above" the graph of g_k (i.e., this line with slope y_k must simply "support" the graph of g_k at the graphical point $(x_k, g_k(x_k)) \in E_2$).

Now, the defining equations (11) and the convexity of the functions g_k imply that

$$\partial g_k(x_k) = E_1, \qquad \text{for } x_k = d_k, \qquad \text{for } k = 1, 2, \cdots, r$$

$$\partial g_k(x_k) = \begin{cases} \{y_k \le c_k(0)\}, & \text{for } x_k = 0, \\ \{c_k(x_k)\}, & \text{for } x_k \in (0, b_k), \end{cases} \qquad \text{for } k = r+1, r+2, \cdots, n.$$

$$\partial g_k(x_k) = \varnothing, \text{ otherwise.}$$

Comparing these equations with the defining equations for the functions γ_k (given in Section 9), we see that

$$\begin{gathered} \text{the domain of } \gamma_k \text{ is identical to the domain of } \partial g_k \\ \text{(i.e., the set of all points } x_k \text{ for which } \partial g_k(x_k) \text{ is not empty),} \end{gathered} \tag{12a}$$

and

$$\gamma_k(x_k) = \partial g_k(x_k) \qquad \text{for each } x_k \text{ in the domain of } \gamma_k. \tag{12b}$$

In fact, the functions γ_k were originally defined so that (12) would be satisfied, that is, so that γ_k would be the subderivative of g_k while g_k is an integral of γ_k.

The optimization problem that serves as the basis for one of our dual variational principles involves the functions g_k in the guise of a function g whose domain

$$C \triangleq \bigtimes_1^n C_k, \tag{13a}$$

and whose function-values

$$g(x) \triangleq \sum_1^n g_k(x_k). \tag{13b}$$

For obvious reasons, this function g is said to be "separable" into the functions g_k. Moreover, it is not difficult to show that *g inherits the strict convexity of the g_k*. From the cone X and the function g we now construct the following optimization problem.

Problem A. *Using the "feasible solution" set*

$$S \triangleq X \cap C,$$

calculate both the "problem infimum"

$$\Phi \triangleq \inf_{x \in S} g(x)$$

and the "optimal solution" set

$$S^* \triangleq \{x \in S \,|\, \Phi = g(x)\}.$$

The fact that X is a cone means that Problem A *is a (generalized) "geometric programming problem."* Moreover, the fact that X is actually a convex polyhedral cone and that g is a strictly convex separable function implies that Problem A *is a strictly convex separable programming problem with (essentially) linear constraints*. It is, of course, obvious that *the set of all feasible total flows x is identical to the feasible solution set S*.

In view of our previous hypotheses concerning the travel-time functions c_k, *each function γ_k* (defined in Section 10) *has a nondecreasing (in this case, point-to-point) inverse function γ_k^{-1} with function-values*

$$\gamma_k^{-1}(y_k) = d_k, \qquad \text{for } y_k \in E_1, \qquad \text{for } k = 1, 2, \cdots, r$$

and

$$\gamma_k^{-1}(y_k) = \begin{cases} 0, & \text{for } y_k \leq c_k(0), \\ c_k^{-1}(y_k), & \text{for } c_k(0) \leq y_k, \end{cases} \qquad \text{for } k = r+1, r+2, \cdots, n,$$

where $c_k^{-1}(y_k)$ is, of course, that x_k for which $y_k = c_k(x_k)$.

For future use, the reader should derive from the specific travel-time formula introduced in Section 6 the correspondingly specific formula

$$c_k^{-1}(y_k) = b_k(1 - t_k/y_k), \qquad \text{for } k = r+1, r+2, \cdots, n.$$

With a judicious choice of the constant of integration, the integral $\int^{y_k} \gamma_k^{-1}(t)\, dt$ clearly produces a function h_k whose domain is

$$D_k \triangleq E_1, \qquad \text{for } k = 1, 2, \cdots, n \tag{14a}$$

and whose function-values are

$$h_k(y_k) \triangleq \begin{cases} d_k y_k, & \text{for } k = 1, 2, \cdots, r \\ \displaystyle\int_0^{y_k} \gamma_k^{-1}(t)\, dt, & \text{for } k = r+1, r+2, \cdots, n. \end{cases} \tag{14b}$$

Each function h_k *is convex* by virtue of both the linearity of h_k for $k = 1, 2, \cdots, r$, and the nondecreasing nature of γ_k^{-1} for $k = r + 1, r + 2, \cdots, n$. Since the (hypothesized) strictly increasing nature of c_k clearly implies that γ_k^{-1} is strictly increasing on $\{y_k | c_k(0) \leq y_k\}$, *the function* h_k *is, in fact, strictly convex on* $\{y_k | c_k(0) \leq y_k\}$ *for* $k = r + 1, r + 2, \cdots, n$.

For future use, the reader should derive from the specific travel-time formula introduced in Section 6 the correspondingly specific formula

$$h_k(y_k) = \begin{cases} 0 & \text{for } y_k \leq t_k, \\ b_k(y_k - t_k[1 + \log(y_k/t_k)]) & \text{for } t_k \leq y_k, \end{cases} \quad \begin{array}{l} \text{for } k = r+1, \\ r+2, \cdots, n. \end{array}$$

Now, the defining equations (14) and the convexity of the functions h_k along with the point-to-point nature of γ_k^{-1} clearly imply that

$$\gamma_k^{-1}(y_k) = \partial h_k(y_k) \qquad \text{for each } y_k \text{ in } E_1. \tag{15}$$

In summary, γ_k^{-1} is the subderivative (in fact, the derivative) of h_k while h_k is an integral of γ_k^{-1}.

The optimization problem that serves as the basis for our other dual variational principle involves the functions h_k in the guise of a function h whose domain is

$$D \triangleq \mathop{\times}_{1}^{n} D_k, \tag{16a}$$

and whose function values are

$$h(y) \triangleq \sum_{1}^{n} h_k(y_k). \tag{16b}$$

Of course this function h is separable, and h *inherits the convexity of the* h_k.

From the cone Y and the function h we now construct the following optimization problem.

Problem B. *Using the feasible solution set*

$$T \triangleq Y \cap D,$$

calculate both the problem infimum

$$\Psi \triangleq \inf_{y \in T} h(y)$$

and the optimal solution set

$$T^* \triangleq \{y \in T | \Psi = h(y)\}.$$

The fact that Y is a cone means that *problem B is a (generalized) geometric programming problem.* Moreover, the fact that Y is actually a convex polyhedral cone and that h is a convex separable function implies that *problem B is a convex separable programming problem with (essentially) linear constraints.*

Problem B has important features not possessed by Problem A. In particular, note that relations (14a) and (16a) imply that

$$D = E_n, \tag{17}$$

which in turn clearly implies that

$$T = Y.$$

Since each cone Y contains at least the zero vector, we infer that *problem B has at least one feasible solution (even when problem A has no feasible solutions).*

We shall eventually see that "geometric programming duality theory" reduces the study of traffic equilibria to a study of either problem A or problem B. Actually, we shall then see that problems A and B should be studied simultaneously in order to provide the most insight into traffic equilibria. In essence, problem A describes traffic equilibria entirely in terms of total flows x, while problem B describes traffic equilibria entirely in terms of quasi-travel time vectors y.

An important ingredient in our simultaneous study of problems A and B is the fact that *the cones X and Y defined by equations* (8) *and* (9), *respectively, constitute a pair of* (*convex polyhedral*) "*dual cones*"; that is

$$Y = \{y \in E_n | 0 \leq \langle x, y \rangle \qquad \text{for each } x \in X\}, \tag{18}$$

while

$$X = \{x \in E_n | 0 \leq \langle x, y \rangle \qquad \text{for each } y \in Y\}. \tag{19}$$

Of course, equation (18) is just a repetition of the defining equation (9a) for Y; and it is obvious from equation (18) that X is a subset of the right-hand side of equation (19). To show in turn that the right-hand side of equation (19) is a subset of X, and hence establish equation (19), simply use the representation formula (9b) for Y in conjunction with both the "Farkas lemma" [7] and the defining equation (8) for X.

Another important ingredient in our simultaneous study of problems A and B is the fact that *the functions g_k and h_k defined by equations* (11) *and* (14), *respectively, constitute a pair of convex "conjugate functions"*; that is,

$$D_k = \{y_k | \sup_{x_k \in C_k} [y_k x_k - g_k(x_k)] \text{ is finite}\} \tag{20a}$$

and

$$h_k(y_k) = \sup_{x_k \in C_k} [y_k x_k - g_k(x_k)], \tag{20b}$$

while

$$C_k = \{x_k | \sup_{y_k \in D_k} [x_k y_k - h_k(y_k)] \text{ is finite}\} \tag{21a}$$

and

$$g_k(x_k) = \sup_{y_k \in D_k} [x_k y_k - h_k(y_k)]. \tag{21b}$$

The reader can readily verify these equations for $k = 1, 2, \cdots, r$. Moreover, using only elementary calculus, he can easily verify them for $k = r + 1$, $r + 2, \cdots, n$ when the preceding specific formulas are employed. However, for verifications in greater degrees of generality, the reader may want to consult [3], [5], [23].

It is also important to know that *the functions g and h defined by equations* (13) *and* (16), *respectively, inherit the conjugacy of the g_k and h_k*; that is,

$$D = \{y \in E_n | \sup_{x \in C} [\langle y, x\rangle - g(x)] \text{ is finite}\} \tag{22a}$$

and

$$h(y) = \sup_{x \in C} [\langle y, x\rangle - g(x)], \tag{22b}$$

while

$$C = \{x \in E_n | \sup_{y \in D} [\langle x, y\rangle - h(y)] \text{ is finite}\} \tag{23a}$$

and

$$g(x) = \sup_{y \in D} [\langle x, y\rangle - h(y)]. \tag{23b}$$

In fact, it is easy to see that equations (13), (16), (20) and (21) imply these equations.

The most fundamental properties of problems A and B are not induced by the special nature of roadway networks, only by the (conical) duality of X and Y and the (functional) conjugacy of g and h. We henceforth concentrate on such properties [20], some of which also depend on the separability of g and h.

Notice how problem B can be obtained directly from problem A: simply replace the convex polyhedral cone X with its (convex polyhedral) "dual" Y, and replace the convex separable function g with its (convex separable) "conjugate transform" h. The symmetry of (conical) duality demonstrated by (18) and (19) together with the symmetry of (functional) conjugacy demonstrated by (22) and (23) imply that the problem obtained by applying the same transformation to B is again A. This symmetry justifies the terminology "(geometric) dual problems" for problems A and B. Needless to say, it also induces a symmetry on the theory that relates A to B, in that *each statement about A and B (whose proof uses only the duality of X and Y together with the conjugacy of g and h) automatically produces an equally* *valid "dual statement" about B and A (obtained by interchanging the symbols X and Y, the symbols C and D, and the symbols g and h).* However, to be concise, each dual statement will be left to the reader's imagination.

Unlike the usual min–max formulations of duality in mathematical programming (e.g., in linear programming), both problem A and its dual problem B are minimization problems. The relative simplicity of this min–min formulation will soon become clear, but the reader who is accustomed to the usual min–max formulation must bear in mind that a given duality theorem generally has slightly different statements depending on the formulation in use. In particular, a theorem that asserts the equality of the min and max in the usual formulation asserts that the sum of the mins is zero (i.e., $\Phi + \Psi = 0$) in the present formulation. (The reader interested in the precise connections between the various formulations of duality in mathematical programming should consult [18] and the references cited therein.)

By definition, the "extremality conditions" (for separable geometric programming) are:

(i) $x \in X$ and $y \in Y$
(ii) $0 = \langle x, y \rangle$
(iii) $y_k \in \partial g_k(x_k)$ for $k = 1, 2, \cdots, n$.

Each vector (x, y) that satisfies these conditions is termed an *extremal vector*.

From equations (12) we immediately see that *each network-equilibrium vector* (x, y) *is an extremal vector, and each extremal vector* (x, y) *is a network-equilibrium vector*. Consequently, *we can henceforth concentrate (without any loss of generality) on the properties of extremal vectors* (x, y). The main reason for doing so is that the following powerful theory of (generalized) geometric programming is particularly suited to the study of extremal vectors.

13. Duality Theory

Proofs for the most fundamental duality theorems actually exploit the conjugacy of g and h indirectly via the following (Young–Fenchel) "conjugate inequality"

$$\langle x, y \rangle \le g(x) + h(y) \qquad \text{for each } x \in C \text{ and each } y \in D$$

This inequality is an immediate consequence of either equations (22) or equations (23); and its equality state can be conveniently characterized in terms of the "subgradient" set

$$\partial g(x) \triangleq \{y \in E_n | g(x) + \langle y, x' - x \rangle \le g(x') \qquad \text{for each } x' \in C\},$$

which is of course a multidimensional generalization of the subderivative set discussed in Section 12.

Now, elementary (algebraic) manipulations of the defining inequality for $\partial g(x)$ show that $y \in D$ and $\langle x, y \rangle = g(x) + h(y)$ when $y \in \partial g(x)$. On the other hand, equally elementary (algebraic) manipulations show that $y \in \partial g(x)$ when $\langle x, y \rangle = g(x) + h(y)$. Hence,

$$\langle x, y \rangle = g(x) + h(y) \text{ if and only if } y \in \partial g(x),$$

from which it follows via the symmetry of (functional) conjugacy that

$$y \in \partial g(x) \text{ if and only if } x \in \partial h(y).$$

Finally, in the event that g and h are separable (e.g., in the present roadway network case), it is clear that

$$y \in \partial g(x) \text{ if and only if } y_k \in \partial g_k(x_k) \qquad \text{for each } k = 1, 2, \ldots, n.$$

It is, of course, equally clear that

$$y_k \in \partial g_k(x_k) \text{ if and only if } x_k \in \partial h_k(y_k) \qquad \text{for } k = 1, 2, \cdots, n.$$

The following duality theorem relates the dual problems A and B directly to the extremality conditions (i)–(iii) and hence to traffic equilibria. This theorem is also basic to all succeeding duality theorems.

Theorem 2. *If x and y are feasible solutions to the dual problems A and B respectively* (*in which case the extremality conditions* (i) *are satisfied*), *then*

$$0 \leq g(x) + h(y),$$

with equality holding if and only if the extremality conditions (ii) *and* (iii) *are satisfied, in which case x and y are optimal solutions to problems A and B respectively.*

PROOF. The defining inequality for Y and the conjugate inequality for h show that $0 \leq \langle x, y \rangle \leq g(x) + h(y)$, with equality holding in both of these inequalities if and only if the extremality conditions (ii) and (iii) are satisfied, in which case a subtraction of the resulting equality $0 = g(x) + h(y)$ from each of the inequalities $0 \leq g(x') + h(y)$ and $0 \leq g(x) + h(y')$ (which are valid for arbitrary feasible solutions x' and y') shows that x and y are actually optimal solutions. □

The fundamental inequality given by Theorem 2 implies important properties of the dual infima Φ and Ψ.

Corollary 2A. *If the dual problems A and B both have feasible solutions, then*

(a) *the infimum Φ for problem A is finite, and*

$$0 \leq \Phi + h(y) \qquad \text{for each } y \in T,$$

(b) *the infimum Ψ for problem B is finite, and*

$$0 \leq \Phi + \Psi.$$

The proof of this corollary is, of course, a trivial application of Theorem 2.

The strictness of the inequality $0 \leq \Phi + \Psi$ in conclusion (b) plays a crucial role in almost all that follows. In fact, dual problems A and B that have feasible solutions and for which $0 < \Phi + \Psi$ are said to have a "duality gap" of $\Phi + \Psi$. Although duality gaps do occur in (generalized) geometric programming, we shall eventually see that they do not occur in the present context of traffic equilibria. This lack of duality gaps is extremely fortunate because of their highly undesirable properties. In particular, we shall soon see that they weaken the bond between the dual problems A and B and imply that the extremality conditions have no solutions (i.e., no "extremal vectors" exist, even though problems A and B may have optimal solutions). They also tend to destroy the possibility of using the inequality $0 \leq g(x) + h(y)$ to provide an "algorithmic stopping criterion".

Such a criterion results from specifying a positive tolerance ε so that the numerical algorithms being used to minimize both $g(x)$ and $h(y)$ are terminated when they produce a pair of feasible solutions x^+ and y^+ for which

$$g(x^+) + h(y^+) \leq \varepsilon.$$

Because conclusion (a) to Corollary 2A along with the definition of Φ shows that

$$-h(y^+) \leq \Phi \leq g(x^+),$$

we conclude from the preceding tolerance inequality that

$$\Phi \leq g(x^+) \leq \Phi + \varepsilon.$$

Hence Φ can be approximated by $g(x^+)$ with an error no greater than ε (and, dually, Ψ can be approximated by $h(y^+)$ with an error no greater than ε). This stopping criterion can be useful even though it does not estimate the proximity of x^+ to an optimal solution x^* (and, dually, even though it does not estimate the proximity of y^+ to an optimal solution y^*). In any event, it is clear that the algorithms being used need not terminate unless $\Phi + \Psi < \varepsilon$, a situation that is guaranteed for each tolerance ε only if $0 = \Phi + \Psi$ (i.e., there is no duality gap).

The equality characterization of the fundamental inequality given by Theorem 2 implies the key relations between extremal vectors (x, y) and optimal solutions $x \in S^*$ and $y \in T^*$.

Corollary 2B. *If* (x^+, y^+) *is an extremal vector (i.e., if* (x^+, y^+) *is a solution to the extremality conditions* (i)–(iii)), *then*

(a) $x^+ \in S^*$ *and* $y^+ \in T^*$.
(b) $S^* = \{x \in X | 0 = \langle x, y^+ \rangle, \text{ and } x_k \in \partial h_k(y_k^+) \text{ for } k = 1, 2, \cdots, n\}$
(c) $T^* = \{y \in Y | 0 = \langle x^+, y \rangle, \text{ and } y_k \in \partial g_k(x_k^+) \text{ for } k = 1, 2, \cdots, n\}$
(d) $0 = \Phi + \Psi$.

On the other hand, if the dual problems A and B both have feasible solutions and if $0 = \Phi + \Psi$, *then* (x, y) *is an extremal vector if and only if* $x \in S^*$ *and* $y \in T^*$.

The proof of this corollary comes from a trivial application of Theorem 2 along with the (previously mentioned and easily established) conjugate transform relations $\partial g_k(x_k) \subseteq D_k$ and $\partial h_k(y_k) \subseteq C_k$.

Corollary 2B readily yields certain basic (nontrivial) information about traffic equilibria. In particular, the first part of Corollary 2B asserts that *if at least one network-equilibrium vector* (x^+, y^+) *exists, then no duality gap occurs.* On the other hand, the second part of Corollary 2B clearly implies that *if no duality gap occurs, then the set* E *of all network equilibrium vectors* (x, y) *is just the Cartesian product* $S^* \times T^*$ *of the optimal solution sets* S^* *and* T^*. Consequently, *when network equilibria exist, each quasi-equilibrium total flow* x^* *(namely, each* $x^* \in S^*$*) can be paired with each quasi-travel time vector* y^* *(namely, each* $y^* \in T^*$*) to produce a network-equilibrium vector* $(x^*, y^*) \in E$. *Moreover, network equilibria exist if and only if our initial hypotheses concerning the travel-time function* c_k, $k = r + 1, r + 2, \cdots, n$, *are sufficiently strong to guarantee the absence of a duality gap along with the existence of optimal solutions* $x^* \in S^*$ *and* $y^* \in T^*$. Actually, the existence of network equilibria for our general model is established in [12] by showing for an even more general model that $0 = \Phi + \Psi$ and neither S^* nor T^* is empty. However, that proof is well beyond the prerequisites for this module.

Computationally, it is important to note that conclusions (a)–(c) of Corollary 2B provide *methods for calculating all network-equilibrium vectors* $(x, y) \in E$ *from the knowledge of a single network-equilibrium vector* $(x^+, y^+) \in E$ (or even from just the knowledge of a single optimal solution $x^* \in S^*$, or a single optimal solution $y^* \in T^*$). Of course, *the (nonempty) set* $Z(x, y)$ *of all quasi-equilibrium circuit flows* z *that generate a given network-equilibrium vector* $(x, y) \in E$ *can be calculated simply by calculating the set of all solutions* z *to the feasibility conditions* (1, 2) *and the complementary slackness conditions* (7′) as explained in the next-to-last paragraph of Section 10.

It is also important to note that conclusions (b)–(c) of Corollary 2B imply that *both the set* S^* *(of all quasi-equilibrium total flows* x^**) and the set* T^* *(of all quasi-travel time vectors* y^**) are convex polyhedral sets, and hence so is the set*

$$E = S^* \times T^*$$

(of all network-equilibrium vectors (x^*, y^*)*).* The reason is that $\partial h_k(y_k^+)$ and $\partial g_k(x_k^+)$ are clearly closed intervals, and convex polyhedralness is preserved under Cartesian products. Of course, *the set* Z *of all quasi-equilibrium circuit flows* z *has the representation formula*

$$Z = \bigcup_{(x,y) \in E} Z(x, y)$$

where each set

$$Z(x, y) \triangleq \{z \in E_m | \text{conditions (1, 2) and (7′) are satisfied}\}$$

is clearly a bounded convex polyhedral set.

14. Uniqueness Theory

It is worth noting that various uniqueness theorems result from the preceding representation formula $E = S^* \times T^*$.

In particular, *there is at most one quasi-equilibrium total flow* x^*. The reason is that for $k = 1, 2, \cdots, r$ the condition $x_k = d_k$ asserts the uniqueness of x_k^*, while for $k = r + 1, r + 2, \cdots, n$ the strict monotonicity of c_k implies that $\gamma_k(x^*)$ and $\gamma_k(x^{**})$ do not intersect when $x_k^* \neq x_k^{**}$ (so a given y_k^* cannot be in both $\gamma_k(x_k^*)$ and $\gamma_k(x_k^{**})$ unless $x_k^* = x_k^{**}$).

On the other hand, *each quasi-travel time vector* y^* *exhibits a unique component* y_k^* *for each* k *for which* $x_k^* > 0$ (*in fact, always for* $k = 1, 2, \cdots, r$). For $k = r + 1, r + 2, \cdots, n$, the reason is that the continuity of c_k implies that $\gamma_k(x_k^*)$ contains a unique element $c_k(x_k^*)$ when $x_k^* > 0$. For $k = 1, 2, \cdots, r$, the reason is that the condition $x_k = d_k > 0$ and the preceding uniqueness along with conditions (6′) and the finiteness of the feasible circuit family [i] clearly imply that y_k^* can have at most a finite number of values, while the convexity of T^* clearly implies that y_k^* can have infinitely many values if it has more than one value.

Finally, even when there is a unique network-equilibrium vector (x^*, y^*), infinitely many quasi-equilibrium circuit flows z^* may exist. In fact, uniqueness of z^* in that case clearly requires a "positive independence" of those circuit vectors δ^j for which $\langle \delta^j, y^* \rangle = 0$ and $z_j^* > 0$. Clearly, such circuit vectors δ^j are less likely to be positively independent the more the corresponding circuits "overlap" (i.e., the more possibility there is for congestion). In any event, the reader can probably construct simply examples where z^* has no unique components.

15. Computational Considerations

At least four different approaches to computing traffic equilibria (x, y) can be found.

The first approach [4], [8], [14] seems to be the only one that has been widely used. In essence, it "simulates" the behavior of drivers by using "shortest path algorithms" (as described, for example, in [10], [13]) to iteratively solve in heuristic ways the defining equations (1–6) for Wardrop-equilibria. However, it is widely known that such algorithms are inadequate, in that they frequently fail to converge.

The second approach has been set forth only recently [12]. It uses any algorithm (possibly those in [11], [15], [25]) that can solve the network-equilibrium conditions (i)–(iii). This approach is analogous to a well-known (rather successful) approach used in analyzing electric and hydraulic networks. In the context of such networks, condition (i) constitutes the "Kirchoff current and potential (conservation) laws", condition (ii) is implied by (i) and hence is redundant, and condition (iii) is just Ohm's law.

The third approach uses any algorithm that can solve the convex separable optimization problem A. Actually, any such algorithm that exploits the absence of nonlinear constraints (for example, [17]) can be used. To do so, simply replace the only nonlinear constraints (4) with the (weaker) linear constraints

$$0 \leq x_k \leq b_k, \qquad \text{for } k = r + 1, r + 2, \cdots, n. \tag{4'}$$

This reformulation does not alter the infimum Φ and the optimal solution set S^* for problem A, because the hypothesis at the beginning of Section 12 clearly implies via the defining equation (13b) for $g(x)$ that $g(x)$ approaches $+\infty$ when at least one component x_q approaches b_q. Of course, once a single optimal solution $x^* \in S^*$ has been computed, the set E of all traffic equilibria (x^*, y^*) can be computed via the technique outlined after Corollary 2B. This approach is analogous to a classical variational approach [26], [6], [2], [16], [1], [23] used in analyzing electric and hydraulic networks. In the context of such networks, problem A consists of minimizing the "content" of the network (which is frequently just the power dissipated in the network), subject to only the Kirchoff current (conservation) law.

The fourth approach uses any algorithm that can solve the convex separable linearly constrained optimization problem B (for example, [17]). Needless to say, once a single optimal solution $y^* \in T^*$ has been computed, the set E of all traffic equilibria (x^*, y^*) can be computed via the technique outlined after Corollary 2B. This approach is also analogous to a classical variational approach [26], [6], [2], [16], [1], [23] used in analyzing electric and hydraulic networks. In the context of such networks, problem B consists of minimizing the "co-content" of the network (which is frequently just the power dissipated in the network), subject to only the Kirchoff potential (conservation) law.

Unlike the first and second approaches, the only nonlinear functions to be reckoned with in the third and fourth approaches (the functions g_k and h_k, respectively) are both continuous and convex. Naturally, the third and fourth approaches can be carried out in unison to provide an algorithmic stopping criterion, as explained after Corollary 2A. However, that criterion may not be of much use because it provides direct information only about the degree of convergence of $g(x)$ and $h(y)$, not the desired degree of convergence of x and y. Nevertheless, the degree of convergence of x and y can frequently be deduced from the degree of convergence of $g(x)$ and $h(y)$ through an analysis of the rates of change of the γ_k and the γ_k^{-1} (e.g., the second derivative of the g_k and the h_k when they exist). In any event, the fact that $D = E_n$ (as asserted by equation (17)) may endow the fourth approach with a numerical advantage over the third approach (because $C \neq E_n$). Moreover, the immediate availability of the data required in the representation formula (9b) for Y (i.e., the generators δ^j for X) may endow the fourth approach with still another numerical advantage over the third approach (because comparable data required in the corresponding (unstated) representation formula for X can be obtained only from the generators for Y, which can be computed only with the aid of a rather intricate linear algebraic algorithm due to Uzawa [27]).

Actually, network planners need only know T^* to uncover the possible "bottlenecks" in a given network. Of course, the same information can also be extracted from S^*. Moreover, it is worth noting that either S^* or T^* can be computed without computing the set Z (of all demand-equilibrium circuit flows) simply by using either the second, third, or fourth approaches. Consequently, the first approach is, no doubt, the least attractive approach for network planning, even though it is the only approach that is now widely used.

Finally, the decomposition principles given in [19]–[21] should probably be used in conjunction with the preceding approaches when the network is large but "sparse" (e.g., when the network results from modeling a metropolitan roadway network in which various identifiable subnetworks are only weakly linked to one another).

16. Model Construction and Solution

To predict the peak travel times that would result if an additional link were constructed between Blairsville and Bolivar, Joe represents the resulting area road map by the graph shown in Figure 9.5. As with the original problem, Blairsville, Bolivar, the intersection of Highbridge and Valley Roads, and the Consolidated Coal Mines are represented by nodes 1, 2, 3, and 4, respectively. Unlike the original problem, we now have only two commodities, commodities 1 and 2, represented by return links 1 and 2, respectively. The reason is that the strengthening of the bridge on Highbridge Road permits the aggregation of the original commodities 2 and 3 into a single commodity 2 (as mentioned at the end of the Appendix to Part I). Needless to say, the loss of commodity 3 (with the corresponding loss of return link 3) changes the numbering of the original real links from 4, 5, 6, and 7 to 3, 4, 5, and 6, respectively. Of course, the additional road whose construction is

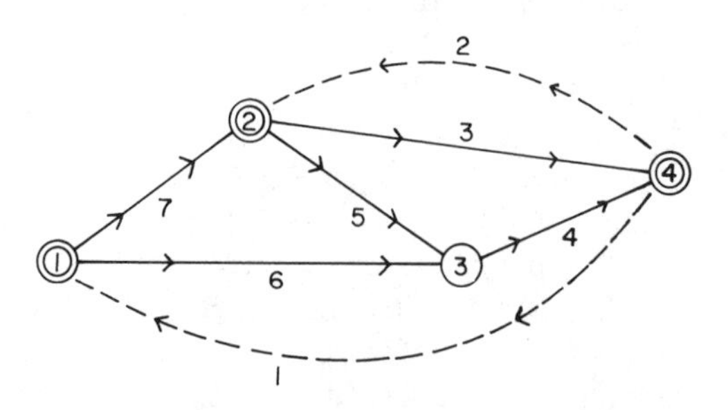

Figure 9.5. Graphical Representation after Aggregation and Link Construction.

being considered is represented by the additional real link 7, which is directed in the expected direction of the morning rush-hour traffic flow.

Now, the feasible circuit vectors for commodities 1 and 2 are

$$\begin{cases} \delta^1 \triangleq (1, 0, 0, 1, 0, 1, 0) \\ \delta^2 \triangleq (1, 0, 0, 1, 1, 0, 1) \\ \delta^3 \triangleq (1, 0, 1, 0, 0, 0, 1) \end{cases}$$

$$\begin{cases} \delta^4 \triangleq (0, 1, 0, 1, 1, 0, 0) \\ \delta^5 \triangleq (0, 1, 1, 0, 0, 0, 0). \end{cases}$$

The corresponding index sets are

$$[1] \triangleq \{1, 2, 3\}$$

$$[2] \triangleq \{4, 5\}.$$

The commodity input flows are

$$d_1 = 10, \qquad d_2 = 16,$$

and the real link capacities are

$$b_3 = 25, \qquad b_4 = 20, \qquad b_5 = 10, \qquad b_6 = 15, \qquad b_7 = 30.$$

The zero flow travel times are

$$t_3 = 15, \qquad t_4 = 1, \qquad t_5 = 2, \qquad t_6 = 5, \qquad t_7 = 1,$$

and the travel-time functions have the formula

$$c_k(x_k) = t_k(1 - x_k/b_k)^{-1}, \qquad \text{for } k = 3, 4, 5, 6, 7.$$

The data for (the additional) link 7 have been estimated by Bill Ivory, with the aid of a detailed description of the road that would be built.

As noted in Section 12, the preceding travel-time formula for $c_k(x_k)$ gives rise to the formula

$$g_k(x_k) = -b_k t_k \log(1 - x_k/b_k), \qquad \text{for } k = 3, 4, 5, 6, 7.$$

The resulting function g has domain

$$C = \{d_1\} \times \{d_2\} \times [0, b_3) \times [0, b_4) \times [0, b_5) \times [0, b_6) \times [0, b_7)$$

and function values

$$g(x) = 0 + 0 + g_3(x_3) + g_4(x_4) + g_5(x_5) + g_6(x_6) + g_7(x_7).$$

The corresponding variational principle is

minimize $g(x)$ subject to both the

condition $x = \sum_{j=1}^{5} z_j \delta^j$ and the

condition $z_j \geq 0, j = 1, 2, 3, 4, 5.$

As noted in Section 12, the preceding travel-time formula for $c_k(x_k)$ gives rise to the formula

$$h_k(y_k) = \begin{cases} 0, & \text{for } y_k \le t_k \\ b_k(y_k - t_k[1 + \log(y_k/t_k)]), & \text{for } t_k \le y_k \end{cases}, \quad \text{for } k = 3, 4, 5, 6, 7.$$

The resulting function h has domain

$$D = E_7$$

and function values

$$h(y) = d_1 y_1 + d_2 y_2 + h_3(y_3) + h_4(y_4) + h_5(y_5) + h_6(y_6) + h_7(y_7).$$

The corresponding variational principle is

$$\text{minimize } h(y) \text{ subject to the condition } 0 \le \langle \delta^j, y \rangle, j = 1, 2, 3, 4, 5.$$

A solution to the preceding dual variational principles, obtained by a computer algorithm [17], is

$$z_1^{***} = 9.49582, \quad z_2^{***} = 0.50418, \quad z_3^{***} = 0,$$

$$z_4^{***} = 7.90965, \quad z_5^{***} = 8.09035$$

$$x_1^{***} = 10, \quad x_2^{***} = 16, \quad x_3^{***} = 8.09035, \quad x_4^{***} = 17.90965,$$

$$x_5^{***} = 8.41382, \quad x_6^{***} = 9.49582, \quad x_7^{***} = 0.50418$$

$$y_1^{***} = -23.19378, \quad y_2^{***} = -22.17669, \quad y_3^{***} = 22.17669,$$

$$y_4^{***} = 9.56776, \quad y_5^{***} = 12.60893, \quad y_6^{***} = 13.62602, \quad y_7^{***} = 1.01709$$

A comparison of (x^{**}, y^{**}, z^{**}) and $(x^{***}, y^{***}, z^{***})$ shows that constructing the additional link between Blairsville and Bolivar would decrease the peak travel time of Blairsville residents by 2.72727 while increasing the peak travel time of Bolivar residents by 0.33458. Joe communicates this information to the Blairsville residents, who must now decide whether this predicted decrease in their peak travel time and other benefits are sufficiently large to justify the cost of building the additional link.

Although x^{***} and y^{***} are uniquely determined (by virtue of the theory in Section 14), the reader should note that z^{***} is not uniquely determined. In fact, since the travel time $y_3^{***} + y_7^{***} = 23.19378$ over path 3 for Blairsville residents is identical to their peak travel time $-y_1^{***}$, the flow z_3 can clearly be increased from $z_3^{***} = 0$ with compensating increases and decreases in z_4 and z_5, respectively. It may now be beneficial for the reader to study (via the analysis in the next-to-last paragraph of Section 10) the whole uniqueness question concerning z^{***}.

17. Final Remarks

A model that is more general than the model given in Section 10 is studied in [12]. In addition to weakening the monotonicity hypotheses on, and allowing discontinuities in, the travel-time functions c_k, it permits the input flows d_i to be monotone nonincreasing point-to-set functions of the corresponding travel times $-y_i$, a flexibility that is needed when the "demand" for travel is "elastic."

Exercises

1. Given that a control light is at the intersection of Highbridge and Valley Roads, determine the conditions under which Joe will not have to perform a separate analysis of the evening rush-hour traffic for Bolivar residents. Also, model those situations in which a separate analysis has to be performed.
2. Model the situation in which the relatively weak bridge on Highbridge Road is not to be strengthened but the new link between Blairsville and Bolivar is to be built.
3. Predict the peak flows and travel times for the hypothetical situation described in Exercise 2. (This exercise may require the use of a computer algorithm, such as the one described in [17].)
4. Given the travel time formula

$$c(x) = t(1 - x/b)^{-p}, \qquad \text{for } 0 \leq x < b$$

with positive parameters b, p, and t, show the following.
(a) Show that

$$\gamma(x) \triangleq \begin{cases} \{y \leq c(0)\}, & \text{for } x = 0 \\ \{c(x)\}, & \text{for } x \in (0, b) \end{cases}$$

has an inverse

$$\gamma^{-1}(y) = \begin{cases} 0, & \text{for } y \leq c(0) \\ c^{-1}(y), & \text{for } c(0) \leq y \end{cases}$$

where

$$c^{-1}(y) = b(1 - [t/y]^{1/p}).$$

(b) Show that when $p \neq 1$, the function $g(x) \triangleq \int_0^x \gamma(s)\, ds$ has a finite value on

$$C = \begin{cases} [0, b], & \text{when } p < 1 \\ [0, b), & \text{when } 1 < p \end{cases}$$

with

$$g(x) = -bt[(1 - x/b)^{(1-p)} - 1]/(1 - p),$$

(c) Show that when $p \neq 1$, the function $h(y) \triangleq \int_0^y \gamma^{-1}(t)dt$ has a finite value on

$$D = (-\infty, +\infty)$$

with

$$h(y) = \begin{cases} 0, & \text{for } y \leq c(0) \\ b(y - t[1 + qt^{-1/q}(y^{1/q} - t^{1/q})], & \text{for } c(0) \leq y, \end{cases}$$

where

$$\frac{1}{q} \triangleq 1 - \frac{1}{p}.$$

References

[1] C. Berge and A. Ghouila-Houri, *Programming, Games, and Transportation Networks.* New York: Wiley, 1963.
[2] G. Birkhoff and J. B. Diaz, "Nonlinear network problems," *Quart. Appl. Math.*, vol. 13, pp. 431–443, 1956.
[3] R. P. Boas and M. Marcus, "Inverse functions and integration by parts," *Amer. Math. Month.*, vol. 81, pp. 760–761, 1974.
[4] Bureau of Public Roads, *Traffic Assignment Manual*, U.S. Gov't. Printing Office, Washington, DC, 1964.
[5] J. B. Diaz and F. T. Metcalf, "An analytic proof of Young's inequality," *Amer. Math. Month.*, vol. 77, pp. 603–609, 1970.
[6] R. J. Duffin, "Nonlinear networks IIa," *Bull. Amer. Math. Soc.*, vol. 53, pp. 963–971, 1947.
[7] R. J. Duffin, E. L. Peterson and C. Zener, *Geometric Programming.* New York: Wiley, 1967.
[8] Federal Highway Administration, *Urban Transportation Planning*, Washington, DC, 1972, III-1 to III-20.
[9] W. Fenchel, "Convex cones, sets and functions," *Mathematics Department Mimeographed Lecture Notes*, Princeton Univ. Princeton, NJ, 1951.
[10] L. K. Ford and D. R. Fulkerson, *Flows in Networks.* Princeton, NJ: Princeton Univ. Press 1962.
[11] G. J. Habetler and A. L. Price, "Existence theory for generalized nonlinear complementarity problems," *J. Opt. Th. Appl.*, vol. 7, 1971.
[12] M. A. Hall and E. L. Peterson, "Traffic equilibria analysed via geometric programming," in *Traffic Equilibrium Methods Proceedings-Montreal*, M. Florian, Ed., Lecture Notes in Economics and Mathematical Systems #118. New York: Springer-Verlag, 1976, pp. 53–105.
[13] T. C. Hu, *Integer Programming and Network Flows* Reading, MA: Addison-Wesley, 1969.
[14] M. Huber *et al.*, "Comparative analysis of traffic assignment techniques with actual highway use," *N.C.H.R.P. Report* 58.
[15] S. Karamardian, "Generalized complementarity problem," *J. Opt. Th. Appl.*, vol. 8, 1971.
[16] G. J. Minty, "Monotone Networks," *Proc. of Royal Soc., London, Ser. A* 257, (1960), 194–212.

[17] B. A. Murtagh and M. A. Saunders, "Nonlinear programming for large, sparse systems," *Systems Optimization Lab. Tech. Rep., SOL* 76–15, Operations Research Dept., Stanford Univ., Stanford, CA, 1976.
[18] E. L. Peterson, "Mathematical foundations of geometric programming," Appendix to "Geometric programming and some of its extensions," in *Optimization and Design*, Avriel, Rijckaert, Wilde, Eds. Englewood Cliffs, NJ: Prentice-Hall, 1973, pp. 244–289.
[19] ——, "The decomposition of large (generalized) geometric programming problems by tearing," *Proc. NATO Inst. Large Scale Systems*, Himmelblau, Ed. Amsterdam: North Holland, 1973, pp. 525–545.
[20] ——, "Geometric programming," *SIAM Rev.*, vol. 18, pp. 1–51, 1976.
[21] ——, *The Mathematical Foundations of Convex and Nonconvex Programming*, in preparation.
[22] R. B. Potts and R. M. Oliver, *Flows in Transportation Networks*. New York: Academic, 1972.
[23] R. T. Rockafeller, "Convex programs and systems of elementary monotonic relations," *J. Math. Anal. Appls.*, vol. 19, pp. 543–564, 1967.
[24] ——, *Convex Analysis*. Princeton, NJ: Princeton Univ. Press, 1970.
[25] R. Saigal, "On the class of complementary cones and Lemke's algorithm," *SIAM J. Appl. Math.*, vol. 23, 1972.
[26] W. Thompson (Lord Kelvin), *Cambridge and Dublin Math. J.* pp. 84–87, 1848.
[27] H. Uzawa, "A theorem on convex polyhedral cones," *Studies in Linear and Nonlinear Programming*, K. J. Arrow, L. Hurwicz, and H. Uzawa, Eds., Stanford, CA: Stanford Univ. Press, 1958, p. 3.
[28] J. G. Wardrop, "Some theoretical aspects of road traffic research," *Proc. Inst. Civ. Eng. Part II*, vol. 1, pp. 325–378, 1952.

ACKNOWLEDGEMENT: The author is indebted to Dr. M. A. Saunders of Stanford University's System Optimization Laboratory for using his computer code to calculate the solutions contained herein.

Notes for the Instructor

Prerequisites. Part I requires a familiarity with the elementary algebra of n-tuples. It also requires previous experience with the network flow concepts usually covered in an introductory physics or electrical engineering course, and preferably with the network flow theory covered in an operations research or combinatories course. Much of Part II requires the mathematical maturity usually not attained until after a first course on real analysis. A complete mastery of Part II requires some exposure to mathematical programming. It also requires a knowledge of the Farkas lemma and other topics for which references are provided.

Time. Part I requires one or two class periods. Some of the sections can be initially withheld while the class attempts its own analyses. Part II could require six or more class periods.

CHAPTER 10

An Optimal Mix Problem: Capacity Expansion in a Power Field

Jacob Zahavi*

1. Introduction

The capacity expansion problem in the power field is indeed a very complex one. In this module we present a static, or one-year, capacity expansion problem that can be solved rather easily using a breakeven point analysis.

The demand to be satisfied in the target year is represented by the temporal distribution of demand known as the load duration curve (LDC) [1]. A typical yearly LDC is exhibited in Figure 10.1. Each point on the abscissa denotes the number of hours in the given year, or the fraction of time, during which the associated demand on the ordinate is equalled or exceeded. The load level which is required throughout the year (8760 hours) is referred to as the base load ($L_{\min}$), the highest level of demand which is required only a small fraction of the time is the peak demand ($L_{\max}$), and all load levels in between are the intermediate demand.

The supply side is given by an array of generating units of different types, e.g., nuclear, coal-fired, oil-fired, and gas turbine units, and different sizes, as measured by MW, which differ from one another with respect to both their capital and operating cost.

The optimal mix problem, then, amounts to finding the optimal mix of units, from amongst the set of available units, to satisfy the demand in the target year at minimum cost. This static, or stationary, optimal mix problem has been widely discussed in the literature for the case the power system in the target year contains new units only [1]. We extend the algorithm in this chapter to the case where the mix of units in the target year also contain some existing units.

* Faculty of Management, Tel-Aviv University, Tel-Aviv, Israel.

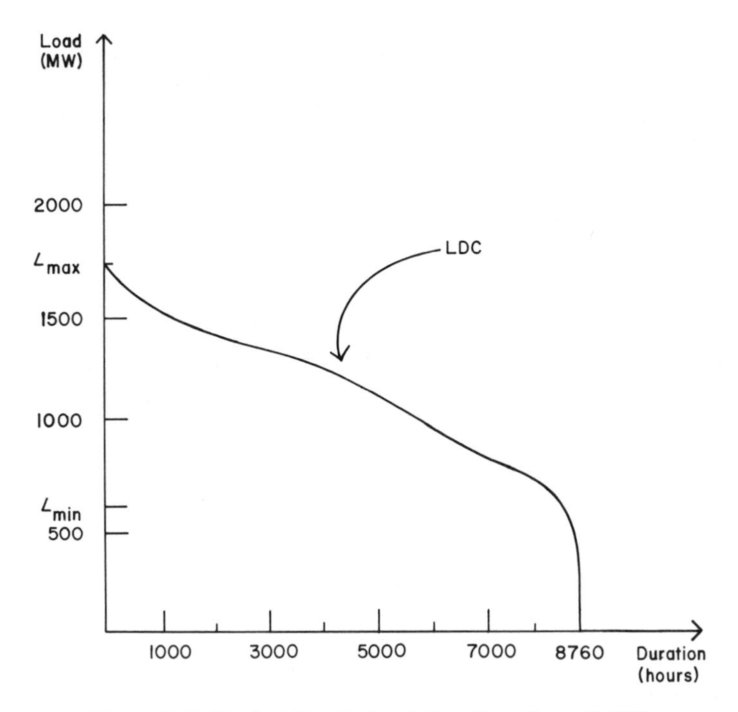

Figure 10.1. Typical Yearly Load Duration Curve (LDC).

2. The Optimization Problem—New Units Only

In this case we assume that the power system in the target year consists of new units only, i.e., we start the power system from scratch. To formulate the optimization problem, we introduce the following notation:

n	number of units in the target year,
t	target year,
x_i	capacity of unit i (MW),
c_i	total fixed cost for unit i (\$/MW/year),
b_i	total variable costs for unit i (\$/MWHr),
E_i	the energy output of unit i (MWHr),
$L(\tau)$	the LDC for the target year,
L_{max}	the peak demand in the target year (MW),
TC	the total cost for the system.

Unless stated otherwise, all the variables will correspond to the given target year t. The index t will be omitted for brevity.

To facilitate the analysis, we assume that c_i is constant, regardless of the unit's capacity, and that b_i is constant, regardless of the unit's output (i.e., no economies of scale exist in both the capacity and the operating costs).

Thus, ignoring the setup and the shutdown costs to bring the generating unit on or off line, the least cost procedure to operate the power system is to load the generating units to production, at their rated capacity, in increasing order of their total variable cost (the merit order loading). Without loss of generality, we arrange the generating units so that:

$$b_1 < b_2 < \cdots < b_i < \cdots < b_n. \tag{1}$$

It is further assumed that

$$c_1 > c_2 > \cdots > c_i > \cdots > c_n \tag{2}$$

because otherwise, if both $b_i < b_j$ and $c_i < c_j$, unit j is inferior and could be discarded from the mix of units to supply the demand in the target year.

The merit order loading is described in Figure 10.2. The point at which we load unit i is referred to as the loading point of the unit. Denoting this point by D_i, we have

$$D_1 \triangleq 0$$

$$D_i \triangleq \sum_{j=1}^{i-1} x_j, \qquad i = 2, \cdots, n.$$

The area under the LDC between the loading points of unit i and unit $i + 1$ is the energy output of unit i (see Figure 10.2).

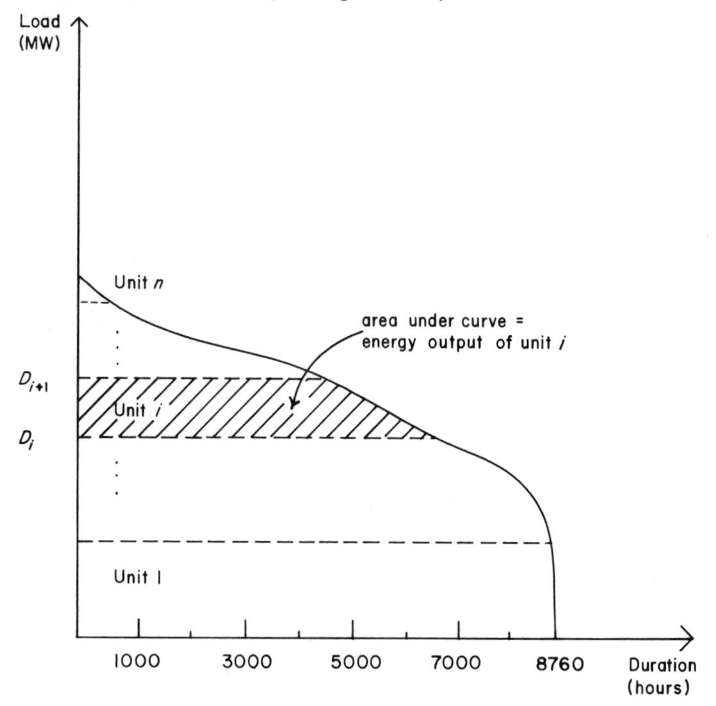

Figure 10.2. The Merit Order Loading.

Given these assumptions, the capacity expansion problem for a given year can be formulated as follows:

$$\min_{x_1, \cdots, x_n} \text{TC} = \sum_{i=1}^{n} c_i x_i + \sum_{i=1}^{n} b_i E_i \tag{3}$$

subject to

$$x_i \geq 0 \tag{4}$$

$$\sum_{i=1}^{n} x_i \geq L_{\max} \tag{5}$$

where $E_i, i = 1, \cdots, n$ is calculated by integrating the LDC over the loading interval of the unit, as implied by Figure 10.2.

We note that the first expression in Equation (3) is the total fixed cost incurred; the second expression is the total operating expenses; constraint (4) is the nonnegativity constraint; and constraint (5) is the peak demand constraint.

The optimization problem (3)–(5) is a convex nonlinear programming problem, that can be solved either graphically or analytically using a simple breakeven point analysis. This procedure can also be shown to satisfy the Kuhn Tucker conditions of the problem, thus yielding a unique optimal solution for the optimization problem (3)–(5).

We pursue the graphical solution to the problem in Figure 10.3 for a three-unit system, containing a coal-fired (unit 1), an oil-fired (unit 2), and a gas turbine unit (unit 3). Typically, the coal-fired unit is relatively cheap to operate (fuel costs are low) but expensive to install, the gas turbine unit is very expensive to operate but very cheap to install, with the oil-fired unit lying in between. Thus the variable and fixed costs of these units satisfy the inequalities (1) and (2). We plot the cost curves for the units in the upper portion of Figure 10.3. Since the costs b_i and c_i are constant for any unit i regardless of its capacity and time of operation, the cost function representing the total cost for unit i as a function of operating time is linear, with the intercept representing the total investment cost per MW per year (c_i) and the slope denoting the total variable cost for a unit of energy (1 MWHr) produced by unit i (b_i). Since the operating and investment costs for the various units are inversely related, the curve describing the minimum total expenses for the system as a function of time, is a piecewise concave function, as shown in the upper portion of Figure 10.3. We refer to this function as the minimum cost polygon. Thus the gas turbine unit is not economical if operated more than t_2 percent of the time, the oil-fired unit is economical only if operated more than t_2 but less than t_1 percent of the time, whereas the coal-fired unit is economical only if operated continuously throughout the time period. It should be noted that the oil-fired unit (unit 2) is indeed more expensive to install than the gas turbine unit (unit 3), but is also cheaper to operate. Thus, when the oil fired unit works more than t_2 percent of the time, the additional

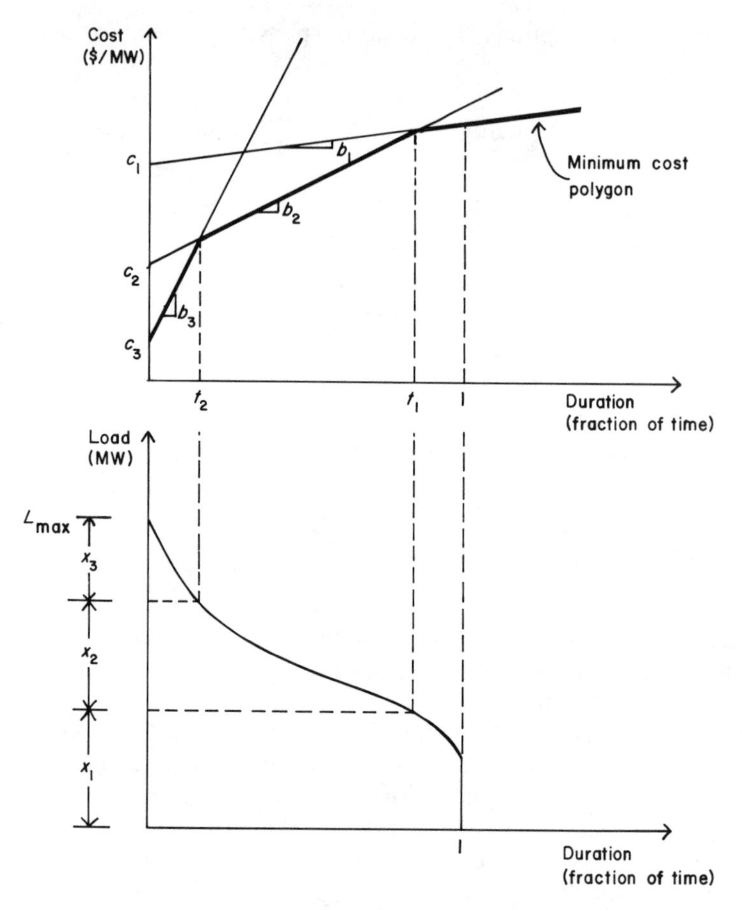

Figure 10.3. Optimal Mix Algorithm—New Units Only.

expenses in capital cost are actually offset by the savings in operating cost, so that the oil-fired unit becomes cheaper in terms of total cost. The breakeven point occurs when the two costs are equal, i.e.,

$$c_2 + b_2 t_2 = c_3 + b_3 t_2$$

from which

$$t_2 = \frac{c_2 - c_3}{b_3 - b_2}$$

or in general

$$t_i = \frac{c_i - c_{i+1}}{b_{i+1} - b_i}, \qquad i = 1, 2, \cdots, n - 1.$$

Then, by projecting the breakeven points onto the LDC, as demonstrated in the lower portion of Figure 10.3, we obtain the optimal capacities to install of each unit, i.e.:

$$x_i = D_i - D_{i-1}, \qquad i = 1, 2, \cdots, n$$
$$D_1 = 0$$
$$D_n = L_{\max}$$

3. Optimization with Existing Units—1 EMU

Clearly, the assumption that the power system in the target year consists of new units only is not realistic, especially when dealing with short and medium range investment planning. We therefore devise an approach to allow for existing units to also participate in the mix of units in the target year, distinguishing between two cases: (a) only efficient existing units are allowed in the optimal mix of units; (b) all existing units, even if inefficient, are required to participate in the mix of units in the target year.[1]

To develop the algorithm, we arrange the new and existing units in a merit order. We refer to any consecutive series of existing units in the merit order as an existing–multi-unit (EMU). To start with, we assume only 1 EMU, consisting of $s - r + 1$ units occupying positions $r, r + 1, \cdots, s$ in the merit order. The optimization problem is the same as in (3)–(5). To allow for efficient existing units to enter the optimal mix of units in the target year, we add the additional set of constraints:

$$x_i \leq y_{i+1-r}, \qquad i = r, \cdots, s \tag{6}$$

where $y_1, \cdots, y_\ell$, $\ell = s - r + 1$, are given constants, expressing the capacities, in MW, of the existing units.

While not required mathematically, it is plausible to assume that the peak demand at the target year $L_{\max}$ is higher than the peak demand at the base year, i.e.:

$$\sum_{i=1}^{\ell} y_i \leq L_{\max}$$

To find a clue for the solution procedure in this case, we evaluate the Kuhn–Tucker condition for the problem (3)–(6) (see note at end of the module).

Letting

- u_i for $i = 1, \cdots, n$, dual variables (Lagrange multipliers) of the nonnegativity constraints (4),
- u dual variable of the peak demand constraint (5),
- λ_{i+1-r} for $i = r, \cdots, s$, dual variables of the existing units constraints (6),
- $g(\cdot)$ inverse of the LDC,[2]

[1] The terms "efficient" and "inefficient" existing units will be made clear later.

[2] We use the inverse of the LDC in formulating the Kuhn–Tucker conditions, to comply with the conventional notation in mathematics where the independent variable (demand) is placed on the x axis and the dependent variable (duration of time) on the y axis.

the Kuhn–Tucker conditions are

$$c_i + R_i - u_i - u = 0, \qquad i < r \text{ or } s < i \leq n \tag{7}$$

$$c_i + R_i + \lambda_{i+1-r} - u_i - u = 0, \qquad r \leq i \leq s \tag{8}$$

where

$$R_i \triangleq \sum_{j=1}^{n} (b_j - b_{j+1})g(D_{j+1}) \tag{9}$$

$$x_i \geq 0, \qquad i = 1, \cdots, n \tag{10}$$

$$\sum_{i=1}^{n} x_i - L_{\max} \geq 0 \tag{11}$$

$$y_{i+1-r} - x_i \geq 0, \qquad i = r, \cdots, s \tag{12}$$

$$u_i \cdot x_i = 0, \qquad i = 1, \cdots, n \tag{13}$$

$$u \cdot \left(\sum_{i=1}^{n} x_i - L_{\max} \right) = 0 \tag{14}$$

$$\lambda_{i+1-r}(y_{i+1-r} - x_i) = 0, \qquad i = r, \cdots, s \tag{15}$$

$$u_i \geq 0, \qquad i = 1, \cdots, n \tag{16}$$

$$u \geq 0 \tag{17}$$

$$\lambda_{i+1-r} \geq 0, \qquad i = r, \cdots, s. \tag{18}$$

We now compare the Kuhn–Tucker conditions (7) and (8), assuming the capital costs for the existing units are actually sunk costs and can thus be set equal to zero. Substituting $c_i = 0$ for all existing units in equation (8), we obtain:

$$R_i + \lambda_{i+1-r} - u_i - u = 0. \tag{19}$$

By comparing (7) to (19), the dual variables $\lambda_1, \cdots, \lambda_\ell$, $\ell = s - r + 1$, can be interpreted as the "effective" fixed cost for the existing unit; i.e., if we were to install a generating unit, in an all-new power system, whose total variable expenses are given by b_i \$/MWHr and total fixed costs by λ_{i+1-r} \$/MW, we would obtain this unit in the optimal mix solution with capacity identical to that of existing unit i. Thus λ_{i+1-r}, $i = r, \cdots, s$, can be interpreted as the marginal cost to the system of increasing the existing unit capacity by 1 unit (1 KW). Since the existing unit is already installed, then if $\lambda_{i+1-r} > 0$, we actually "earn" the capital cost by keeping the existing unit in the optimal solution. By the Kuhn–Tucker condition (15), the existing unit will be utilized at its rated capacity ($x_i = y_{i+1-r}$) in the optimal mix solution. We refer to this unit as an "efficient" unit. If the dual variable is zero, we actually gain nothing by keeping the unit in the optimal mix solution at its rated capacity. On the contrary, by the Kuhn–Tucker condition (15), the unit should be either discarded from the optimal mix solution or utilized at a reduced capacity. We refer to this type of existing unit as an "inefficient" unit.

Incidentally, the other dual variables can also be interpreted as marginal costs or shadow prices. Thus u_i, the dual variable of the nonnegativity constraint i is the marginal cost to the system of installing an increment of unit i, whereas u, the dual variable of the peak demand constraint, is the additional cost incurred by increasing the peak demand ($L_{\max}$) by 1 unit (1 KW).

We use this interpretation of the dual variable λ_{i+1-r} in order to devise an approach to solve the optimal mix problem in this case. We again demonstrate the procedure graphically, starting with a power system which contains, in the target year, two new units, say one coal-fired and one gas turbine unit, and an EMU consisting of one oil-fired unit, whose merit order location lies in between the merit order location of the coal-fired and the gas turbine units (i.e., $r = 2$, $s = 2$). The solution is depicted in Figure 10.4. We start by

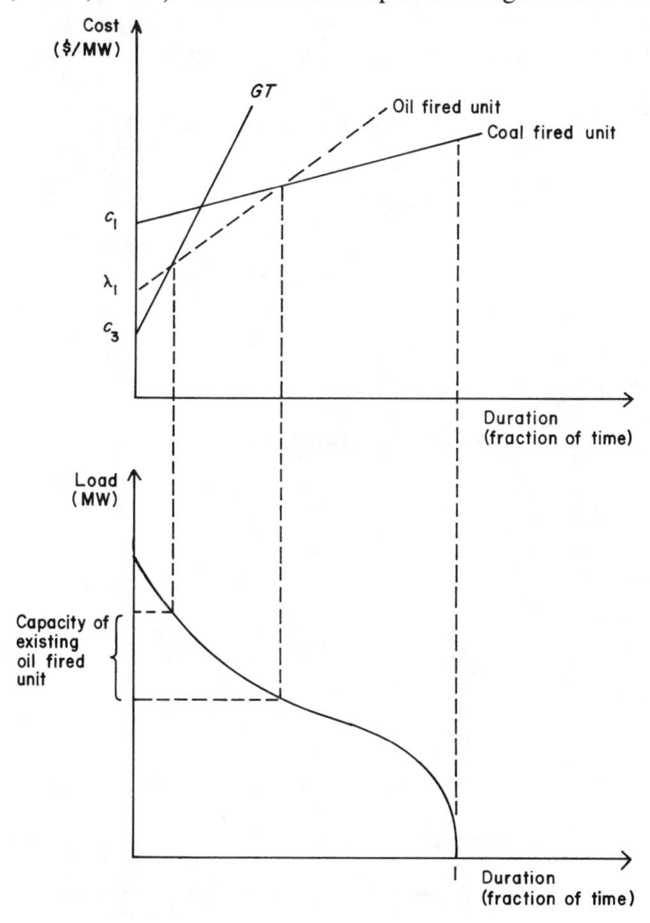

Figure 10.4. Graphical Solution, 1 EMU Consisting of One Efficient Existing Oil-fired Unit.

plotting the minimum cost polygon for the new units only, as shown by the solid lines in the top portion of Figure 10.4. We then slide the cost curve for the existing oil-fired unit, represented by the broken line in the upper portion of Figure 10.4, up and down, until the projection of the breakeven points onto the LDC yields the capacity of the existing oil-fired unit. The capacities of the new coal-fired and gas turbine units to be installed in the system, are then read directly off the LDC. Since the effective cost for the existing oil-fired unit λ_1 is positive, the oil fired unit is "efficient" and is utilized at the optimal solution at its rated capacity.

Next we demonstrate the algorithm for a case of two new units (a coal-fired and a gas turbine unit) and an EMU consisting of two oil-fired units, denoted as units 1 and 2, respectively. We again assume that the merit order locations of the two existing units lies in between the merit order location of the new coal-fired and gas turbine units, (i.e., $r = 2$, $s = 3$). We describe the algorithm in Figure 10.5 for the case where both existing units are efficient, i.e., their corresponding dual variables are positive.

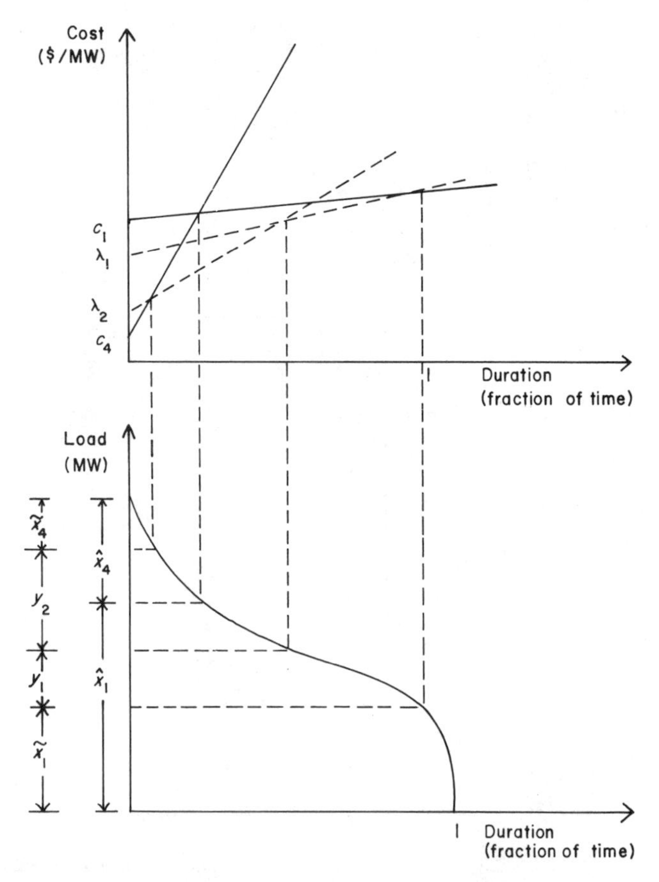

Figure 10.5. Optimal Mix Algorithm, 1 EMU Containing Two Existing Oil-fired Units, All Existing Units Are Efficient.

As in the previous case, we start by finding the minimum cost polygon for the new units only (solid lines in the upper portion of Figure 10.5). We then shift the minimum cost polygon for the EMU (broken lines in the upper portion of Figure 10.5) up and down, right or left, until the projections of the breakeven points onto the LDC yield the capacities of the existing units. As can be seen from Figure 10.5, the two dual variables λ_1 and λ_2 are positive. Thus, by definition, both existing units are efficient and appear in the optimal mix of units at their rated capacities. Given the loading points of the existing units, the capacities of the new coal-fired and gas turbine units are then read directly off the LDC. Clearly, these capacities are smaller than those obtained when no existing units are allowed in the optimal mix of units (denoted with a "hat" in Figure 10.5).

The case of inefficient existing units is illustrated in Figure 10.6. Repeating the procedure described above, we obtain that in this case λ_2, the dual

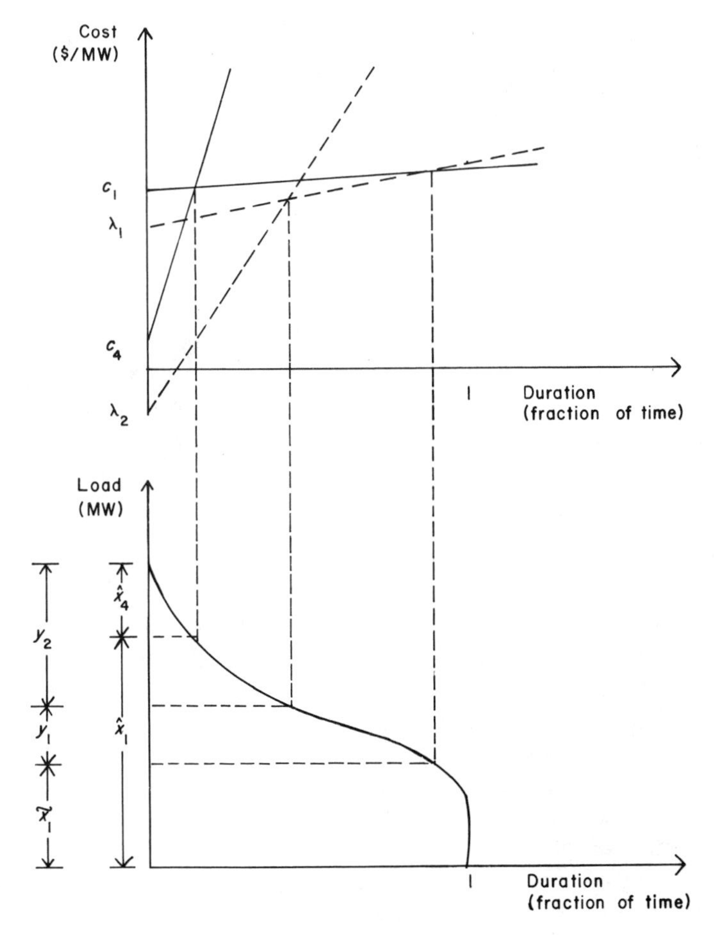

Figure 10.6. Optimal Mix Algorithm, 1 EMU, Containing Two Existing Oil-fired Units, All Existing Units Required To Operate at Target Year.

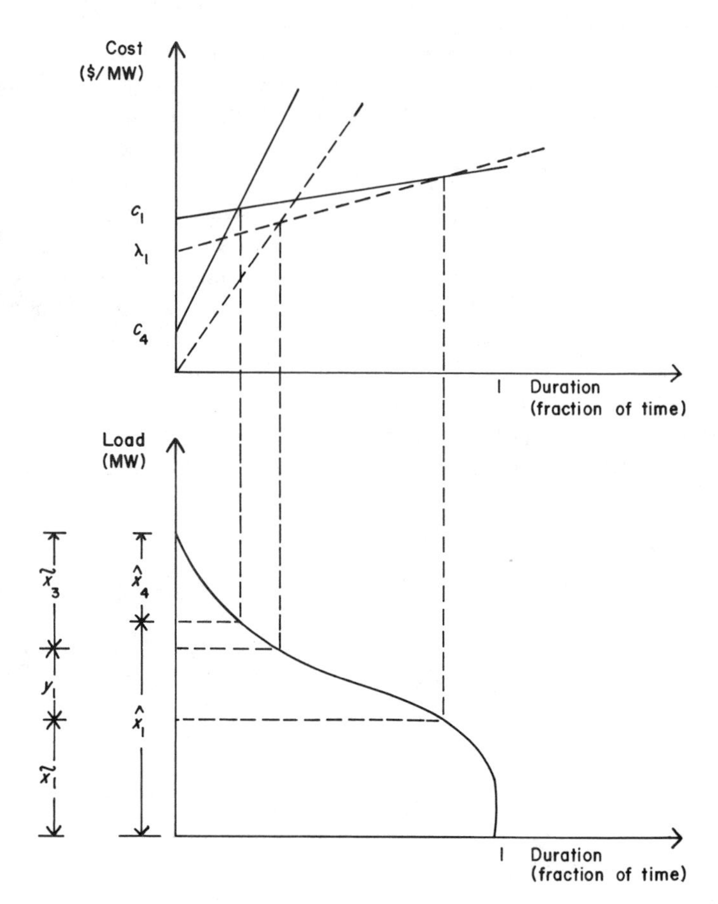

Figure 10.7. Optimal Mix Algorithm, 1 EMU, Containing Two Existing Oil-fired Units, Only Efficient Existing Units Are Allowed in Optimal Mix.

variable for the second existing unit, is negative. Since by Kuhn–Tucker condition (18) λ_2 cannot be negative, we slide the minimum cost polygon of the EMU upwards, until λ_2 becomes zero, as demonstrated in Figure 10.7, thus obtaining the new breakeven points. The optimal capacities to install of each unit are then obtained by projecting the breakeven points onto the LDC. We note that the second existing unit is utilized, in the optimal mix of units, at a capacity which is less than its rated capacity.

Finally, we consider the case where all existing units, even if inefficient, are required to participate in the mix of units to meet the demand in the target year. This situation might arise as a result of financial squeeze that limits the amount of money to be invested in power system expansion. The optimization problem in this case is given again by (3)–(5). Constraint (6), however, takes the form of an equality, i.e.

$$x_i = y_{i+1-r}, \qquad i = r, \cdots, s \tag{20}$$

The solution procedure for this case was demonstrated in Figure 10.6. As can be seen from the figure, existing unit 2 appears in the optimal mix of units at its rated capacity, in compliance with constraint (20). We note that since (20) is an equality constraint, the dual variables of the existing units could be negative, as is the case with λ_2. The negative dual variables yield, in this case, the losses incurred to the system by the requirement to include the inefficient existing units in the optimal mix of units in the target year.

4. Sensitivity Analysis

The graphical solution can also be used to study the sensitivity of the optimal solution to changes in the total variable costs (b_i) and the investment cost (c_i). We demonstrate the procedure graphically for the case of an all new power system, in Figures 10.8 and 10.9, respectively. Similar analysis can be

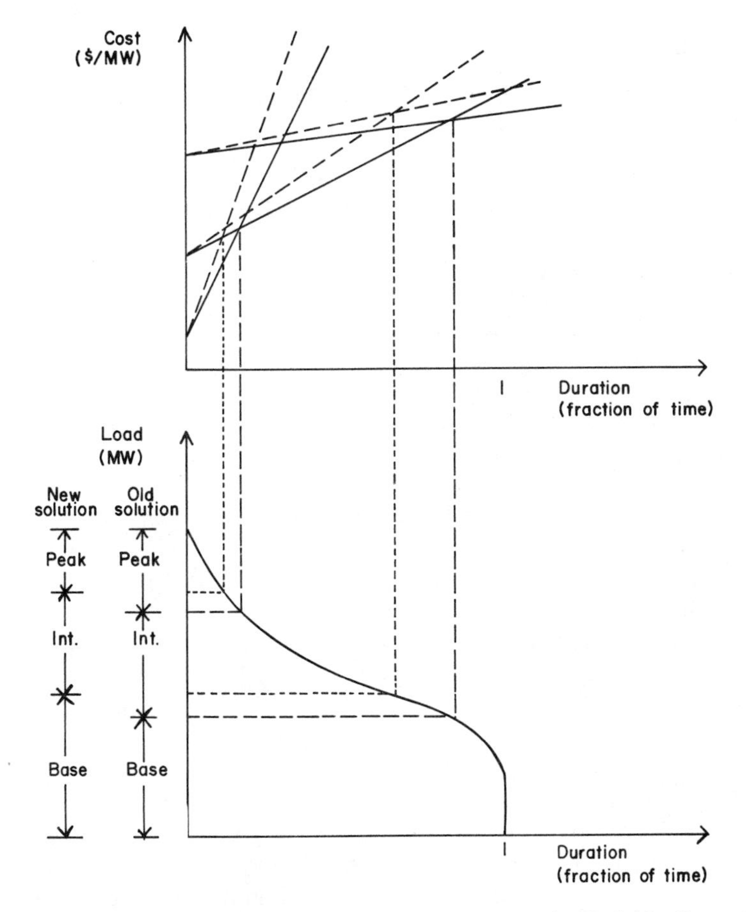

Figure 10.8. Sensitivity Analysis with Respect to Changes in Variable Costs, for a System Containing New Units Only.

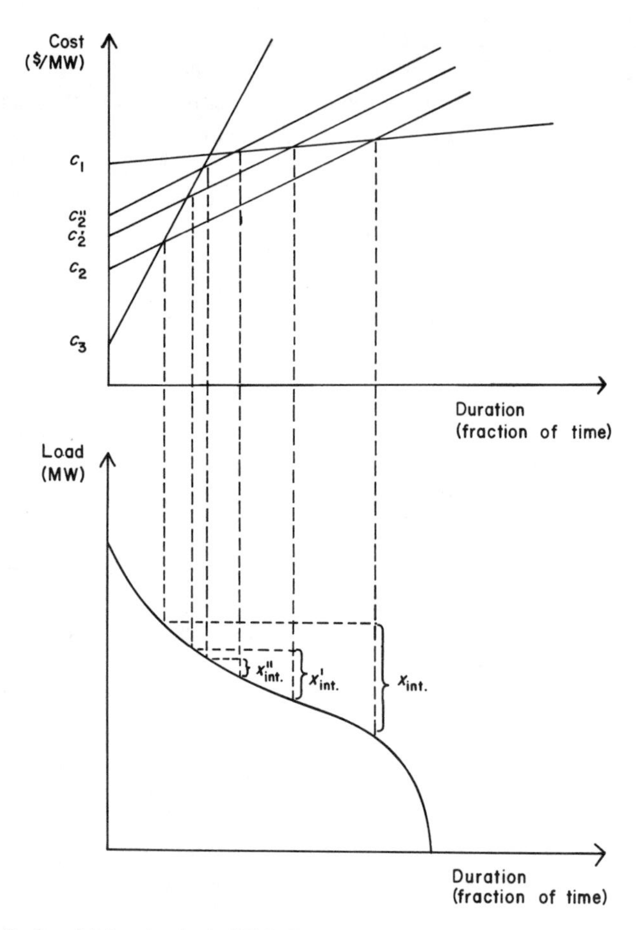

Figure 10.9. Sensitivity Analysis With Respect to Changes in the Capital Cost of the Intermediate Loaded Unit, for a System Containing New Units Only.

applied for the case where existing units are allowed to participate in the mix of units in the target year.

Changes in the demand for power are reflected in the LDC. Thus, for any change in either the peak demand or the temporal distribution of demand, one can obtain the resulting LDC and apply the above procedures to determine the resulting impact on the optimal mix solution and the total cost of the system.

5. Extensions

The above algorithms can also be extended to the case of multiple EMUs. However, the analysis becomes more complex and will not be presented here. The analytical solution of these problems is also very complex and falls beyond the scope of this chapter.

Note—The Kuhn–Tucker Conditions. Let

$\boldsymbol{X}$	$(x_1, \cdots, x_n)$, n-dimensional vector of variables,
$\boldsymbol{b}$	$(b_1, \cdots, b_{m+n})$, $m + n$ − dimensional vector of constants,
A	$(m + n) \times n$ coefficient matrix,
$F(\boldsymbol{x})$	convex, differentiable function.

Then $\boldsymbol{X}^*$ is an optimal solution to the problem:

$$\min_{X} \mathrm{F}(\boldsymbol{X}) \tag{21}$$

subject to

$$\boldsymbol{b} - A\boldsymbol{X} \geq 0 \tag{22}$$

where (22) contains m system constraints and n nonnegativity constraints, if there exists an $(n + m)$ dimensional vector of dual variables $\boldsymbol{U}$ such that:

$$\nabla F(\boldsymbol{X}^*) + A'\boldsymbol{U} = 0 \tag{23}$$

$$\boldsymbol{b} - A\boldsymbol{X} \geq 0 \tag{24}$$

$$\boldsymbol{U}'(\boldsymbol{b} - A\boldsymbol{X}) = 0 \tag{25}$$

$$\boldsymbol{U} \geq 0. \tag{26}$$

$\nabla F(\boldsymbol{X}^*)$ is an n-dimensional gradient vector with elements $(\partial F/\partial x_j)|_{x_j = x_j^*}$. Equation (25) is a scalar containing $(m + n)$ products of the form $u_i(b_i - \Sigma_{j=1}^n A_{ij}x_j)$. By (26), the first term in the product is nonnegative; by (24) the second term is nonnegative. Hence (25) implies

$$u_i\left(b_i - \sum_{j=1}^{n} A_{ij}x_j\right) = 0, \qquad i = 1, \cdots, m + n. \tag{27}$$

Reference

[1] T. W. Berrie, "The economics of system planning in bulk electricity supply," in *Public Enterprise*, R. Turvey, Ed. Penguin Modern Economics, 1968.

Notes for the Instructor

Objective. To present a complex decision problem which under simplifying assumptions can be solved rather easily. Several aspects of nonlinear optimization are presented, including minimization of total cost, the Kuhn–Tucker conditions, and sensitivity analysis.

Prerequisites. Basic familiarity with nonlinear programming and the Kuhn–Tucker conditions.

Time. Three to four class meetings.

CHAPTER 11

Hierarchies, Reciprocal Matrices, and Ratio Scales[1]

Thomas L. Saaty*

1. Introduction

We are interested in the problem of finding a scale which reflects the relative intensity of a property shared by n objects. The objects may be n stones and the properties may be their weights. What is needed is a theory that would enable us to conduct measurement which produces not only known results but is useful and amenable to generalization to the social and behavioral fields. The first problem we have to face is how to assign numbers to all manifestations of a given property in any of the objects so that ratios give a faithful reflection of variation in this property from object to object.

Suppose we are now looking for a way to compare objects using ratios not just with regard to one property but several taken together. Is this a meaningful idea and how do we carry out the calculations to derive these numbers called the priorities of the objects? The tool needed for this purpose is a hierarchy. We shall give a formal definition of a hierarchy and apply it to derive the desired priorities.

This note falls into three parts, the first two of which are totally distinct. We begin by recovering ratio scales from reciprocal pairwise comparison matrices. The problem is to assume that an individual has an internal ratio scale on which a number of items are located; we wish to determine his scale by eliciting cardinal pairwise comparisons. Therefore, we must consider what questions to ask in order to elicit information from the individual; how the individual's responses should be represented; if his responses are "con-

[1] Since the writing of this chapter in 1976, a book has been published on this subject called *The Analytic Hierarchy Process*, McGraw-Hill, 1980.

* 2025 C. L., University of Pittsburgh, Pittsburgh, PA 15260.

sistent" (i.e., if they conform exactly with his internal scale), how his scale can be determined, and if his responses are not consistent, how his scale can be approximately determined. The last is, of course, the central question. The answer provides an excellent illustration of a mathematical approach to many such problems. We analyze the properties of consistent matrices and discover how the properties change as consistency is gradually sacrificed. Then, we use these results as the foundation for a computational method of approximate scale construction. Finally, we examine how the likely accuracy of the reconstructed scale can be measured. The answer here is that it can be done only indirectly, by looking at the consistency of the elicited information and making some assumption relating consistency of the information to the accuracy of the same information. We illustrate with examples.

Next we study the problem of representing complex systems as hierarchies. Assume that a system is "hierarchically" structured, with items at each level influencing those items directly "below" them. Then it is natural to ask how the item(s) at the lowest level are influenced by the "fundamental" items (those at the highest level). We consider what, precisely, is a hierarchy, and what properties we want the hierarchies and priority functions to have. We offer a way to determine "indirect" influence across several levels of the hierarchy from direct influence. Again, we illustrate with examples.

Finally, a synthesis of the ideas is studied through the following kind of questions. From subjective judgments, how might one choose to cluster the various items of a system into levels? What clusterings are "efficient"? If the "direct influences" are estimated by the method of ratio scales and the indirect influences are then computed using hierarchies, how sensitive are the final results to the consistency (accuracy) of the data which yielded the direct influences? Some brief illustrations are provided. A more complete mathematical version is in [43′].

2. Ratio Scales from Reciprocal Pairwise Comparison Matrices

We need a scale of numbers to assign which can be used to form ratios. We want the scale to reflect accurately the strength of the property being measured. Thus an object that is twice as heavy as another should have a scale value twice that of the first. Most weight scales have this property. So does the Kelvin scale of temperature. However, the Fahrenheit and Celsius scales do not. A day that registers 30°F temperature is not 3/2 warmer than a day which registers 20°F. More than one scale may exist which replaces ratios as indicated, and therefore the scale may not be unique. The question then is to find all scales which yield the same ratio for any pair of objects to which measurement is applied.

Suppose we are comparing n objects. If one scale gives the values $(x_1, \cdots, x_n)$ for the weights of these objects (e.g., in pounds) and if another scale assigns these objects the weights $(y_1, \cdots, y_n)$ (e.g., in kilograms), what we want is that ratios should be preserved
i.e.,

$$\frac{x_i}{x_j} = \frac{y_i}{y_j}, \qquad i, j = 1, \cdots, n.$$

Thus we seek a transformation f from the first scale to the second such that $y_i = f(x_i)$, $i = 1, \cdots, n$, and $x_i/x_j = f(x_i)/f(x_j)$. From this we have $f(x_i)/x_i = f(x_j)/x_j$ for all i and j and hence we may conclude that for any x, $f(x)/x$ is independent of the object and has the same value $\boldsymbol{a}$ for all objects. Consequently, we have $f(x) = ax$. Thus a ratio scale must be transformable in this way in order to yield the same ratios before and after the transformation. Now $\boldsymbol{a}$ must be positive to preserve the relative order of the objects because a negative constant would result in the reversal of the rank. In other words, if $x_i > x_j$ and if $a < 0$, then $ax_i < ax_j$ or $y_i < y_j$, contrary to an important basic property of any scale, i.e., that it should preserve rank or be a monotone increasing function of the ranked objects. A function of the form $y = ax$, $a > 0$ is called a positive or order preserving similarity transformation. Thus we characterize a ratio scale by its invariance under such a transformation; it is unique to within such a transformation. Clearly, a ratio scale has an origin or starting point, and all scale values fall above it (or else they all fall below the origin).

By contrast, the weaker scale (called an interval scale) which is invariant under a positive linear transformation of the form $y = ax + b, b \neq 0, a > 0$, requiring ratios of differences to be the same, has no origin. Transformed values could fall on either side of any point. However, both scales have a unit of measurement and all other values are given as multiples of it.

Suppose we wish to compare a set of n objects in pairs according to their relative weights. Denote the objects by $A_1, \cdots, A_n$ and their weights by $w_1, \cdots, w_n$. The pairwise comparisons may be represented by a matrix of underlying ratios (assumed to exist) as follows:

$$A = \begin{array}{c|cccc} & A_1 & A_2 & \cdots & A_n \\ \hline A_1 & \frac{w_1}{w_1} & \frac{w_1}{w_2} & \cdots & \frac{w_1}{w_n} \\ A_2 & \frac{w_2}{w_1} & \frac{w_2}{w_2} & \cdots & \frac{w_2}{w_n} \\ \vdots & \vdots & \vdots & & \vdots \\ A_n & \frac{w_n}{w_1} & \frac{w_n}{w_2} & \cdots & \frac{w_n}{w_n} \end{array}.$$

This matrix has positive entries everywhere and satisfies the reciprocal property $a_{ji} = 1/a_{ij}$. It is called a reciprocal matrix. We note that if we multiply this matrix by the column vector $(w_1, \cdots, w_n)$, we obtain the vector nw. That is,

$$Aw = nw.$$

We started out with the assumption that w was given, but if we only had A and wanted to recover w, we would have to solve the system $(A - nI)w = 0$ in the unknown w. This has a nonzero solution if and only if n is an eigenvalue of A, i.e., it is root of the characteristic equation of A. However, A has unit rank since every row is a constant multiple of the first row. Thus all the eigenvalues λ_i, $i = 1, \cdots, n$ of A are zero except one. Also we know that

$$\sum_{i=1}^{n} \lambda_i = \operatorname{tr}(A) \equiv \text{sum of the diagonal elements} = n.$$

Therefore only one of λ_i, we call it $\lambda_{\max}$, equals n, and

$$\lambda_i = 0, \qquad \lambda_i \neq \lambda_{\max}.$$

The solution w of this problem is any column of A. These solutions differ by a multiplicative constant. However, it is desirable to have this solution normalized so that its components sum to unity. The result is a unique solution no matter which column is used. We have recovered the scale from the matrix of ratios by, for example, simply normalizing the first column.

The matrix A satisfies the "cardinal" consistency property $a_{ij}a_{jk} = a_{ik}$ and is called consistent. For example, if we are given any row of A, we can determine the rest of the entries from this relation. This also holds for any set of n entries no two of which fall in the same row or column.

Now suppose that we are dealing with a situation in which the scale is not known, but we have estimates of the ratios in the matrix. In this case the cardinal consistency relation need not hold, nor need an "ordinal" relation of the form: $A_i > A_j$, $A_j > A_k$ imply $A_i > A_k$ hold (where the A_i are rows of A).

As a realistic representation of the situation in preference comparisons, we wish to account for inconsistency in judgments because, despite their best efforts, people's feelings and preferences remain inconsistent and intransitive.

We know that in any matrix, small perturbations in the coefficients imply small perturbations in the eigenvalues. Thus the problem $Aw = nw$ becomes $A'w' = \lambda_{\max} w'$. We also know from the theorem of Perron–Frobenius that a matrix of positive entries has a real positive eigenvalue (of multiplicity 1) whose modulus exceeds those of all other eigenvalues. The corresponding eigenvector solution has nonnegative entries, and when normalized it is unique. Some of the remaining eigenvalues may be complex.

Suppose then that we have a reciprocal matrix. What can we say about an overall estimate of inconsistency for both small and large perturbations of its entries? In other words how close is $\lambda_{\max}$ to n and w' to w? If they are not close, we may either revise the estimates in the matrix or take several matrices from which the solution vector w' may be improved. Note that improving consistency does not mean getting an answer closer to the "real" life solution. It only means that the ratio estimates in the matrix, as a sample collection, are closer to being logically related than to being randomly chosen.

From here on we shall use $A = (a_{ij})$ for the estimated matrix and w for the eigenvector. Dropping the primes should cause no confusion.

A reciprocal matrix A with positive entries turns out to be consistent if and only if $\lambda_{\max} = n$. This can be seen by writing

$$\lambda_{\max} = \sum_{j=1}^{n} a_{ij} w_j,$$

putting $a_{ji} = 1/a_{ij}$, and reducing this equation to the form

$$(\lambda_{\max} - 1) = \frac{1}{n} \sum_{1 \le i < j \le n} \left(y_{ij} + \frac{1}{y_{ij}} \right), \qquad \text{where } y_{ij} = a_{ij} \frac{w_i}{w_j}.$$

Each term in parentheses has the minimum value 2 at $y_{ij} = 1$, and hence $\lambda_{\max} \ge n$. If $\lambda_{\max} = n$, we can easily show that the sum in parentheses must attain its minimum at $y_{ij} = 1$, i.e., $a_{ij} = w_i/w_j$ the consistent case. With inconsistency $\lambda_{\max} > n$ always. One can also show that ordinal consistency is preserved, i.e., if $A_i \ge A_j$ (or $a_{ik} \ge a_{jk}$, $k = 1, \cdots, n$), then $w_i \ge w_j$. This follows from

$$\lambda_{\max} w_i = \sum_{k=1}^{n} a_{ik} w_k \ge \sum_{k=1}^{n} a_{jk} w_k = \lambda_{\max} w_j.$$

We now establish

$$\frac{\lambda_{\max} - n}{n - 1} = \frac{-\sum_{i=2}^{n} \lambda_i}{n - 1} \text{ (the average of } \lambda_i\text{)}$$

as a measure of the consistency or reliability of judgments (supplied by an individual) whose true value is w_i/w_j. We assume that, because of possible error, the estimate has the observed value $(w_i/w_j)\varepsilon_{ij}$ where $\varepsilon_{ij} > 0$. First we note that to study the sensitivity of the eigenvector to perturbations in a_{ij}, we cannot make a precise statement about a perturbation $dw = (dw_1, \cdots, dw_n)$ in the vector $w = (v_1, \cdots, v_n)$ because everywhere we deal with w, it appears in the form of ratios w_i/w_j or with perturbations (mostly multiplicative) of this ratio. Thus we cannot hope to obtain a measure of the absolute error in w.

From general considerations one can show that the larger the matrix, the less significant are small perturbations or a few large perturbations on the eigenvector. If the order of the matrix is small, the effect of a large array perturbation on the eigenvector can be relatively large. We may assume that when the consistency index shows that perturbations from consistency are large and hence the result is unreliable, the information available cannot be used to derive a reliable answer. If the consistency can be improved to a point where its reliability indicated by the index is acceptable, i.e., the value of the index is small (as compared with its value from a randomly generated reciprocal matrix of the same order), we can carry out the following type of perturbation analysis.

The choice of perturbation most appropriate for describing the effect of inconsistency on the eigenvector depends on what is thought to be the psychological process which goes on in the individual. Mathematically, general perturbations in the ratios may be reduced to the multiplicative form mentioned above. Other perturbations of interest can be reduced to the general form $a_{ij} = (w_i/w_j)\varepsilon_{ij}$. For example,

$$\frac{w_i}{w_j} + \alpha_{ij} = \frac{w_i}{w_j}\left(1 + \frac{w_j}{w_i}\alpha_{ij}\right).$$

Starting with the relation

$$\lambda_{\max} = \sum_{j=1}^{n} a_{ij}\frac{w_j}{w_i}$$

from the ith component of $Aw = \lambda_{\max} w$,
we consider the two real valued parameters $\lambda_{\max}$ and μ, the negative average of λ_i, $i \geq 2$, i.e.,

$$\mu = -\frac{1}{n-1}\sum_{i=2}^{n}\lambda_i = \frac{\lambda_{\max} - n}{n-1} \geq 0, \qquad \lambda_{\max} = \lambda_1.$$

We wish to have μ near zero, thus also to have $\lambda_{\max}$ which is always $\geq n$ near its lower bound n, and thereby obtain consistency. Now we show that $(\lambda_{\max} - n)/(n-1)$ is related to the statistical root mean square error. To see this, we have from

$$\lambda_{\max} - 1 = \sum_{j \neq i} a_{ij}\frac{w_j}{w_i}$$

that

$$n\lambda_{\max} - n = \sum_{1 \leq i < j \leq n} a_{ij}\frac{w_j}{w_i} + a_{ji}\frac{w_i}{w_j},$$

and therefore

$$\mu = \frac{\lambda_{\max} - n}{n-1} = \frac{1}{n-1} - \frac{n}{n-1} + \frac{1}{n(n-1)}\sum_{1 \leq i < j \leq n} a_{ij}\frac{w_j}{w_i} + a_{ji}\frac{w_i}{w_j}.$$

Let $a_{ij} = (w_i/w_j)\varepsilon_{ij}$, $\varepsilon_{ij} > 0$. Clearly, we have consistency at $\varepsilon_{ij} = 1$. Now

$$\mu = -1 + \frac{1}{n(n-1)}\sum_{1 \leq i < j \leq n}\left(\varepsilon_{ij} + \frac{1}{\varepsilon_{ij}}\right)$$

which $\to 0$ as $\varepsilon_{ij} \to 1$. Also, μ is convex in the ε_{ij} since $\varepsilon_{ij} + 1/\varepsilon_{ij}$ is convex (and has its minimum at $\varepsilon_{ij} = 1$), and the sum of convex functions is convex.

Thus μ is small or large depending on ε_{ij} being near or far from unity, respectively; i.e., near or far from consistency.

If we write $\varepsilon_{ij} = 1 + \delta_{ij}$, $|\delta_{ij}| < 1$, we have

$$\mu = \frac{1}{n(n-1)} \sum_{1 \le i < j \le n} \delta_{ij}^2 - \frac{\delta_{ij}^3}{1 + \delta_{ij}}.$$

Now $\mu \to 0$ as $\delta_{ij} \to 0$.

Multiplication by two gives the variance of the δ_{ij}. Thus 2μ is this variance. To test for how acceptable μ is for a given matrix, we require that its ratio with its average value for a randomly generated matrix of the same order be about 10% or less. Random values of μ for different matrices are known. The eigenvalue method has several advantages in developing a ratio scale as compared with direct estimates of the scale or with least square methods. For example, compared with the former, it captures more information through redundancy of information obtained from pairwise comparisons and the use of reciprocals. When compared with either method, it addresses the question of the consistency by a single numerical index and points to the reliability of the data and to revisions in the matrix.

L. Vargas has developed a formula which relates w to A and hence can be used to analyze the sensitivity of w to changes in the a_{ij}. The principal eigenvector of the perturbed matrix A' is given by $w'_{ij} = C(w_i y_i)$, $i = 1, \cdots, n$, where C is a normalizing constant and y_i is the principal eigenvector of the matrix $(p_{ij}\varepsilon_{ij})$ where $a'_{ij} = p_{ij}a_{ij}$.

As already mentioned, it is easy to prove that the solution of the problem $Aw = nw$ when A is consistent is given by the normalized row sums or any normalized column of A. In addition, the solution to $AW = \lambda_{\max} w$ when $\lambda_{\max}$ is close to n may be approximated by normalizing each column of A and taking the average of each of the components of the resulting vector. This yields a vector $\bar{w}$; in this case one can readily obtain an estimate for $\lambda_{\max}$ by computing $A\bar{w}$, dividing each of the components of the resulting vector by the corresponding component of $\bar{w}$ and averaging the results.

3. The Scale

The scale we recommend for use which has been successfully tested and compared with other scales will now be discussed. The judgments elicited from people are taken qualitatively and corresponding scale values assigned to them. In general, we do not expect cardinal consistency to hold everywhere in the matrix because people's feelings do not conform to an exact formula. Nor do we expect ordinal consistency as people's judgments may not be transitive. However to preserve sanity in the numerical judgments, whatever value a_{ij} is assigned in comparing the ith activity with the jth one, the reciprocal value is assigned to a_{ji}. Thus we put $a_{ji} = 1/a_{ij}$. Usually, we first record whichever value represents dominance greater than unity. Roughly speaking, if one activity is judged to be α times stronger than another, then we record the latter as only $1/\alpha$ times as strong as the former. We can easily see that when we have consistency, the matrix has unit rank, and knowing one row of the matrix is sufficient to construct the remaining entries. For example, if we know the first row then $a_{ij} = a_{1j}/a_{1i}$ (under the rational assumption, of course, that $a_{1i} \neq 0$ for all i).

Let us repeat that reported judgments need not be even ordinally consistent, and hence they need not be transitive; i.e., if the relative importance of C_1 is greater than that of C_2 and the relative importance of C_2 is greater than that of C_3, then the relation of importance of C_1 need not be greater than that of C_3, a common occurrence in human judgments. An interesting illustration is afforded by tournaments regarding inconsistency or lack of transitivity of preferences. A team C_1 may lose against another team C_2 which has lost to a third team C_3; yet C_1 may have won against C_3. Thus team behavior is inconsistent—a fact which has to be accepted in the formulation, and nothing can be done about it. Of course, our model may require transitivity internally, but the "reporting" procedure may scramble things.

We now turn to the question of what numerical scale to use in the pairwise comparison matrices. Whatever problem we deal with, we must use numbers that are sensible. From these the eigenvalue process would provide a scale. As we said earlier, the best argument in favor of a scale is if it can be used to reproduce results already known in physics, economics, or in whatever area is a scale. The scale we propose is useful for small values of $n < 10$.

Our choice of scale hinges on the following observation. Roughly, the scale should satisfy these requirement.

(1) It should be possible to represent people's differences in feelings when they make comparisons. It should represent as much as possible all distinct shades of feeling that people have.
(2) If we denote the scale values by $x_1, x_2, \cdots, x_p$, then it would be desirable that $x_{i+1} - x_i = 1, i = 1, \cdots, p - 1$.

Since we require that the subject must be aware of all gradations at the same time and since we agree with the psychological [36] experiments which show that an individual cannot simultaneously compare more than seven objects (plus or minus two) without being confused, we are led to choose a $p = 7 + 2$. Using a unit difference between successive scale values is all that we allow, and using the fact that $x_1 = 1$ for the identity comparison, it follows that the scale values will range from one to nine.

As a preliminary step towards the construction of an intensity scale of importance for activities, we have broken down the importance ranks as

Table 1

Intensity of Importance	Definition	Explanation
1[†]	Equal importance	Two activities contribute equally to the objective
3	Moderate importance of one over another	Experience and judgment slightly favor one activity over another
5	Essential or strong importance	Experience and judgment strongly favor one activity over another
7	Demonstrated importance	Activity is strongly favored and its dominance demonstrated in practice
9	Absolute importance	Evidence favoring one activity over another is of the highest possible order of affirmation
2, 4, 6, 8	Intermediate values between the two adjacent judgments	When compromise is needed
Reciprocals of above nonzero	If activity i has one of the above nonzero numbers assigned to it when compared with activity j, then j has the reciprocal value when compared with i.	See below
Rationals	Ratios arising from the scale	If consistency were to be forced by obtaining n numerical values to span the matrix

[†] On occasion in 2×2 problems, we have used $1 + \varepsilon$, $0 < \varepsilon \leq 1/2$ to indicate very slight dominance between two nearly equal activities.

shown in the scale in Table 1. In using this scale, the reader should recall that we assume that the individual providing the judgment has knowledge about the relative values of the elements being compared whose ratio is ≥ 1 and that the numerical ratios he forms are nearest-integer approximations scaled in such a way that the highest ratio corresponds to nine. We have assumed that an element with weight zero is eliminated from comparison. This, of course, need not imply that zero may not be used for pairwise comparison. Reciprocals of all scaled ratios that are ≥ 1 are entered in the transpose positions (not taken as judgments).

Note that the eigenvector solution of the problem remains the same if we multiply the unit entries on the main diagonal, for example, by a constant greater than one. (If $w_i/w_j > 9$, one may need to imbed the problem in a hierarchical framework to obtain reasonable answers (discussed later).) In practice, one way or another, the numerical judgments will have to be approximations, but how good is the question at which our theory is aimed?

A typical question to ask in order to fill in the entries in a matrix of comparisons is, "Consider two properties i on the left side of the matrix and another j on the top; which of the two has the property under discussion more, and how much more (using the scale values 1 to 9)? This gives us a_{ij}. The reciprocal value is then automatically entered for a_{ji}.

EXAMPLE (Distance Estimation through Air Travel Experience). The eigenvalue method was used to estimate the relative distance of six cities from Philadelphia by making pairwise comparisons between them as to which was how strongly farther from Philadelphia.

It is interesting to note that the cities cluster into three classes—those nearest to Philadelphia (Montreal and Chicago), those which are intermediate (San Francisco and London), and those farthest (Cairo and Tokyo). The last, because of relatively large value due to errors of uncertainty, cause the values of the others to be perturbed from where we want them to be. Thus if their eigenvector components change comparatively little, and the increment is distributed among the others, the relative values of these can be altered considerably.

Table 2 gives numerical values to the perceived remoteness from Philadelphia for each pair of cities. The rows indicate the strength of dominance. The question to ask is, Given a city (on the left) and another (on top) how much strongly further is the first one from Philadelphia than the second? We then put the reciprocal value in the transpose position. Compare the solution of the eigenvalue problem with the actual result (Table 3).

Considerable experimental data have been gathered to compare the scale 1–9 with 25 other different scales all used within the eigenvalue formulation. Various statistical measures were calculated. The evidence strongly favors

Table 2

Comparison of Distances of Cities from Philadelphia	Cairo	Tokyo	Chicago	San Francisco	London	Montreal
Cairo	1	1/3	8	3	3	7
Tokyo	3	1	9	3	3	9
Chicago	1/8	1/9	1	1/6	1/5	2
San Francisco	1/3	1/3	6	1	1/3	6
London	1/3	1/3	5	3	1	6
Montreal	1/7	1/9	1/2	1/6	1/6	1

Table 3

City	Distance to Philadelphia in Miles	Normalized Distance	Eigenvector
Cairo	5729	0.278	0.263
Tokyo	7449	0.361	0.397
Chicago	660	0.032	0.033
San Francisco	2732	0.132	0.116
London	3658	0.177	0.164
Montreal	400	0.019	0.027

$\lambda_{max} = 6.45$.

the use of the scale 1–9 as a reflection of our mental ability to discriminate between different degrees of strengths of dominance between a few objects.

The problem may be reformulated by forming clusters of cities according to the order of magnitude of their perceived distance from Philadelphia. The elements in each cluster are first compared among themselves and then the clusters would be themselves compared to obtain the appropriate weights for each cluster and finally the entire set of values normalized (discussed later).

We performed many similar experiments. Among them are one in optics and one in estimating the wealth of seven nations. The optics experiment was to determine relative illumination from a single source. Judgments were made by children whose ages are five and seven. The following result was obtained from comparing the relative brightness of four similar objects placed at 9, 15, 21, and 28 yards from a light source:

Eigenvector	0.62	0.23	0.10	0.05
Actual	0.61	0.22	0.11	0.06.

The "actual" vector is obtained from the inverse square law in optics.

The relative wealth of nations was estimated by the author and an economist, based on feelings, not data. The results are

	U.S.	USSR	China	France	U.K.	Japan	W. Germany
Eigenvector	0.427	0.230	0.021	0.052	0.052	0.123	0.094
Relative GNP's 1972	0.413	0.225	0.043	0.069	0.055	0.104	0.091.

In both of these cases we can see remarkable agreement. One is led to believe that the mind has much greater powers of discrimination than we would have thought, particularly in areas where there is some knowledge and experience. This type of systematic evaluation seems to bring out this ability of the mind.

4. Hierarchies—General Considerations

Although most people have an idea of what a hierarchy is, few use the concept in their thinking. Fewer still realize how important and powerful a hierarchy is as a model of reality when viewing a complex system of interacting components.

First, we are going to give an overview of the idea of a hierarchy, its application to real systems and to thought processes, and its usefulness as a general model. We then go on to present a formal theory for the analysis of hierarchies and their stability and what to do with the model after it is obtained. Although the latter research is relatively new, it has been put to use in a large number of practical applications.

Among these we mention 1) a development of a theory of priorities, 2) a theory of two-point boundary planning (forward and backward processes), and 3) a new method in conflict resolution.

Any system can be represented by a large interaction matrix whose rows and columns are components of the system. When component i and component j interact, strongly the i, jth entry is near ± 1. When they do not interact, the entry is near zero. In a large system, most of the entries are close to zero. Using the concept of a reachability matrix and its power, a distinct hierarchic structure [70] is often discerned. In fact, this arrangement of the elements of a system in an incidence type matrix can be used to identify the levels of a hierarchy. We shall not describe this well-known process here because it usually produces results in line with one's intuition about what falls in which levels. In the simplest type of hierarchy an upper level dominates the neighboring lower level.

Different levels of a hierarchy are generally characterized by differences in both structure and function. The proper functioning of a higher level depends on the proper functioning of the lower levels. The basic problem with a hierarchy is to seek understanding at the highest levels from interactions of the various levels of the hierarchy rather than directly from the elements of the levels. Rigorous methods for structuring systems into hierarchies are gradually emerging in the natural and social sciences and, in particular, in general systems theory as it relates to the planning and design of social systems.

Hierarchies are order presserving structures. They involve the study of order among partitions of a set. The partitions are called the levels of the hierarchy. Conceptually, the simplest hierarchy is linear, rising from one level to an adjacent level. The complexity of the arrangement of the elements in each level may be the same or it may increase from level to level. This also applies to the depth of analytical detail. A hierarchy may emerge gradually from one root (e.g., the development of the human race from a first ancestor) or it may descend in rank from one boss as in an organization. It may grow by adding parts like a snowball, or it may be a simple gradual arrangement of the levels according to a pattern. The structure of each level may take the form of a general network representing the appropriate connections among its elements. This last subject is of considerable interest to us in developing a mathematical theory of hierarchies as we obtain a method for evaluating the impact of a level on an adjacent level from the interactions of the elements in that level.

A simple comprehensive example of a hierarchy begins with the entire universe as one level, galactic clusters as the next level, then successively to galaxies, constellations, solar systems, planets, clumps of matter, crystals, compounds, molecular chains, molecules, atoms, nucleii, protons and electrons. Another example of a hierarchy is that representing the structure of living organism, and a third example would be one which represents the functions of an organizational hierarchy. Two illustrations are provided in Figures 11.1 and 11.2.

Hierarchies may be used to represent both the *structural* and *functional relations* of a system. Because of the close identification of hierarchies, i.e., autonomous, dynamic, etc., each element of a given hierarchy may belong functionally to several other different hierarchies. A spoon may be arranged with other spoons of different sizes in one hierarchy or with knives and forks in a second hierarchy. For example, it may be a controlling component in a level of one hierarchy or it may simply be an unfolding of higher or lower order functions in another hierarchy.

A hierarchical structure may not be reversible. We can see this by looking at processes of planning. This type of planning is simply and graphically demonstrated in a relevant way in space travel. Launching a manned craft and returning it to its starting point is a two-point boundary problem. Different

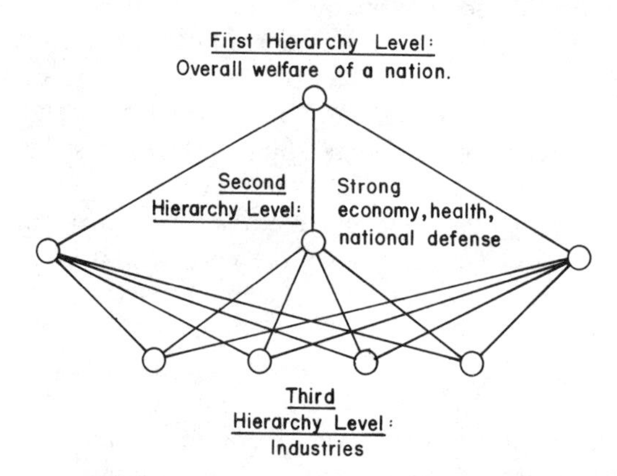

Figure 11.1. A tree of prioritization. *First hierarchy* level has a single objective used as a criterion below. Its priority value is assumed to equal unity. *Second hierarchy* level objectives have priorities derived from the matrix of comparison of their impact on the objective given in the first level. *Third hierarchy* level objectives derive their priorities from their comparison matrix with respect to each objective of the second level. Then to obtain the overall priority of each objective, a weighted sum of the priorities is taken, using the priorities of the second level objectives as weights.

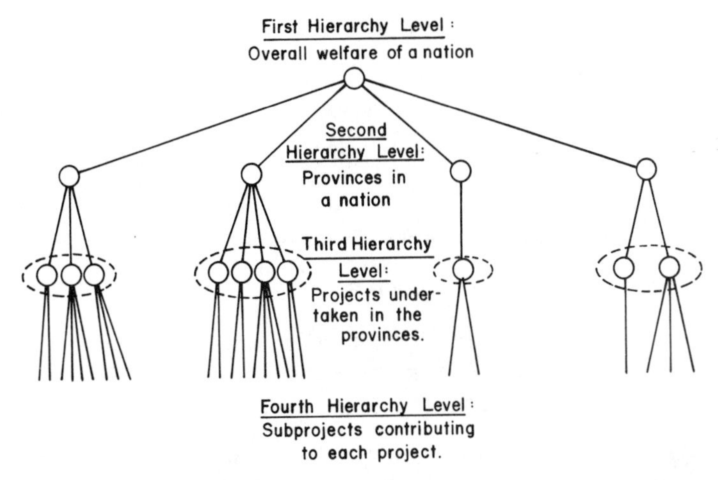

Figure 11.2. Another tree of prioritization. *First hierarchy* level has a single objective used as a criterion below. Its priority value is assumed to equal unity. *Second hierarchy* level objectives have priorities derived from the matrix of comparison of their impact on the objective given in the first level. *Third hierarchy* level objectives fall in groups according to which element of the second hierarchy they contribute. Their overall priority is obtained from the matrix of comparison of their impact on the relevant objective of the second level multiplied by n_i/m (the fraction of elements in their group indicated by dotted lines with $\Sigma n_i = m$, where i is the index of groupings) and by the priority of the element from the second level class they fall under. The sum of the priorities in each level must add up to unity. This is done by a normalization process of dividing the priority of each objective by the sum of all priorities in the level. *Lower hierarchy* level priorities are similarly derived based on the immediately preceding level.

considerations are involved in getting it from one point to the second than from second to first. In the forward process, high velocity and the effect of gravity are the critical factors. How many G's are exerted on the body is important to know. In the backward process, air resistance and the need for parachutes or other deceleration devices, heat transfer, and the tolerance of the heat shield material are the important factors. These factors are also found in the launching process, but they are not the critical ones. The two sets of factors must be taken into consideration to solve the entire problem.

Another feature of the two hierarchies of the forward and backward planning processes is that they can be easily linked. To see this we note that both processes can be represented by "exploding trees." The theory we briefly describe here enables us to contain or limit the size of the hierarchy to its essential components and still get an effective representation of a system. The result is to facilitate the process of interaction between the two hierarchies.

A dichotomy exists as to whether a hierarchy is a convenient tool of the mind or whether nature is actually endowed with hierarchical structures and functions. Considerable evidence has been put together to support the first idea. Here are brief eloquent expressions in defense of each point of view.

> "The immense scope of hierarchical classification is clear. It is the most powerful method of classification used by the human brain–mind in ordering experience, observations, entities and information. Though not yet definitely established as such by neurophysiology and psychology, hierarchical classification probably represents the prime mode of coordination or organization (i) of cortical processes, (ii) of their mental correlates, and (iii) of the expression of these in symbolisms and languages. The use of hierarchical ordering must be as old as human thought, conscious and unconscious . . . " (L. L. Whyte)

Direct confrontation of the large and the small is avoided in nature through the use of a hierarchical linkage. Bigness is avoided by functionally bounding the ratio between the size of the hierarchy and that of its levels.

> "In the more than fifty years of my intimate preoccupation, with the phenomena and problems of morphogenesis . . . I have been unable to find a way of deriving, free from all preoccupations, a comprehensive and realistic description of the developmental process otherwise than by reference to a dualistic concept, according to which the discrete units are enmeshed in, and interplay with, an organized reference system of unified dynamics of the collective of which they are the members." (Paul Weiss)

In either case, our present purpose is to assist our understanding of the interrelations that at least the model claims exist.

We now present some properties of hierarchical structures.

1) A significant observation is that they usually consist of a few kinds of subsystems in various combinations and arrangements—a multitude of proteins from about twenty amino acids, a very large variety of molecules from about a hundred elements.

2) They are nearly decomposable; i.e., connections between levels are far simpler than the connections between the elements in a level. Thus only the aggregate properties of the level determine the interactions between levels and not the properties of the individual elements.

3) Regularities in the interactions between levels may themselves be classified and coded, taking advantage of redundancy in complex hierarchical structures to obtain greater simplicity in explanation. Thus, for example, the trajectory of a system over an entire period of time may be simply described in terms of the differential law generating that trajectory at individual instants of time.

Some advantages of hierarchies are as follows.

1) They provide meaningful integration of a system. Integrated behavior or function of a hierarchical organization accounts for how complicated changes in a large system can influence a single component. It is the opposite of what we generally expect.

2) They use aggregates of elements in the form of levels to accomplish tasks.

3) They provide greater detail down the hierarchy levels and thus greater

depth in understanding its purpose up the hierarchy levels. From the upper level the constraints of the hierarchy are taken for granted, and the question is, "How could the constraints arise?"

4) Hierarchies are efficient and will evolve in natural systems much more rapidly than nonhierarchic systems having the same number of elements. This will be demonstrated later.

5) They possess reliability and flexibility. Local perturbation does not perturb the entire hierarchy. The overall purpose of the hierarchy is divided among the levels whereby each solves a partial problem and the totality meets the overall purpose. The units on the higher level are not concerned with the overall purpose but with specific goals of that level. Explanation of the functioning of a hierarchical system should be attempted not in terms of the overall goal but in terms of specific goals of each level.

5. Formal Hierarchies

The laws characterizing different levels of a hierarchy are generally different. The levels differ in both structure and function. The proper functioning of a higher level depends on the proper functioning of the lower levels. The basic problem with a hierarchy is to seek understanding at the highest levels from interactions of the various levels of the hierarchy rather than directly from elements of the levels.

First we introduce lattices, which we will use in the definition of a hierarchy.

Definition 1. An ordered set is any set S with a binary relation $\leq$ which satisfies the reflexive, antisymmetric, and transitive laws:

Reflexive: for all x, $x \leq x$;
Antisymmetric: if $x \leq y$ and $y \leq x$, then $x = y$;
Transitive: if $x \leq y$ and $y \leq z$, then $x \leq z$.

For any relation $x \leq y$ (read, y includes x) of this type, we may define $x < y$ to mean that $x \leq y$ and $x \neq y$. y is said to cover (dominate) x if $x < y$ and if $x < t < y$ is possible for no t.

Ordered sets with a finite number of elements can be conveniently represented by a directed graph. Each element of the system is represented by a vertex so that an arc is directed from a to b if $b < a$.

Definition 2. A simply or totally ordered set (also called a chain) is an ordered set with the additional property that if $x, y \in S$, then either $x \leq y$ or $y \leq x$.

Definition 3. A subset E of an ordered set S is said to be bounded from above if an element $s \in S$ exists such that $x \leq s$ for every $x \in E$. The element s

is called an *upper bound* of E. We say E has a supremum or least upper bound in S if E has upper bounds and if the set of upper bounds U has an element u_1 such that $u_1 \leq u$ for all $u \in U$. The element u_1 is unique and is called the supremum of E in S. The symbol sup is used to represent a supremum. (For finite sets largest elements and upper bounds are the same.)

Similar definitions may be given for sets bounded from below, a *lower bound* and *infimum*. The symbol inf is used.

A hierarchy may be defined in many ways. The one which suits our needs best here is the following.

We use the notation $x^- = \{y|x \text{ covers } y\}$ and $x^+ = \{y|y \text{ covers } x\}$, for any element x in an ordered set.

Definition 4. Let H be a finite partially ordered set with largest element b.

H is a *hierarchy* if it satisfies the conditions

(a) H is partitioned into sets L_k, $k = 1, \cdots, h$ where $L_1 = \{b\}$;
(b) $x \in L_k$ implies $x^- \subset L_{k+1}$, $k = 1, \cdots, h-1$.
(c) $x \in L_k$ implies $x^+ \subset L_{k-1}$, $k = 2, \cdots, h$.

For each $x \in H$, there is a suitable weighting function (whose nature depends on the phenomenon being hierarchically structured):

$$w_x : x^- \to [0, 1] \text{ such that } \sum_{y \in x^-} w_x(y) = 1.$$

The sets L_i are the *levels* of the hierarchy, and the function w_x is the *priority function* of the elements in one level with respect to the objective x. We observe that even if $x^- \neq L_k$ (for some level L_k), w_x may be defined for all of L_k by setting it equal to zero for all elements in L_k not in x^-.

The weighting function, we feel, is a significant contribution towards the application of hierarchy theory.

Definition 5. A hierarchy is *complete* if, for all $x \in L_k$ $x^+ = L_{k-1}$, for $k = 2, \cdots, h$.

Basic Problem. We can state the central question as follows. Given any element $x \in L_\alpha$ and subset $S \subset L_k$, $(\alpha < \beta)$, how do we define a function $w_{x,S}: S \to [0, 1]$ which reflects the properties of the priority functions w_y on the levels L_k, $k = \alpha, \cdots, \beta - 1$? Specifically, what is the function w_{b,L_h} : $L_h \to [0, 1]$?

In less technical terms, this can be paraphrased as follows. Given a social (or economic) system with a major objective b, and the set L_h of basic activities, such that the system can be modeled as a hierarchy with largest element b and lowest level L_h, what are the priorities of the elements of L_h with respect to b?

From the standpoint of optimization, to allocate a resource among the elements any interdependence must also be considered. Analytically, interdependence may take the form of input–output relations such as, for example, the interflow of products between industries. A high priority industry may depend on flow of material from a low priority industry. In an optimization framework, the priority of the elements enables one to define the objective function to be maximized, and other hierarchies supply information regarding constraints, e.g., input–output relations.

Solution. We now present our method of solving the basic problem. Assume that $Y = \{y_1, \cdots, y_{m_k}\} \in L_k$ and that $X = \{x_1, \cdots, x_{m_{k+1}}\} \in L_{k+1}$. (Observe that, according to the remark following Definition 5, we may assume that $X = L_{k+1}$.) Also assume that an element $z \in L_{k-1}$ exists such that $Y \subset z^-$. We then consider the priority functions

$$w_z : Y \to [0, 1] \quad \text{and} \quad w_y : X \to [0, 1], \qquad j = 1, \cdots n_k.$$

We construct the "priority function of the elements in X with respect to z," denoted w, $w: X \to [0, 1]$, by

$$w(x_i) = \sum_{j=1}^{n_k} w_{y_j}(x_i) w_z(y_j), \qquad i = 1, \cdots, n_{k+1}.$$

Obviously, this is no more than the process of weighting the influence of the element y_j on the priority of x_i by multiplying it with the importance of y_i with respect to z.

The algorithms involved will be simplified if one combines the $w_{y_j}(x_i)$ into a matrix B by setting $b_{ij} = w_{y_j}(x_i)$. If we further set $w_i = w(x_i)$ and $w'_j = w_z(y_j)$, then the above formula becomes

$$w_i = \sum_{j=1}^{n_k} b_{ij} w'_j, \qquad i = 1, \cdots, n_{k+1}.$$

Thus we may speak of the *priority vector* w and, indeed, of the *priority matrix* B of the $(k + 1)$st level; this gives the final formulation

$$w = Bw'.$$

The following is easy to prove.

Theorem. *Let H be a complete hierarchy with largest element b and h levels. Let B_k be the priority matrix of the kth level, $k = 1, \cdots, h$. If w' is the priority vector of the pth level with respect to some element z in the $(p - 1)$st level, then the priority vector w of the qth level $(p < q)$ with respect to z is given by*

$$w = B_q B_{q-1} \cdots B_{p+1} w'.$$

Thus the priority vector of the lowest level with respect to the element b is given by

$$w = B_h B_{h-1} \cdots B_2 b_1.$$

If L_1 has a single element, $b_1 = 1$. Otherwise, b_1 is a prescribed vector.

The following observation holds for a complete hierarchy but is also useful in general. The priority of an element in a level is the sum of its priorities in each of the comparison subsets to which it belongs; each weighted by the fraction of elements of the level which belong to that subset and by the priority of that subset. The resulting set of priorities of the elements in the level is then normalized by dividing by its sum. The priority of a subset in a level is equal to the priority of the dominating element in the next level.

EXAMPLE 1 (School Selection). Three highschools, A, B, and C, were analyzed from the standpoint of a candidate according to their desirability. Six characteristics were selected for the comparison. They are learning, friends, school life, vocational training, college preparation, and music classes. The pairwise judgment matrices were as follows:

- comparison of characteristics with respect to overall satisfaction with school—

	Learning	Friends	School Life	Vocational Training	College Preparation	Music Classes
Learning	1	4	3	1	3	4
Friends	1/4	1	7	3	1/5	1
School Life	1/3	1/7	1	1/3	1/5	1/6
Vocational Training	1	1/3	3	1	1	3
College Preparation	1/3	5	5	1	1	3
Music Classes	1/4	1	6	1/3	1/3	1

- comparison of schools with respect to the six characteristics—

Learning

	A	B	C
A	1	1/3	1/2
B	3	1	3
C	2	1/3	1

Friends

	A	B	C
A	1	1	1
B	1	1	1
C	1	1	1

School Life

	A	B	C
A	1	5	1
B	1/5	1	1/5
C	1	5	1

Vocational Training

	A	B	C
A	1	9	7
B	1/9	1	1/5
C	1/7	5	1

College Preparation

	A	B	C
A	1	1/2	1
B	2	1	2
C	1	1/2	1

Music Classes

	A	B	C
A	1	6	4
B	1/6	1	1/3
C	1/4	3	1

The eigenvector of the first matrix is given by:

$$(0.32, 0.14, 0.03, 0.13, 0.23, 0.14),$$

and its corresponding eigenvalue is $\lambda = 7.49$ which is far from the consistent value 6. No revision of the matrix was made. Normally, such inconsistency would indicate that we should reconsider the arrangements.

The eigenvalues and eigenvectors of the other six matrices are

$\lambda = 3.05$	$\lambda = 3$	$\lambda = 3$	$\lambda = 3.21$	$\lambda = 3.00$	$\lambda = 3.05$
Learning	Friends	School Life	Vocational Training	College Preparation	Music Classes
0.16	0.33	0.45	0.77	0.25	0.69
0.59	0.33	0.09	0.05	0.50	0.09
0.25	0.33	0.46	0.17	0.25	0.22.

To obtain the overall ranking of the schools, we multiply the last matrix on the right by the transpose of the vector of weights of the characteristics. This yields

$$A = 0.37$$

$$B = 0.38$$

$$C = 0.25.$$

The individual went to school A because it had almost the same rank as school B, yet school B was a private school charging close to \$1600 a year and school A was free. This is an example where we were able to bring in a lower priority item, e.g., the cost of the school to add to the argument that A is favored by the candidate. The actual hierarchy is shown in Figure 11.3.

EXAMPLE 2 (Psychotherapy). The hierarchical method of prioritization may be used to provide insight into psychological problem areas in the following manner. Consider an individual's overall well-being as the single top level entry in a hierarchy. Conceivably, this level is primarily affected by childhood, adolescent, and adult experiences. Factors in growth and maturity which impinge upon well-being may be the influences of the father and the mother separately, as well as their influences together as parents, the socio-

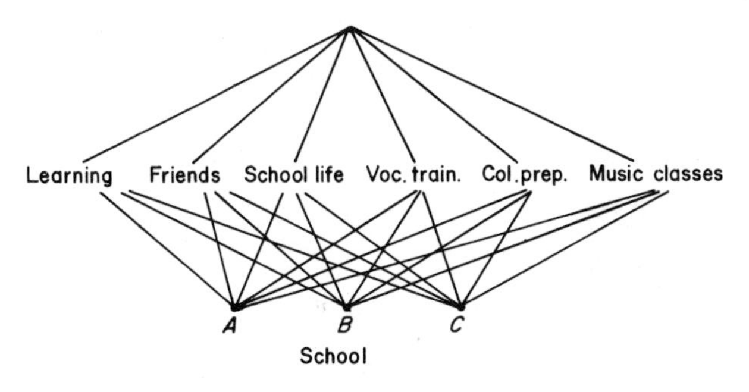

Figure 11.3

economic background, sibling relationships, one peer group, schooling, religious status, and so on.

The factors above, which comprise the second level in our hierarchy, are further affected by criteria pertinent to each. For example, the influence of the father may be broken down to include his temperament, strictness, care, and affection. Sibling relationships can be further characterized by the number, age differential, and sexes of siblings; peer pressure and role modeling provide a still clearer picture of the effects of friends, schooling and teachers.

As an alternative framework for the second level, we might include self-respect, security, adaptability to new people and new circumstances, and so on, influencing or as influenced by the elements above.

A more complete setting for a psychological history might include several hundreds of elements at each level, chosen by trained individuals and placed in such a way as to derive the maximum understanding of the subject in question.

Here we will consider a highly restricted form of the above, where the individual in question feels his self-confidence has been severely undermined and his social adjustments impaired by a restrictive situation during childhood. He is questioned about his childhood experiences only and asked to relate the following elements pairwise on each level:

- Level I overall well-being (OW),
- Level II self-respect, sense of security, ability to adapt to others (R, S, A)
- Level III visible affection shown for subject (V), ideas of strictness, ethics (E), actual disciplining of child (D), emphasis on personal adjustment with others (O),
- Level IV influence of mother, father, both (M, F, B).

The replies in the matrix form were as follows:

OW	R	S	A
R	1	6	4
S	1/6	1	3
A	1/4	1/3	1

R	V	E	D	O
V	1	6	6	3
E	1/6	1	4	3
D	1/6	1/4	1	1/2
O	1/3	1/3	2	1

S	V	E	D	O
V	1	6	6	3
E	1/6	1	4	3
D	1/6	1/4	1	1/2
O	1/3	1/3	2	1

A	V	E	D	O
V	1	1/5	1/3	1
E	5	1	4	1/5
D	3	1/4	1	1/4
O	1	5	4	1

V

	M	*F*	*B*
M	1	9	4
F	1/9	1	8
B	1/4	1/8	1

E

	M	*F*	*B*
M	1	1	1
F	1	1	1
B	1	1	1

D

	M	*F*	*B*
M	1	9	6
F	1/9	1	1/4
B	1/6	4	1

O

	M	*F*	*B*
M	1	5	5
F	1/5	1	1/3
B	1/5	3	1.

The eigenvector of the first matrix, *a*, is given by

	OW
R	0.701
S	0.193
A	0.106.

The matrix *b* of eigenvectors of the second row of matrices is given by

	R	*S*	*A*
V	0.604	0.604	0.127
E	0.213	0.213	0.281
D	0.064	0.064	0.120
O	0.119	0.119	0.463.

The matrix *c* of eigenvectors of the third row of matrices is given by

	V	*E*	*D*	*O*
M	0.721	0.333	0.713	0.701
F	0.210	0.333	0.061	0.097
B	0.069	0.333	0.176	0.202.

The final composite vector of influence on well-being obtained from the product *cba* is given by

Mother	0.635
Father	0.209
Both	0.156.

EXAMPLE 3 (Choosing a Job). A student who had just received his Ph.D. degree was interviewed for three jobs. His criteria for selecting the jobs and their pairwise comparison matrix are as follows:

• overall satisfaction with job—

	Research	Growth	Benefits	Colleagues	Location	Reputation
Research	1	1	1	4	1	1/2
Growth	1	1	2	4	1	1/2
Benefits	1	1/2	1	5	3	1/2
Colleagues	1/4	1/4	1/5	1	1/3	1/3
Location	1	1	1/3	3	1	1
Reputation	2	2	2	3	1	1

The pairwise comparison matrices of the jobs with respect to each criterion are as follows:

Research

	A	*B*	*C*
A	1	1/4	1/2
B	4	1	3
C	2	1/3	1

Growth

	A	*B*	*C*
A	1	1/4	1/5
B	4	1	1/2
C	5	2	1

Benefits

	A	*B*	*C*
A	1	3	1/3
B	1/3	1	1
C	3	1	1

Colleagues

	A	*B*	*C*
A	1	1/3	5
B	3	1	7
C	1/5	1/7	1

Location

	A	*B*	*C*
A	1	1	7
B	1	1	7
C	1/7	1/7	1

Reputation

	A	*B*	*C*
A	1	7	9
B	1/7	1	5
C	1/9	1/5	1

The eigenvalue and eigenvector of the first matrix are, respectively,

$$
\begin{aligned}
\lambda_{\max} = {} & 6.35 \\
& 0.16 \\
& 0.19 \\
& 0.19 \\
& 0.05 \\
& 0.12 \\
& 0.30.
\end{aligned}
$$

The remaining eigenvalues and eigenvectors are given by

	Research	Growth	Benefits	Colleagues	Location	Reputation
$\lambda_{\max} =$	3.02	3.02	3.56	3.06	3	3.21
	0.14	0.10	0.32	0.28	0.47	0.77
	0.63	0.33	0.22	0.65	0.47	0.17
	0.24	0.57	0.46	0.07	0.07	0.05

The composite vector for the jobs is given by

$$A = 0.40$$

$$B = 0.34$$

$$C = 0.26.$$

The differences were sufficiently large for the candidate to accept the offer to job A.

EXAMPLE 4 (Selecting a Plan for Vacation). With a view to spend a week for vacation, four places were evaluated in terms of the following criteria:

- F_1 cost of the trip from Philadelphia,
- F_2 sight-seeing opportunities,
- F_3 entertainment (doing things),
- F_4 way of travel,
- F_5 eating places.

The places considered were:

- S short trips (i.e., New York, Washington, Atlantic City, New Hope, etc.),
- Q Quebec,
- D Denver,
- C California.

The matrices of pairwise comparison of these vacation places for each criterion together with their eigenvectors, are as follows:

Cost

	S	Q	D	C	Eigenvector
S	1	3	7	9	0.58
Q	1/3	1	6	7	0.30
D	1/7	1/6	1	3	0.08
C	1/9	1/7	1/3	1	0.04

$\lambda_{max} = 4.21$

Sightseeing

	S	Q	D	C	Eigenvector
S	1	1/5	1/6	1/4	0.06
Q	5	1	2	4	0.45
D	6	1/2	1	6	0.38
C	4	1/4	1/6	1	0.12

$\lambda_{max} = 4.34$

Entertainment

	S	Q	D	C	Eigenvector
S	1	7	7	1/2	0.36
Q	1/7	1	1	1/7	0.06
D	1/7	1	1	1/7	0.06
C	2	7	7	1	0.52

$\lambda_{max} = 4.06$

Way of Travel

	S	Q	D	C	Eigenvector
S	1	4	1/4	1/3	0.21
Q	1/4	1	1/2	3	0.19
D	4	2	1	3	0.41
C	3	1/3	1/3	1	0.18

$\lambda_{max} = 5.38$

Eating Places

	S	Q	D	C	Eigenvector
S	1	1	7	4	0.43
Q	1	1	6	3	0.38
D	1/7	1/6	1	1/4	0.05
C	1/4	1/3	4	1	0.14

$\lambda_{max} = 4.08.$

The criterion matrix is

	Cost	Sight Seeing	Entertainment	Way of Travel	Eating Places	Eigenvector
Cost	1	1/5	1/5	1	1/3	0.09
Sightseeing	5	1	1/5	1/5	1	0.13
Entertainment	5	5	1	1/5	1	0.23
Way of Travel	1	5	5	1	5	0.43
Eating Places	3	1	1	1/5	1	0.13

$\lambda_{max} = 6.78.$

The ranking of the four places obtained by the prioritization is

$$\begin{array}{ll} S & 0.29 \\ Q & 0.23 \\ D & 0.25 \\ C & 0.24. \end{array}$$

This shows that the four are almost equally preferred with short trip places having a slight edge over the others.

6. Remarks on Conflict and Planning Applications [44]

In an interesting application to conflict analysis, the theory has been used to construct a hierarchy whose levels represent the actors who influence or control the outcome of the conflict, the objectives of the actors, their policies, their strategies, and the set of plausible outcomes which can result from their actions. The analysis leads to weights or priorities for the outcomes. The analysis offers ground for approaching the parties on what may work best when all their combined interests are taken into consideration or to show them where to modify their positions to obtain a jointly more desirable outcome.

An outcome in a planning problem is often referred to as a scenario. To insure taking into account the widest "plausible" set of possible outcomes (or scenarios), adopting an outcome for each actor which he would like to pursue by himself is desirable. The set of outcomes is then hierarchically weighted by the weights of the actors composed with the weights of their objectives and finally with those of their strategies.

A composite outcome, the resultant of all the influences on the set of outcomes is obtained. This is the likely or composite future. The likely future is characterized in detail by a set of state variables. The values of these variables are calibrated by weighting the corresponding values of these variables from each individual future considered. These are usually determined on a difference scale according to the strength of differences of each variable from its value in the present outcome taken as the zero reference point. The present is assumed to be the best known outcome with which other outcomes may be compared. The purpose of such an analysis is to examine the attitudes of the actors about the future within a hierarchical framework which they can help define and offer them an opportunity to bargain and change their position, hopefully to obtain a more favorable outcome.

7. Decomposition and Aggregation or Clustering

There are essentially two fundamental ways in which the idea of a hierarchy can be used. The first, by now clear, has to do with modeling the real world hierarchically. The second is probably even more fundamental than the first and points to the real power of hierarchies in nature. It is to break things down into large groupings or clusters and then break each of these into smaller clusters and so on. The object would then be to obtain the priorities of all the elements by means of clustering. This is by far the more efficient process than treating all the elements together. Thus whether we think of hierarchies as intrinsic in nature, as some have maintained, or whether we simply use them because of our limited capacity to process information is immaterial. In either case, they are an efficient way of looking at complex problems.

To decompose a hierarchy into clusters, one must first decide on which elements to group together in each cluster. This is done according to the proximity or similarity of the elements with respect to the function they perform or property they share and regarding which we need to know the priority of these elements. One must then conduct comparisons on the clusters and on the subclusters and then recompose the clusters to obtain a true reflection of the overall priorities. If this process works, the result after the decomposition should be the same as if there were no decomposition. Let us illustrate with the example of distance from Philadelphia of the six cities mentioned earlier.

Example (A Distance Hierarchy). The example of distances between cities will now be structured into a hierarchy.

If we group the cities into clusters according to their falling in nearly equivalent distances from Philadelphia, we have three classes compared in the following matrix.

	Chicago Montreal	London San Francisco	Cairo Tokyo	Eigenvector
Chicago Montreal	1	1/7	1/9	0.056
London San Francisco	7	1	1/4	0.26
Cairo Tokyo	9	4	1	0.68

If we now compare the cities in each cluster separately according to their relative distance from Philadelphia, we have on using for the 2×2 case the scale $1 + \varepsilon$:

	Chicago	Montreal	Eigenvector
Chicago	1	2	0.67
Montreal	1/2	1	0.33

	Cairo	Tokyo	Eigenvector
Cairo	1	1/1.5	0.4
Tokyo	1.5	1	0.6

	San Francisco	London	Eigenvector
San Francisco	1	1/1.3	0.43
London	1.3	1	0.57

Now we multiply the first eigenvector by 0.056, the second by 0.26, and the third by 0.68 to obtain the overall relative distance vector:

Recall that the actual result is

Cairo	Tokyo	Chicago	San Fran.	London	Montreal
(0.27	0.41	0.037	0.11	0.15	0.019)
(0.278	0.361	0.032	0.132	0.177	0.019).

Let us assume that we have a set of n elements. If we wish to compare the elements in pairs to obtain a ratio scale ranking by solving the eigenvalue problem, $(n^2 - n)/2$ judgments would be necessary. Suppose now that seven is the maximum number of elements which can be compared with any reasonable (psychological) assurance of consistency. Then n must be first decomposed into equivalence classes of seven clusters or subsets, each of these decomposed in turn to seven new clusters and so on down generating levels of a hierarchy until we obtain a final decomposition each of whose sets has no more than seven of the original elements. Let $\{x\}$ denote the smallest integer greater than or equal to x. We have the following theorem.

Theorem 1. *The maximum number of comparisons obtained from the decomposition of a set of $n > 1$ elements into a hierarchy of clusters (under the assumption that no more than seven elements are compared simultaneously), is bounded by* $(7/2)\,(7^{\{\log n/\log 7\}} - 1)$, *and this bound is sharp.*

PROOF. We have the following for the *number of comparisons* in each level where we must have in the hth or last level at most seven elements in each cluster.

1. 0
2. $\dfrac{7^2 - 7}{2}$
3. $7 \times \dfrac{7^2 - 7}{2}$

$\vdots$

k. $7^{h-2} \times \dfrac{7^2 - 7}{2}$ where $7^{h-2} \times 7 = n$, $h = \left\{\dfrac{\log n}{\log 7}\right\} + 1$, $h > 2$.

The sum of these comparisons is

$$21 \times \frac{7^{h-1} - 1}{7 - 1} = \frac{7}{2}(7^{\{\log n/\log 7\}} - 1).$$

To show that the bound is sharp it is sufficient to put $n = 7^m$. □

Remark. It looks as if the Saint Ives conundrum finds its solution in hierarchies.

The efficiency of a hierarchy may be defined to be the ratio of the number of direct pairwise comparisons required for the entire set of n elements involved in the hierarchy, as compared with the number of pairwise comparisons resulting from clustering as described above.

Theorem 2. *The efficiency of a hierarchy is of the order of* $n/7$.

PROOF. To prove the theorem we must compare

$$\frac{n^2 - n}{2} \text{ with } \frac{7}{2}(7^{\{\log n/\log 7\}} - 1).$$

Let $n = 7^{m+\varepsilon} 0 \leq \varepsilon < 1$. Then we clearly have

$$7^{2m+2\varepsilon} - \frac{7^{m+\varepsilon}}{7} \cdot (7^m - 1) \geq \frac{7^{m+\varepsilon}}{7} = \frac{n}{7}.$$

Thus $n/7$ is equal to the efficiency. □

One might naturally ask why we do not use two in place of seven for even greater efficiency. We note that in using a hierarchy we seek both consistency and good correspondence to reality. The former is greater the smaller the size of each matrix; the latter is greater the larger the size of the matrix, due to the use of redundant information. Thus we have a trade-off. Actually, we have shown, using the consistency index, that the number seven is a good practical bound on n, a last outpost so far as consistency is concerned.

Remark. The exponential efficiency is $\log_7 n$ or, in general, if we replace 7 by s we have $\log_s n$.

Suppose we have a set of 98 elements to which we want to assign priority. We decompose the problem into seven sets, each having on the average 14 elements. Now, we cannot compare 14 elements, so we decompose each of these sets into two sets, each having no more than seven elements. We then compare the elements among themselves.

To look at the efficiency of this process closely, we note that if it were possible to compare 98 elements among themselves, we would require $((98)^2 - 98)/2 = 4753$ comparisons. On the other hand, if we divide them into seven clusters of 14 elements each, then do pairwise comparisons of the seven clusters, we need $(7^2 - 7)/2 = 21$ comparisons. Each cluster can now be divided into two clusters each with seven elements. Comparing two clusters falling under each of the 14 element clusters requires one comparison, but there are seven of these; hence we require seven comparisons on this level; then we need $14 \times 21 = 294$ comparisons on the lowest level. The total number of comparisons in this hierarchical decomposition is $21 + 7 + 294 = 322$ as compared with 4753 comparisons without clustering. Indeed, the theorem is satisfied since $322 \ll 4753/7$.

Clustering a complex problem into hierarchical form has two advantages:

(1) great efficiency in making pairwise comparisons,
(2) greater consistency under the assumption of a limited capacity of the mind to compare more than 7 ± 2 elements simultaneously.

The efficiency of a hierarchy has been illustrated by Simon with an example of two men assembling watches, one by constructing modular or component parts from elementary parts and using these to construct higher order parts and so on, and the other by assembling the entire watch piece by piece from beginning to end. If the first man is interrupted, he only has to start reassembling a small module, but if the second man is interrupted, he has to start reassembling the watch from the beginning. If the watch has 1000 components and the components at each level have 10 parts, the first man will, of course, have to make the components and then make subassemblies in a total of 111 operations. If p is the probability of an interruption while a part is being added to an incomplete assembly, then the probability that the first man completes a piece without interruption is $(1 - p)^{10}$ and that for the second is $(1 - p)^{1000}$. For the first man, an interruption would cost the time required to assemble five parts. The cost to the second man will, on the average, be the time needed to assemble $1/p$ parts which is approximately the expected number of parts without interruption. If $p = 0.01$ (a chance in a hundred that either man would be interrupted in adding any one part), the cost to the first man is five and to the second man 100. The first man will assemble 111 components while the second would make just one component. However, the first man will complete an assembly in $(1 - 0.01)^{-10} = 10/9$ attempts, whereas the second man will complete an assembly in $e^{10} = (1 - 0.01)^{-1000} = (1/44) \times 10^6$ attempts. Thus the efficiency of the first man to that of the second man is given by

$$\frac{100/0.99^{1000}}{111\{[(1/0.99^{10}) - 1]5 + 10\}} \simeq 2000.$$

In man-made systems, the task of managing a complex enterprise is, in general, considerably simplified when it is broken down into subsystems or levels that are individually more tractable, i.e., a manager having a limited span of management. The steps of solving a large-scale problem are simplified and efficiently accomplished when they are modularized, e.g., by taking n sets of m variables each, than by taking mn variables simultaneously.

8. The Consistency of a Complete Hierarchy

We have generalized the measurement of consistency to a complete hierarchy. What we do is to multiply the index of consistency obtained from a pairwise comparison matrix by the priority of the property with respect to which the comparison is made and add all the results for the entire hierarchy. This is then compared with the corresponding index obtained by taking randomly generated indices, weighting them by the priorities, and adding. The ratio should be in the neighborhood of 0.10 in order not to cause concern for improvements in the judgments.

Let n_j, $j = 1, 2, \cdots, h$, be the number of elements in the jth level of the hierarchy. Let w_{ij} be the composite weight of the ith activity of the jth level, and let $\mu_{i,j+1}$ be the consistency index of all elements in the $(j + 1)$st level compared with respect to the ith activity of the jth level.

The consistency index of a hierarchy is defined by

$$C_H = \sum_{j=1}^{h} \sum_{i=1}^{n_{ij}} w_{ij}\mu_{i,j+1}$$

where $w_{ij} = 1$ for $j = 1$, and $n_{i_{j+1}}$ is the number of elements of the $(j + 1)$st level with respect to the ith activity of the jth level. When this has been applied to the school selection example, we have

second level priority vector: (0.32, 0.14, 0.03, 0.13, 0.23, 0.14),

second level μ_2: $\mu_2 = \dfrac{7.49 - 6}{5} = 0.298$,

third level vector of μ's: (0.025, 0, 0, 0.105, 0, 0.025).

Hence

$$C_H = 0.298 + (0.32, 0.14, 0.03, 0.13, 0.24, 0.14)\begin{bmatrix}0.025\\0\\0\\0.105\\0\\0.025\end{bmatrix} = 0.323$$

and using the corresponding random indices we have

$$\bar{C}_H = 1.24 + (0.32, 0.14, 0.03, 0.13, 0.24, 0.14)\begin{bmatrix}0.58\\0.58\\0.58\\0.58\\0.58\\0.58\end{bmatrix} = 1.82.$$

The consistency ratio of the hierarchy is therefore $C_H/\bar{C}_H = 0.18$, which is not very good, because it reflects the high inconsistency arising from ($\lambda_{\max} = 7.49$ for $n = 6$).

In this module we have presented a theory of measurement which appears to work in subjective areas and applied it to develop a hierarchy theory for both objective (tangible) and subjective (intangible) variables. We have shown this to be an efficient and reasonable approach. Currently, we are concentrating on a more abstract theory of impact matrices for general systems with the view of studying the stability of systems by these matrices. The method is also being applied to planning—the forward process to measure attitudes and the backward process to measure desired [44] futures and finding policies to attain these futures. Together, the forward and backward processes define a two-point boundary value problem to which the theory has been applied.

I am grateful to Robert J. Weber and to William W. Kuhn for careful reading of the paper and for making a number of valuable suggestions.

Since this chapter was first written, the material has been considerably elaborated in a number of papers and books on the subject.

Appendix

Interval Scale from Pairwise Comparisons

Suppose we have n activities $A_1, \cdots, A_n$ weighted on a difference scale, and we wish to study their numerical differences on a certain attribute by pairwise comparisons. Let their actual weights be $w_1, \cdots, w_n$, respectively. We form the matrix of pairwise comparisons using differences.

$$\begin{array}{cccccc} & A_1 & A_2 & \cdots & A_n \\ A_1 & w_1 - w_1 & w_1 - w_2 & & w_1 - w_n \\ A_2 & w_2 - w_1 & w_2 - w_2 & & w_2 - w_n \\ \vdots & & & & \\ A_n & w_n - w_1 & w_n - w_2 & & w_n - w_n. \end{array}$$

This matrix is consistent if we can construct it from n elements which form a spanning tree, such as the elements of one row. This leads to the fundamental consistency condition

$$a_{ik} - a_{ij} = a_{jk}$$

or

$$a_{ij} + a_{jk} = a_{ik}.$$

On putting $k = i$ and noting that $a_{ii} = 0$ we have $a_{ij} = -a_{ji}$ which says that the matrix $A = (a_{ij})$ is skew-symmetric. Since

$$(w_i - w_k) - (w_i - w_j) = (w_j - w_k),$$

the above matrix is consistent.

Let us assume that the matrix (a_{ij}) comprises small perturbations of the consistent matrix $(w_i - w_j)$. We can then write

$$a_{ij} - (w_i - w_j) = \varepsilon_{ij}, \qquad i, j = 1, \cdots, n. \tag{1}$$

We note that this system and the system obtained from it by replacing w_i with $w_i - \alpha$ where α is an arbitrary constant are the same and thus any solution has the form $w_i - \alpha$, $i = 1, \cdots, n$.

Now (1) is a system of n^2 equations in the $n^2 + n$ unknowns w_i and ε_{ij}. First we sum with respect to j, and we have

$$\sum_{j=1}^{n} a_{ij} - nw_i + \sum_{j=1}^{n} w_j = \sum_{j=1}^{n} \quad {}_{ij}, \qquad i = 1, \cdots, n. \tag{2}$$

We note that when we have consistency, $\Sigma_{j=1}^{n} \varepsilon_{ij} = 0$. For the general case (without consistency we impose the condition $\Sigma_{j=1}^{n} \varepsilon_{ij} = \lambda$. We also normalize the solution by writing $\Sigma_{j=1}^{n} w_j = 0$. We have on summing with respect to i, $\lambda = (1/n)\Sigma_{i,j=1}^{n} a_{ij}$. Finally, we get

$$w_i = \frac{1}{n}\sum_{j=1}^{n} a_{ij} - \frac{1}{n^2}\sum_{i,j=1}^{n} a_{ij}.$$

With consistency, we have $a_{ij} = -a_{ji}$, and hence $\lambda = 0$ and

$$w_i = \frac{1}{n}\sum_{j=1}^{n} a_{ij}.$$

Inconsistency is measured by

$$\frac{1}{n^2}\sum_{i,j=1}^{n} a_{ij}.$$

Clearly, our estimate of w may be improved by obtaining information regarding both a_{ij} and a_{ji} and using λ to evaluate the overall inconsistency rather than forcing a_{ji} by taking its value as $-a_{ij}$.

We shall use the following numerical scale:

Value	To Represent
0	the objects being compared are the same with regard to the property in the comparison;
2	the first object has the property weakly more than the second;
4	the first object has the property strongly more than the second;
6	the first object has the property demonstratedly more than the second;
8	the first object has the property absolutely more than the second.

Values 1, 3, 5, 7, are used for compromise between the adjacent values. The numbers -1 to -8 are used to indicate "less" than rather than "more" than for the corresponding difference in the comparison.

EXAMPLE. Given four individuals J, R, D, T whose heights are respectively 4′6″, 5′9″, 4′, and 6′. We wish to compare the difference between their heights in pairs. This gives the following matrix and its solution. The solution vector gives the deviation of the height of each individual from the mean height 5′3/4″. This is in fact exactly what is obtained directly, i.e., the respective differences from the mean are $-6\ 3/4''$, $8\ 1/4''$, $-12\ 3/4''$, $11\ 1/4''$:

	J	R	D	T	Solution Row average
J	0	-15	6	-18	$-\frac{27}{4}$
R	15	0	21	-3	$\frac{33}{4}$
D	-6	-21	0	-24	$-\frac{51}{4}$
T	18	3	24	0	$\frac{45}{4}$.

We next constructed a matrix of pairwise comparisons of the individuals using the scale 0–8 according to perceived height difference described above with $a_{ji} = -a_{ij}$. The matrix of differences and the corresponding solution are given below. If we multiply the solution vector by 3 we obtain $-6\ 3/4''$, $8''$, $-11\ 1/4''$, $11''$, which is in striking correspondence with the actual values:

	J	R	D	T	Solution
J	0	-5	2	-6	$-2\frac{1}{4}$
R	5	0	6	-1	$2\frac{2}{4}$
D	-2	-6	0	-7	$-3\frac{3}{4}$
T	6	1	7	0	$3\frac{2}{4}$.

References

[1] R. Beals, D. H. Krantz, and A. Tversky, "Foundations of multidimensional scaling," *Psychological Rev.*, vol. 75, no. 2, pp. 127–142, 1968.
[2] C. Berge, *Theory of Graphs and Its Applications*. New York: Wiley, 1962.
[3] P. Bogart, "Preference structures I: Distances between transitive preference relation," *J. Mathematical Sociology*, vol. 3, pp. 49–57, 1973.
[4] ——, "Preference structure II: Distances between intransitive preference relations," to appear in *SIAM J. Applied Mathematics*.
[5] G. Bronson, "The hierarchical organization of the central nervous system," in *International Politics and Foreign Policy: A Reader in Research and Theory*, rev. ed., James A. Rosenau, Ed. New York: Free Press, 1969.
[6] R. C. Buck, and D. L. Hull, "The logical structure of the linnaean hierarchy," *Systematic Zoology*, vol. 15, pp. 97–111, 1966.
[7] R. Busacker and T. L. Saaty, *Finite Graphs and Networks*. New York: McGraw-Hill, 1965.
[8] G. Cant, "An X-ray analysis of doctors' bills," *Money*, August 1973.
[9] Census of Manufacturers, Bureau of the Census, U.S. Department of Commerce, 1967.
[10] H. Chenery and P. Clark, *Interindustry Economics*. New York: Wiley, 1959.
[11] R. T. Eckenrode, "Weighting multiple criteria," *Management Science*, vol. 12, no. 3, pp. 180–192, November 1965.
[12] Environmental Protection Agency, Office of Research and Monitoring, *The Quality of Life Concept: A Potential New Tool for Decision-Makers*, 1973.
[13] G. Fechner, *Elements of Psychophysics*, vol. 2, Helmut E. Adler, Tr. New York: Holt, Rinehart & Winston, 1966.
[14] J. N. Franklin, *Matrix Theory*. Englewood Cliffs, NJ: Prentice-Hall, 1968.
[15] R. A. Frazer, W. J. Duncan, and A. R. Collar, *Elementary Matrices*. Cambridge University Press, 1963.
[16] F. R. Gantmacher, *The Theory of Matrices*, vol. II. New York, Chelsea Publishing, 1960, pp. 53 and 63.
[17] M. Gardner, "The hierarchy of infinites and the problems it spawns," *Scientific American*, vol. 214, pp. 112–118, Mar. 1966.
[18] J. R. Gillett, "The football league eigenvector," *Eureka*, Oct. 1970.
[19] E. E. Harris, "Wholeness and hierarchy," in *Foundations of Metaphysics in Science*, New York: Humanities, 1965, ch. 7.
[20] I. N. Herstein, *Topics in Algebra*, Blaisdell Publishing, 1964.
[21] N. N. Jardine and R. Sibson, *Mathematical Taxonomy*. New York: Wiley, 1971.
[22] S. C. Johnson, "Hierarchical clustering schemes," *Psychometrica*, vol. 32, no. 3, pp. 241–252, Sept. 1967.
[23] J. G. Kemeny and J. Laurie Snell, *Mathematical Models in the Social Sciences*, Blaisdell Publishing, 1962.
[24] A. J. Klee, "The role of decision models in the evaluation of competing environmental health alternatives," *Management Science*, vol. 18, no. 2, pp. 53–67. Oct. 1971.
[25] A. Koestler and J. R. Smythies, Eds., *Beyond Reductionism: New Perspectives in the Life of the Sciences*. New York: MacMillan, 1970.
[26] D. H. Krantz, "A theory of magnitude estimation and cross modality matchings," *J. Mathematical Psychology*, vol. 9, pp. 168–199, 1972.
[27] ——, R. D. Luce, P. Suppes, and A. Tversky, *Foundations of Measurement*. New York: Academic, 1971.
[28] H. H. Landsberg and S. H. Schurr, *Energy in the United States: Sources, Uses and Policy Issues*. New York: Random House for Resources for the Future, Inc., 1963.

[29] A. Lindenmayer, "Life cycles as hierarchical relations," *Form and Strategy in Science*, J. R. Gregg and F. T. C. Harris, Eds. Dordrecht: D. Reidell, 1964.
[30] Logistics Management Institute, *Identification of War Reserve Stock*, Task 72–04, Washington, DC, June 1972.
[31] M. L. Manheim, *Hierarchical Structure: A Model of Planning and Design Processes*. Cambridge, MA: M.I.T. Press, 1966, p. 222.
[32] R. Mariano, "The study of priorities in electrical energy allocation," University of Pennsylvania, 1975.
[33] E. S. Mason and the staff on the NPA Project on the Economic Aspects of the Productive Uses of Nuclear Energy, *Energy Requirements and Economic Growth*, Washington, National Planning Association, 1955.
[34] M. D. Mesarovic and D. Macko, "Scientific theory of hierarchical systems," in *Hierarchical Structures*, L. L. Whyte, A. G. Wilson, D. Wilson, Eds. New York: American Elsevier, 1969.
[35] J. L. Moreno, *Fondements de la sociometrie*, Lesage-Maucorps, Tr. Presses Universitaires, Paris, 1954.
[36] G. A. Miller, "The magical number seven, plus or minus two: Some limits on our capacity for processing information," *The Psychological Review*, vol. 63, pp. 81–97, Mar. 1956.
[37] H. H. Pattee, "The problem of biological hierarchy," *Towards a Theoretical Biology*, vol. III, C. H. Waddington, Ed. Edinburgh: Edinburgh Univ. Press, 1969.
[38] ——, Ed., *Hierarchy Theory, The Challenge of Complex Systems*. New York: George Braziller, 1973.
[39] J. Pfanzagl, *Theory of Measurement*. New York: Wiley, 1968.
[40] T. L. Saaty, "An eigenvalue allocation model in contingency planning," University of Pennsylvania, 1972.
[41] ——, "Hierarchies and priorities—Eigenvalue analysis," University of Pennsylvania, 107 pages, 1975.
[42] ——, "Measuring the fuzziness of sets," *J. Cybernetics*, vol. 4, no. 4, pp. 53–61, 1974.
[43] T. L. Saaty and M. Khouja, " A measure of world influence," *Peace Science*, June 1976.
[43′] T. L. Saaty, *The Analytic Hierarchy Process*, McGraw Hill, 1980.
[43″] ——, and L. Vargas, *The Logic of Priorities*, Kluwer Nijhoff, 1981.
[43‴] ——, *Decision Making for Leaders*, Wadsworth, 1982.
[44] T. L. Saaty and P. C. Rogers, "The future of higher education in the United States (1985–2000)", *Socio-Economic Planning Sciences*, vol. 10, no. 6, pp. 251–264, Dec. 1976.
[45] T. L. Saaty, Project Director, *The Sudan transport study*, 5 volumes, The Democratic Republic of the Sudan in association with the Kuwait Fund for Arab Economic Development, 1975.
[46] T. L. Saaty, F. Ma, and P. Blair, "Hierarchical theory and operational gaming for energy policy analysis," study done for ERDA, 1975.
[47] A. Sankaranarayanan, "On a group theoretical connection among the physical hierarchies," Res. Communication no. 96, Douglas Advanced Research Laboratories, Huntingdon Beach, Ca.
[48] C. W. Savage, "Introspectionist and Behaviorist Interpretations of Ratio Scales of Perceptual Magnitudes," *Psychological Monographs: General and Applied*, 80, No. 19, Whole No. 627, 1966.
[49] D. Scott, "Measurement structures and linear inequalities," *J. Mathematical Psychology*, vol. 1 pp. 233–247, 1964.
[50] R. N. Shepard, "A taxonomy of some principal types of data and of multidimen-

sional methods for their analysis," in *Multidimensional Scaling: Theory and Applications in the Behavioral Sciences*, vol. 1, R. N. Shepard, A. K. Romney, S. B. Nerlove, Eds. New York: Seminar Press, pp. 21–47. 1972.
[51] R. R. Sokal and P. H. A. Seneath, *Principles of Numerical Taxonomy*. Freeman, 1963.
[52] S. S. Stevens, "On the psychophysical law," *Psychological Review*, vol. 64, pp. 153–181, 1957.
[53] ——, "To honor Fechner and repeal his law," *Science*, vol. 13, Jan. 13, 1961.
[54] P. Suppes and J. L. Zinnes, *Handbook of Mathematical Psychology*, vol. 1. New York: Wiley, 1963.
[55] L. L. Thurston, "A law of comparative judgment," *Psychological Review*, vol. 34, pp. 273–286, 1927.
[56] W. S. Torgerson, *Theory and Methods of Scaling*. New York: Wiley, 1958.
[57] T. H. Wei, "The algebraic foundations of ranking theory," Thesis, Cambridge, 1952.
[58] P. A. Weiss, *Hierarchically Organized Systems in Theory and Practice*. New York: Hafner, 1971.
[59] H. Weyl, "Chemical valence and the hierarchy of structures," Appendix D in *Philosophy of Mathematics and Natural Science*. Princeton, NJ: Princeton Univ. Press, 1949.
[60] L. L. Whyte, "Organic structural hierarchies," in *Unity and Diversity in Systems*, Essays in honor of L. von Bertalanffy, R. G. Jones and G. Brandl, Eds. New York: Braziller, 1969.
[61] ——, "The structural hierarchy in organisms," from *Unity and Diversity in Systems*, Jones and Brandl, Eds. New York: Braziller, in press.
[62] L. L. Whyte, A. G. Wilson, and D. Wilson, Eds., *Hierarchical Structures*, New York: American Elsevier 1969.
[63] H. Wielandt, "Unzerlegbare, Nicht Negative Matrizen," *Math. Z.*, vol. 52, pp. 642–648, 1950.
[64] H. S. Wilf, *Mathematics for the Physical Sciences*. New York: Wiley, 1962.
[65] J. H. Wilkinson, *The Algebraic Eigenvalue Problem*. Clarendon Press, 1965, ch. 2,
[66] A. G. Wilson, "Hierarchical structure in the cosmos," in *Hierarchical Structures*. New York: American Elsevier, 1969.
[67] D. R. Woodall, "A criticism of the football league eigenvector," *Eureka*, Oct. 1971.
[68] P. A. Julien, P. Lamonde, and D. Latouche, *La Methode des Scenarios*, Groupe de Recherches sur le Futur Universite du Quebec, Nov. 1974.
[69] A. Ando and H. Simson, "Aggregation of variables in dynamic systems," *Econometrica*, vol. 29, no. 2, Apr. 1961.
[70] H. Simon, "The architecture of complexity," *Proc. American Philosophical Society*, vol. 106, pp. 467–482, Dec. 1962.

CHAPTER 12
Multiple-Choice Testing

Robert J. Weber*

1. Introduction

The multiple-choice questionnaire has become an integral part of modern society. It is used for such diverse purposes as testing knowledge, estimating ability, polling opinions, and indicating most-suitable careers to individuals. Some of the richness of variation possible within the multiple-choice format will be studied in this module.

The general multiple-choice questionnaire consists of (one or) several items, each of which is followed by several options. For the sake of specificity, we shall primarily concern ourselves with the common achievement/ability test in which the items are questions, and each has precisely one correct answer among its options. Only occasionally shall we refer to other types of questionnaires, such as those in which preferences are elicited (no particular options are "correct") or in which more than one option may be correct. Therefore, we begin with the assumption that we possess a test which presents with each item several options, all but one of which are incorrect (distractors).

The assumption underlying most testing is that the particular subject trait being investigated (for example, knowledge of a specific topic) may be quantitatively measured on a single-dimensional scale. It is often further assumed that a procedure for obtaining the value of this trait, called the "criterion score", is available. (For example, an extremely long, comprehensive examination administered as a series of parallel tests yields an average score per test which is, by the law of large numbers, a good estimate of the subject's "true score.") We shall assume that we have available an effective,

* J. L. Kellogg Graduate School of Management, Northwestern University, Evanston, IL 60201.

although perhaps inefficient, method for obtaining each subject's criterion score. Clearly, extensive testing may be an extremely uneconomical method for generally measuring criterion scores. The ultimate goal of test designers is to fashion an efficient method for computing a "good" estimate of individual criterion scores.

In order to evaluate any proposed method, two measures of "goodness" have been popularly accepted. These are validity and reliability. The validity of a test is the correlation between test score and criterion score. This indicates how well the test accomplishes its primary objective of quantifying the chosen trait. Validity is usually computed for a representative subsample of subjects, for each of whom the extensive direct measurement of criterion scores is carried out. The reliability of a test is an internal (non-criterion-related) measure, the correlation between parallel, but independent, administrations of the same test to the same group of subjects. Since this ideal measurement is not practically obtainable, various estimates are used. A popular one is the "split-half" coefficient. If the test can be split into two comparable, independent halves, then $1 - (v_d/v_t)$ is a lower bound on reliability, where v_d is the variance of the difference in scores on the two halves and v_t is the variance of the sum. A test of high (close to 1) validity must be reliable, as indicated by the relationship

$$2v^2 - 1 \leq r \leq 1.$$

On the other hand, a highly reliable test may totally fail to measure the criterion being studied, and so lack validity.

We have discussed the original motivation for test design, and the final evaluation of tests, but central questions of specific form remain to be treated. What types of responses are to be elicited? How are these responses on each item to be scored, and how are the item scores to be combined into a final score which, one might hope, correlates well with the criterion score?

The most common type of instruction to a subject is to "indicate the answer believed most likely." We shall refer to items with such instructions as "one-choice" items. Other possibilities include: "indicate the one or several answers considered most likely" (equivalently, "indicate the one or several answers considered least likely"), "rank the answers in order of likelihood," and "assign personal (subjective) probabilities to the answers." We should realize that, in order for a subject to be able to reasonably respond to the instructions, he must have some idea of the method of scoring to be used.

The weighting of item scores to obtain an overall score for a subject is an interesting topic on which extensive literature is available. However, we shall restrict our consideration of scoring to the question of establishing item scores, with the assumption that some affine combination of item scores (i.e., a sum of positive multiples of the item scores and an additive constant) will yield the overall score for a subject. We make the further assumptions that a subject's knowledge on an item is completely described by the subjective probabilities he attaches to the correctness of the various options, and that each subject's sole objective is to maximize his overall expected score. With these assumptions, we can view the objective of each subject, when faced with a particular item, as maximization of his expected score on that item (independent of the remainder of the test).

The possibilities for item-scoring methods generally fall into two classes: response-dependent methods (formula scoring) and option-dependent methods. The response-dependent methods score a subject solely on the type of response given and on the position of the correct option relative to that response. These methods might score a subject by the certainty he expresses in his response, or by how far his response differs from the "ideal" response (the response a subject with perfect knowledge and absolute confidence would give). The option-dependent methods are based on the premise that some of the distractors in a set of options may be "less incorrect" than others.

Option-dependent scoring methods can be further classified as *a priori* methods (in which the option scores are based on "expert" assessments) and *a posteriori* methods (in which information about the degree of "correctness" of certain distractors is gleaned from subject response data). Posterior weights are subject to two kinds of usage. The weights can be established by a test group of subjects, and then used as *a priori* weights for a subsequent (larger) group. Alternatively, the weights can be used in conjunction with the data from which they were derived, in an attempt to increase the validity of the original test results. The former use is suitable for the construction of tests to be widely administered, while the latter is relevant to the small-scale (frequently one-shot) type of multiple-choice examination used in many college courses.

Two related observations on the use of scoring systems deserve mention. Clearly, a response-dependent scoring system is meant to be fully described to the subjects in order that their responses be based on their knowledge in the desired manner. On the other hand, specific option-weighting information must be concealed, because such information bears on the correctness particular options. Yet in this latter case, if the subjects know the method by which the option weights are derived, their optimal response strategies involve complicated decision-theoretic or game-theoretic considerations. Our simplifying assumption in such a case is that the subjects follow the (usually near-optimal) strategy of trying to maximize expected "raw" score (the score on which the option weights are/were subsequently based). This observation relates to an important issue treated rarely in the testing literature, namely, the best scoring system for eliciting desired responses may not be the best scoring system for evaluating subject knowledge. Whether it is ethical to promise the use of one scoring system to subjects and then use another for evaluating their responses is debatable. At the least, perhaps the question of eliciting responses in order to derive weights should be kept separate from the question of estimating criterion score in subsequent test administrations.

In Section 2 of this module we study the question of option weighting in more detail. This study is carried out in the context of the traditional examination in which "one-choice" responses are elicited. In the third section, we will examine several response-dependent scoring systems and will investigate optimal subject response strategies. The final section will cover response-dependent scoring systems designed to obtain directly a subject's subjective probabilities as his response.

2. Option Weighting

The primary purpose of multiple choice testing is to gain information about subjects. Consider the scoring of a test in which each item has a single correct option. Assigning $+1$ on each question correctly answered, and 0 otherwise, captures much of the information contained in a subject's pattern of responses. However, other information is available which this system of scoring does not exploit. For instance, information about the subjects' knowledge is also contained in the "types" of wrong answers they give.

An example may illuminate this idea. Consider the following item from an actual test given in an elementary mathematics course.

Q: If a coin is so weighted that "heads" is twice as likely as "tails," what is the probability that a "tail" results from a particular flip?
A: (a) 1/4 (b) 1/3 (c) 1/2 (d) 2/3 (e) none of these.
(51) (83) (40) (66) (43)

The number under each option indicates the average overall test score of subjects responding with that option. As one would hope, the highest group-average score was associated with the correct answer. Let us consider the other options. The response which attracted the second-best group of students was (d). Upon reflection, we find this not surprising. Solution of this problem may proceed in two steps. First, it is realized that 1/3 and 2/3 are the probabilities of the two possible outcomes, and then these probabilities are associated with the outcomes in either the correct or the incorrect order. The incorrect response (d) may indicate that the subject successfully completed the first step before slipping on the second.

In case the example above seems unconvincing due to the simplicity of the item, we present two more items which may discriminate between subjects on the basis of their partial knowledge.

Q: Select the pair which best expresses a relationship similar to that expressed in the original pair. Color : spectrum : :
A: (a) tone : scale (b) sound : waves (c) verse : poem
(d) dimension : space (e) cell : organism.

Q: Select the word most nearly opposite in meaning to "promulgate".
A: (a) distort (b) demote (c) suppress
(d) retard (e) discourage.

On each of these items, a knowledgeable subject will probably choose his response from a more restricted set of alternatives than will a less knowledgeable subject (or at least, the first may have a less uniform subjective probability distribution on the likely correctness of the various answers). (The correct answers are (a) and (c).)

These examples suggest that the use of option weights other than +1 and 0 may capture more of the information available about the subjects' knowledge. Although one would (probably) still want the greatest item score to be associated with the correct option, allowing differential scores for the other options may yield a more accurate test.

In order to investigate the possibilities of differential option weights further, we begin by assuming that we have available the criterion scores of a group of subjects. In this case we shall seek to determine, for each item independently, a set of option weights for which the subjects' scores on the item most closely match their criterion-scores. Specifically, we will minimize the squared deviation of item scores from criterion scores.

Let a q-item examination be administered to a group of p subjects. Let

$$r_{ijk} = \begin{cases} 1, & \text{if subject } i \text{ selects option } j \text{ on item } k, \\ 0, & \text{otherwise.} \end{cases}$$

Throughout this discussion, we assume that each subject answers every item (that is, the examination is a "power", rather than "speed," test). Assume that the test has a total of t options (for example, if each item has l options, $t = lq$). Let R be the $p \times t$ matrix with entry r_{ijk} in the ith row and $\boldsymbol{jk}$th column, where $\boldsymbol{jk}$ indexes the column corresponding to the jth option of the kth item. Let w_{jk} be the weight associated with the $\boldsymbol{jk}$th option.

Let s_i be the criterion-score of the ith subject. The sum of the squared deviations for item k is

$$\sum_i \left[\sum_j w_{jk} r_{ijk} - s_i \right]^2. \qquad (\#1)^1$$

Equating partial derivatives with zero, we find that the minimum is uniquely attained when w_{jk} is just the average criterion score of all subjects choosing option j in answer to item k. (Actually, w_{ij} is not uniquely determined if $r_{ijk} = 0$ for all i, but in this case, the jkth option might as well be removed from the questionnaire.) Further work along these lines has been carried out by Guttman (1941). (#2)

Notice that we make no specific mention of the correctness of particular options in this scoring method. Indeed, it is possible for a greater weight to be associated with an incorrect option than with the correct option on the same item. Therefore, this method is not generally used on examinations in which certain options are *a priori* "better" than others. Instead, a common application of weights similar to these is to personality testing. For example,

[1] The notation (#n) indicates the related exercise to be found at the end of this chapter.

suppose we wish to discover the degree to which an individual's preferences are compatible with various careers. A group of subjects, perhaps involved in these careers, is extensively tested, and a set of career-related criterion scores is developed for each subject. These subjects are then administered a less extensive multiple-choice test, which typically consists of a number of items on which preferences are elicited. A different set of option weights corresponding to each criterion is derived. Subsequently, subjects about which criterion information is not available can be given the test, and their various scores (corresponding to the different sets of weights) can be used as estimates of their criterion scores (for example, for counseling purposes).

If criterion scores independent of the test under consideration are not available, the preceding method is not directly applicable. One could use *a priori* option weights to derive "raw" test scores, and then these scores could be used as criterion scores.

From these, new option weights could be derived, but then, the new weights would yield new test scores, which would yield still another set of weights. We shall investigate the end result of this iterative procedure for adjusting weights and scores.

Let $w^{(0)}$ be the column t-vector of initial option weights. For example, in a test of knowledge, we would have

$$w_{jk}^{(0)} = \begin{cases} 1, & \text{if } j \text{ is the correct option on item } k \\ 0, & \text{otherwise.} \end{cases}$$

For notational convenience, we compute a subject's overall score as his average score per item (of course, this does not affect any use of the scores for ranking purposes). Let $s^{(0)}$ be the column p-vector of "raw" scores. Then $s^{(0)} = (R/q)w^{(0)}$. Let $D = \text{diag}\,(\Sigma_i r_{ijk})$, the diagonal $t \times t$ matrix with the number of subjects responding with option j on item k as the ***jk***th diagonal entry. We assume that any option chosen by no subjects has been eliminated from consideration, so D is nonsingular.

The average score of all subjects choosing the ***jk***th option is

$$\frac{\sum_i r_{ijk} s_i^{(0)}}{\sum_i r_{ijk}}.$$

If these averages are used in a new vector of option weights $w^{(1)}$, then $w^{(1)} = D^{-1}R^T s^{(0)}$, where R^T denotes the transpose of R. Iterating the construction of new scores and weights, we have

$$s^{(n)} = (R/q)w^{(n)}, \qquad w^{(n)} = D^{-1}R^T s^{(n-1)}. \qquad (\#3)$$

Therefore,

$$s^{(n)} = (R/q)D^{-1}R^T s^{(n-1)} = \cdots = [(R/q)D^{-1}R^T]^n s^{(0)}$$

and

$$w^{(n)} = D^{-1}R^T(R/q)w^{(n-1)} = \cdots = [D^{-1}R^T(R/q)]^n w^{(0)}.$$

If $s^{(n)}$ and $w^{(n)}$ approach limits as n becomes large, these limits will be stable under further iterations and might be reasonable scores for the current group of subjects, and weights for use on subsequent administrations of the test.

Let us consider the matrix $A = (R/q)D^{-1}R^T$. It is symmetric and nonnegative; moreover, it is doubly stochastic (all rows and columns sum to one). (#4) The entries of this $p \times p$ matrix indicate the degree of similarity of response patterns of different subjects, giving greatest weight to instances in which only a small number of subjects agree on an option. (#5a) (As an aside, we note that, if a subject is suspected of having cheated on the test, his most likely accomplice is the subject with greatest (nondiagonal) entry in his row of A.)

We require a weak assumption about the responses of the subjects. Call subjects i and $\bar{i}$ *directly related* if they give the same response on at least one item (that is, if $A_{i\bar{i}} \neq 0$) (#5b); simply call them *related* if there is some sequence of subjects $i_1, i_2, \cdots, i_n$ for which i is directly related to i_1, who is directly related to $i_2, \cdots$, who is directly related to $\bar{i}$. Our assumption is that every pair of subjects is related. For example, if one "easy" item exists on which all subjects agree, the assumption holds.

From the preceding assumption, it follows that some power of A (and therefore, every subsequent power) is strictly positive, (#6) We can now easily show that $\lim A^n = J/p$, where J is the $p \times p$ matrix with all entries equal to one. (#7)

Of course, this is a distressing result. It implies that $s^{(n)}$ has as its limit the vector of constant entries, in which each entry is the average of the original scores in $s^{(0)}$. Clearly, these limiting scores do not help to discriminate between subjects.

Next, consider the limit of the sequence of weight vectors. Let $B = D^{-1}R^T(R/q)$. (#8) Then

$$B^n = D^{-1}R^TA^{n-1}(R/q) \to D^{-1}R^T(J/p)(R/q). \qquad (\#9a)$$

Therefore, as expected from the previous result, the limiting weights are all equal. Their common value is an average of the original weights, weighted by the number of subjects who chose each option. (#9b)

This approach to option weighting has been studied by several authors. These authors have generally considered the iterative method as a technique for deriving selfconsistent weights when no natural set of weights exists (when the options do not classify, for example, simply as "correct" and "incorrect"). We mention three specific approaches. Guttman (1941) discards the system of identical weights (which constitute the principal right eigenvector of B) as "extraneous," and uses instead the right eigenvector corresponding to the second-largest eigenvalue of B. This also yields a system of weights (some necessarily negative) and scores which is stable under iteration. Shiba (1965) treats the more general type of examination in which subjects are allowed to indicate as many options as desired. His approach differs from ours only in the more general permitted form of the response matrix R, and therefore our results are quite similar. He selects the principal right eigenvector of a suitably scaled analog of B for his weight vector. Hence he gets equal weights for all subjects in the case in which they each indicate the same number of options on their questionnaires. Perhaps the most surprising work is that of Lawshe and Harris (1958). They treat the same case as Guttman but adopt the principal eigenvector as their weight vector. Due to an obscure computational procedure (designed for hand-computation), they apparently fail to realize that their procedure leads to equal weights on all options.

Naturally, the essential independence of the limiting weights from the original weights makes these approaches unsuitable for scoring examinations in which each item has just one correct option. A natural approach (which has apparently not appeared in the literature) is to combine the original scores and weights with the scores and weights from subsequent iterations. For any $0 \leq r < 1$, let

$$\begin{aligned} s(r) &= (1 - r)(s^{(0)} + rs^{(1)} + r^2 s^{(2)} + \cdots) \\ &= (1 - r)(I + rA + r^2 A^2 + \cdots)s^{(0)} \\ &= (1 - r)(I - rA)^{-1} s^{(0)}. \qquad (\#10) \end{aligned}$$

The factor of $(1 - r)$ is included solely for purposes of normalization and allows us to also define

$$s(1) = \lim_{r \to 1} s(r). \qquad (\#11\text{a})$$

The vector $s(0)$ is simply the original score vector $s^{(0)}$. Disregarding the normalization factor, for small values of r, $s(r)$ is $s^{(0)}$ perturbed primarily in the direction of $s^{(1)}$. On the other end of the interval of convergence, large values of r place proportionately large weight on the structure of R, to the exclusion of "raw" score considerations. Indeed, $s(1)$ is just the vector of identical entries derived previously. (#11b) Note that we can similarly define a parametrized weight vector

$$\begin{aligned} w(r) &= (1 - r)(w^{(0)} + rw^{(1)} + r^2 w^{(2)} + \cdots) \\ &= (1 - r)(I + rB + r^2 B^2 + \cdots)w^{(0)} \\ &= (1 - r)(I - rB)^{-1} w^{(0)}. \end{aligned}$$

We now have a spectrum of scoring methods available. A question remains. How should a specific value of r be chosen? If we are involved in the construction of a large-scale test, a reasonable approach is first to test extensively a group of subjects in order to determine their criterion scores. These subjects can then be given the test under development, and the value of $r = r_{\max}$, which maximizes the correlation of the scores in $s(r)$ with the criterion scores, can be determined. The option weight vector $w(r_{\max})$ can finally be used in subsequent test administrations. If a group of subjects has response matrix R, their score vector will be $s = Rw(r_{max})$. As long as r is relatively small (as one would expect), $s(r)$ and $w(r)$ can be computed efficiently by truncation of the infinite series.

If a test will only be administered to a single group of subjects, to assume that any criterion scores will be available is unrealistic. In such a case, the question of choosing a suitable value of r becomes quite difficult. Either detailed data analysis, or heuristics developed from experience, or both, may help to determine a good choice.

This completes our presentation of a technique for "intensifying" scores on multiple-choice examinations in which each item has a single "correct" option. A philosophical point deserves discussion before we close. Namely, is it fair to the subjects to differentially score the "incorrect" options? On the one hand, "partial credit" is commonly given (and accepted) on subjective examinations in an attempt to reward those subjects who evidence "partial knowledge" on some topics. However, we are going a step further in allowing peer data to influence our assignment of partial credit. Indeed, we are proposing to reward some subjects, and penalize others, for "the intellectual company they keep." Should a subject have his score influenced by the behavior of his fellow subjects? We have no convincing answer to this question. (#12)

3. Response-Determined Scoring

We turn our attention to scoring systems which differentiate between options only to the extent of recognizing one as correct and the others as (equally) incorrect. The systems we consider will be *a priori*, in the sense that they will not depend on data concerning the responses of either individuals or groups of subjects.

Under these restrictions, what freedom remains? Two aspects of test administration still have not been specified. First, we must say precisely what kinds of responses will be allowed. In addition, we must determine the scores to be assigned to these various responses (respective of which option is correct). Before treating these problems in detail, an intermediate question must be answered. Given a response system (a collection of permitted responses) and an associated scoring system, how does a subject choose a particular response on an item? Since our primary goal is to measure the abilities of the subject in terms of his responses to the items, the answer to this question is central to our work.

Our basic assumption is that a subject has, on a given n-option item, subjective probabilities assigned to the options. We view the item as a chance event for the subject. Each option is an "urn," and a "truth-ball" is to be dropped into precisely one of the urns. The subject anticipates the location of the ball according to a probability distribution $(p_1, \cdots, p_n)$ over the n urns, where each $p_i \geq 0$ and $\Sigma\, p_i = 1$.

We also require a behavioral assumption, which is open to some question. We assume that each subject chooses his response on each item so as to maximize his (subjective) expected score. Consider a particular response to an item (which might be more general than the mere specification of an option), and let $r_1(r_2, \cdots, r_n)$ be the score for that response in the event that option $1(2, \cdots, n)$ is correct. Then a subject's expected score, if he gives this response, is $\Sigma\, p_i r_i$, where $(p_1, \cdots, p_n)$ is his probability distribution over the options. Our assumption is that the subject is "rational" and therefore chooses the response which maximizes this quantity. Given a response system and a scoring system, an *optimal strategy* assigns a maximizing response to each probability distribution over the options.

With these assumptions, the behavior of subjects faced with particular systems can be compared. Our general approach will be to present a response system, consider a variety of associated scoring systems, and determine optimal subject strategies for each scoring system. The final choice among systems must be a matter of taste (related to the types of responses desired), or of experimental comparison.

A familiar type of response involves simply indicating one of the options. On a forced-choice item (on which a subject is required to select an option), exactly n possible responses exist. As long as the score for indicating the correct option is greater than the score for indicating an incorrect option, the only optimal strategy is to indicate the option considered most likely to be correct. Assume that the response scores are, respectively, $+1$ and 0, and that a subject has subjective probabilities $p_1 \geq \cdots \geq p_n$. Then this strategy yields an expected score of p_1. In the case in which the subject has no knowledge $(p_1 = \cdots = p_n = 1/n)$, his expectation is $1/n$.

If the subject is allowed to refrain from indicating an option, an additional response is available. Optimal strategies will depend on the scoring system employed. Assume (without loss of generality up to affine transformations) that a score of $+1$ is given if the correct option is chosen, and a score of 0 if no choice is made. Commonly, the score for an incorrect response is then some nonpositive number $-w$. If a response is to be made, clearly it should be the option considered most likely to be correct. Therefore, the maximum achievable expected score is $\max(p_1 - w(1 - p_1), 0)$. The first term is larger when $p_1 > w/(w + 1)$, and this is precisely the case in which an optimal strategy must involve making a response. If $0 \leq w \leq 1/(n - 1)$, a rational subject can always simply respond with his most likely choice. (#13) However, for example, if the test designer wants responses only when subjects believe the most likely choice to have probability $p_1 \geq 1/2$, he must take $w = 1$ (that is, a score of -1 for an incorrect answer). (#14) This scoring system, designed to elicit subject responses of a particular type, should not be confused with systems which attempt, when an examination is over, to compute a "true score" for each subject by subtracting off the fraction of a subject's score believed to be due to guessing.

The preceding system is not particularly sensitive. For example, if two subjects have subjective distributions $(1/2, 1/2, 0, \cdots, 0)$ and $(1/2, 1/2(n-1), \cdots, 1/2(n-1))$, they will have the same expectation and the same optimal response for every value of w. However, it seems as though the first subject is much more certain in his beliefs than the second. To seek a response and scoring system in which these two subjects are differentiated might be reasonable.

Consider a response system in which subjects are allowed to indicate or mark any number of options. The purpose of such a system is to allow the subjects to indicate several of their most likely options. There are $2^n - 1$ essentially different responses (since the two responses of marking all options or marking none are identical for all practical purposes, we rule out the response of marking none). The scoring system depends on $2n - 1$ parameters. Let $r_1, \cdots, r_n$ be the scores for indicating $1, \cdots, n$ options, one of which is the correct one, and let $w_1, \cdots, w_{n-1}$ be the scores for indicating $1, \cdots, n-1$ options, but not the correct one. Setting $r_1 \geq \cdots \geq r_n$ seems reasonable since the marking of a large number of options implicitly expresses little certainty about the position of the correct option. It also seems reasonable that all w_i be less than r_n, since any responses failing to indicate the correct option indicate misinformation on the part of the subject. A natural ordering of the w_i is not as apparent. On the one hand, a subject marking many options and let failing to locate the correct one displays serious misinformation. On the other hand, a subject marking only a few options seems to be expressing confidence in his response, a confidence which is badly misplaced. Which subject should be penalized more? We continue our investigation without taking a stand on this issue.

In nonforced one-choice systems, a common value used for the score of a wrong answer is $-1/(n-1)$. When this value is used, a subject with no knowledge is indifferent between guessing at an option, and refraining from choosing an option. We shall seek for the many-choice system a scoring system in which a subject similarly neither gains or loses from "pure" guessing. More specifically, assume that a subject knows with (subjective) certainty that the correct option is one of a set of k options. Also assume that he has no information to help him further discriminate among these k options. Thus his subjective distribution is $(l/k, \cdots, 1/k, 0, \cdots, 0)$. We want a scoring system which gives him the same expectation from the response of marking the k most likely options, as from randomly choosing some m of these k as his response. Hence we want

$$r_k = \frac{m}{k} r_m + \frac{k-m}{k} w_m, \qquad \text{for all } 1 \leq m \leq k.$$

Without loss of generality, assume the scores to be normalized so that $r_1 = 1$ and $r_n = 0$. Then, in particular,

$$r_k = \frac{1}{k}r_1 + \frac{k-1}{k}w_1 = \frac{1}{k} + \frac{k-1}{k}w_1$$

and

$$r_n = 0 = \frac{1}{n} + \frac{n-1}{n}w_1.$$

This implies that $w_1 = -1/(n-1)$. Therefore,

$$r_k = \frac{n-k}{k(n-1)}, \qquad \text{for all } 1 \le k \le n,$$

$$w_k = -\frac{1}{n-1}, \qquad \text{for all } 1 \le k \le n-1. \qquad (\#15)$$

Notice that the w_i are all equal, neatly freeing us from our earlier quandary concerning the proper direction of their monotonicity. Furthermore, the expected score from a guess on an item on which a subject has no knowledge remains $-1/(n-1)$, as in the one-choice case.

What is an optimal strategy for a subject with subjective probabilities $p_1 \ge \cdots \ge p_n$, when faced with this scoring system? If he marks the first k options, his expected score will be

$$\begin{aligned} E_k &= (p_1 + \cdots + p_k)\frac{n-k}{k(n-1)} + (p_{k+1} + \cdots + p_n)\left(-\frac{1}{n-1}\right) \\ &= \frac{1}{n-1}\left[\frac{n}{k}(p_1 + \cdots + p_k) - 1\right] \\ &= \frac{n-k}{n-1}\left[\frac{(p_1 + \cdots + p_k)}{k} - \frac{(p_{k+1} + \cdots + p_n)}{n-k}\right]. \qquad (\#16a) \end{aligned}$$

Therefore,

$$E_{k+1} - E_k = \frac{n}{n-1}\cdot\frac{1}{k+1}\cdot\left[p_{k+1} - \frac{(p_1 + \cdots + p_k)}{k}\right]. \qquad (\#16b)$$

This is always nonpositive and is equal to zero only if $p_1 = \cdots = p_k = p_{k+1}$. Therefore, a subject can never do better than he does by just marking one option, and indeed, he can hurt himself by marking any options of unequal probability. Thus we can never expect to see the wide variety of responses which we allow, and the only scores we will give will be $+1$, 0, and $-1/(n-1)$, as in the one-choice system!

Of course, this may not be a practical difficulty. It is unlikely that most subjects would perform the computations we have just presented. In consequence, they may feel as though rationality is compatible with the marking of several alternatives. However, the result of the use of this system must be left to be decided by experiment.

An alternative scoring system for many-choice response systems was proposed by Coombs *et al.* (1956). This system results when the scores which we have just derived are weighted by the number of options marked in each response. The score for a correct response involving k options is $(n - k)/(n - 1)$, and the score for an incorrect response involving k options is $-k/(n - 1)$. If we normalize these scores by multiplying by the constant $n - 1$, we have $r_k = n - k$ and $w_k = -k$. This system can be described as "charging" a subject one point for each option marked, but paying a reward of n points for the marking of the correct answer. We can verify that guessing, in the sense described previously, becomes increasingly expensive (in terms of expectation) as a subject becomes more knowledgeable (that is, is able to rule out, on the basis of certain belief, more and more options). (#17)

Assume that a subject has probabilities $p_1 \geq \cdots \geq p_n$. If he marks the first k options, his expected score is

$$E_k = n(p_1 + \cdots + p_k) - k.$$

If follows that

$$E_{k+1} - E_k = np_{k+1} - 1,$$

and therefore, an optimal strategy for the subject is to mark precisely those options to which he ascribes a probability of at least $1/n$. (#18, 19)

Coombs *et al* compared this response and scoring system with the traditional one-choice system. A test was administered to two groups of subjects, one of which responded under the new system and the other under the traditional one. The scores obtained under the two systems were of similar validity, but the new system yielded an increase in reliability equal to that expected from a 20% extension in the length of the test given under the traditional system.

All of the response systems considered so far involve the classification of the options into two categories, marked and unmarked. An approach taken by Willey (1960) requires classification of the options of a five-option item into three categories. One option is to be marked "most likely," and two more options are to be marked "next-most likely". Willey recommended the use of scores of 1, 2/3, or 0, respectively, in the cases in which the correct option is placed in the first, second, or third (unmarked) category. Subsequent investigation indicated that this method does not yield any increase in validity over standard one-choice methods (see Bernhardson (1967)) and is possibly less reliable. This might be accounted for by the fact that a subject with no knowledge on an item has a relatively greater expected score of 7/15 from Willey's method than his expectation of only 1/5 from the earlier method.

Another type of response system requires a punchboard, or computer, or other mechanical device. Subjects choose successive options until the device indicates that the correct option has been chosen, and their scores depend (monotonically) on the number of options chosen. At each stage, choosing the most likely of the remaining options is clearly optimal. Hence the test designer is free to consider a wide variety of scoring systems in terms of validity and reliability. An interesting additional feature is the possible educational benefit of providing the subject with instant feedback concerning the correctness of his responses.

Rather than continuing through the infinite catalog of variations on the response theme, let us turn our attention to test items of a slightly different character. Assume that an item has a continuum of associated options (for the sake of specificity, the points of a real interval), only one of which is correct. One might permit a subject to respond to the item by indicating a (measurable) subset of the continuum, or less generally, a subinterval. A typical item of this type might be as follows:

Q: What fraction of the 1975 U.S. budget was allocated to the payment of interest on the national debt?

A: Give a subinterval (of the unit interval) which contains the correct answer. [10.009 %].

Observe that such items are just limiting cases of the items with many-choice responses treated earlier. Both of the scoring systems previously developed can be extended to this new situation. Assume that a fraction x of the interval of options is given as a response, and let $r(x)$ (respectively, $w(x)$) be the score if the correct answer is (is not) included in the response. The first scoring method extends to yield scoring functions $r(x) = (1 - x)/x$ and $w(x) = -1$. Notice that r is discontinuous at zero. This leads to the unreasonable result that a subject maximizes his expected score by responding with extremely small intervals. We therefore reject this scoring system, and consider instead the method of Coombs *et al.* Their scoring method extends to yield functions $r(x) = 1 - x$ and $w(x) = -x$.

Assume that the option interval is the unit interval, and consider a subject with a subjective probability density function p on this interval. Without loss of generality, assume that p is monotone decreasing. Then, an optimal strategy for the subject involves choosing some interval $[0, t]$ as his response. In this case, the expected score of the subject is

$$E(t) = r(t)\int_0^t p(x)\,dx + w(t)\int_t^1 p(x)\,dx$$

$$= w(t) + (r(t) - w(t))\int_0^t p(x)\,dx.$$

Assuming that p is continuous, we find that $E(t)$ is maximal when

$$E'(t) = w'(t) + (r'(t) - w'(t))\int_0^t p(x)\,dx + (r(t) - w(t))p(t)$$

$$= 0.$$

For example, consider the scoring system of Coombs *et al.* $E(t)$ will be maximized when $E'(t) = -1 + 0 + p(t) = 0$. Therefore, the subject should respond with the interval of all x for which $p(x) \geq 1$. This is equivalent to responding with all x for which $p(x)$ is at least as great as the (constant) value of the uniform density function, a result analogous with the result in the finite-option case.(#20)

We can investigate, in this context, the existence of scoring systems which encourage other types of response strategies. For example, is there any scoring system which leads a subject to respond with the interval $[0, T]$ for which

$$\int_0^T p(x)\,dx = \frac{1}{2} \tag{$*$}$$

(in other words, can we get the subject to "split" his options)? If so, we would have

$$\begin{aligned} E'(t) &= w'(T) + (r'(T) - w'(T))\tfrac{1}{2} + (r(T) - w(T))p(T) \\ &= 0. \end{aligned}$$

We could then solve for $p(t)$ in terms of r, r', w, and w' at T, as

$$p(T) = -\frac{r'(T) + w'(T)}{2(r(T) - w(T))}.$$

The right-hand side of this expression depends only on T, but the left-hand side can vary, for fixed T, as different densities p satisfying $(*)$ are considered. This contradictory situation leads us to conclude that no scoring system of the desired type exists.

We finally return to finite-option tests and consider the possibility of quite general responses. After all, if we assume that a subject has a subjective probability distribution over the alternatives, why not simply let him give this distribution as his response?

The elicitation of such responses has received much attention in the last decade. Let us consider the immediately obvious scoring system: give each subject a score equal to the subjective probability he assigns to the correct option. This system gives the highest scores to subjects who express great condifence in the correct option, a most desirable feature. Assume that a subject has subjective probabilities $p_1 \geq \cdots \geq p_n$. He can, of course, give any distribution $(q_1, \cdots, q_n)$ as a response. His expected score will be

$$E(q_1, \cdots, q_n) = p_1 q_1 + \cdots + p_n q_n.$$

The subject will wish to choose q so as to maximize this expression. We can easily show that the maximum is attained when $q_1 = 1$ and $q_2 = \cdots = q_n = 0$. (#21) Therefore, the optimal response is to express *certainty* in the option which he merely considers most likely!

This distressing result may not be disastrous. Indeed, in practice it has been found that subjects presented with the scoring system just discussed will often attempt to accurately report their subjective probabilities, either in ignorance or in disregard of the nonoptimality of this behavior. Clearly, though, it would be nice to determine scoring systems which lead rational (expectation-maximizing) subjects to accurately report their true subjective probabilities. Such scoring systems are the topic of the next section.

In closing, we return to a point addressed earlier in this module. There are two purposes in the design of a scoring system. We may seek to elicit responses of a particular nature; indeed, in some studies this is the primary objective. The approaches in this section are directly applicable to such a goal, but if our goal is to estimate accurately the scores of the subjects on some external (to the test) criterion, only careful experimentation can tell if a specific scoring system meets that goal. (#22a, b)

4. Reproducing Scoring Systems

A direct approach to measuring a subject's knowledge concerning a test item is to attempt to elicit from him his subjective probability distribution over the options. We shall investigate scoring systems designed for this purpose. After developing some of the mathematical properties of such systems, we will briefly review some empirical studies of them.

Assume that a subject has a (personal) probability distribution $p = (p_1, \cdots, p_n)$ over the n options of a particular item. Let $r = (r_1, \cdots, r_n)$ be any probability distribution. A *scoring system* is a collection of functions $\{f_1, \cdots, f_n\}$, for which $f_i(r)$ is the score awarded to the subject if he indicates that r is his subjective distribution, and then option i develops to be the correct option. The expected score of the subject, from giving the vector r as his response to the item, is

$$E(p, r) = \sum p_i f_i(r).$$

The collection of scoring functions $\{f_i\}$ is a *reproducing scoring system* (RSS) if it encourages the subject to report his true probabilities. That is, $\{f_i\}$ is an RSS if

$$\max_r E(p, r) = E(p, p), \qquad \text{for all } p,$$

where the maximization is over all probability vectors r. Particular conditions might be placed on the scoring functions; we might ask that the indicated maximum be uniquely attained when $r = p$, or that each f_i be nondecreasing (in an appropriate sense) in r_i. (#23) We shall discuss these restrictions as the context demands.

In the preceding section, a specific scoring system was presented, in which each $f_i(r) = r_i$. For this system,

$$E(p, p) = \sum p_i^2 \leq \sum p_{\max} p_i = p_{\max} \sum p_i = p_{\max},$$

where $p_{\max} = \max p_i$, and the inequality is almost always strict. (#24) However, $p_{\max} = E(p, r)$, where r is the distribution assigning probability 1 to the option corresponding to p_{max}. Therefore, this scoring system is not reproducing. Other "natural" scoring systems may similarly fail to be reproducing. (#25) Clearly, a general method for constructing RSS's is needed.

Consider the case in which only two options exist. The expectation of a subject can then be described by two parameters p and r, corresponding to the subjective distribution $(p, 1 - p)$ and the reported distribution $(r, 1 - r)$. If we define $g_1(r) = f_1(r, 1 - r)$ and $g_2(r) = f_2(1 - r, r)$, his expectation is

$$E(p, r) = pg_1(r) + (1 - p)g_2(1 - r).$$

Assume that g_1 and g_2 are both differentiable and nondecreasing on $[0, 1]$. In order for $\{f_1, f_2\}$ to be an RSS, $E(p, r)$ must be is maximized at $r = p$; therefore, it is necessary that $\partial E/\partial r|_{r=p} = 0$. This is equivalent to the requirement that

$$pg_1'(p) = (1 - p)g_2'(1 - p). \tag{$*$}$$

It develops that this condition is also sufficient. Assume that $(*)$ holds. Then

$$\frac{\partial}{\partial r}E(p, r) = pg_1'(r) - (1 - p)g_2'(1 - r) = (p - r)\frac{g_1'(r)}{1 - r}.$$

Since $g_1'(r) \geq 0$, we have

$$\frac{\partial}{\partial r}E(p, r) \begin{matrix} \geq \\ = \\ \leq \end{matrix} 0 \text{ if } r \begin{matrix} < \\ = \\ > \end{matrix} p.$$

Hence $E(p, r)$ is maximized at $r = p$. If g_1 (and therefore g_2) is strictly increasing, the maximum is unique.

Rather than being merely an abstract characterization, this result enables us to construct RSS's. Consider any differentiable, increasing function g_2 on $[0, 1]$ satisfying $g_2'(1) = 0$ (actually, we require slightly more structure on g_2' near 1, namely, that the integral in the following definition converges on intervals of the form $[0, r]$). Let

$$g_1(r) = \int \frac{(1 - r)}{r} g_2'(1 - r)\, dr.$$

The corresponding functions $f_1(r, 1 - r) = g_1(r)$ and $f_2(1 - r, r) = g_2(r)$ form an RSS. Notice that g_1 is determined only up to an additive constant. This is to be expected, since the awarding of an additional fixed score (independent of response) when a particular option is correct does not affect optimal response strategies. As an example, take $g_2(r) = 1 - (1 - r)^k$ (with $k > 1$). Then $g_1(r) = k/(k - 1)r^{k-1} - r^k + C$.

In most testing situations, there is no reason to use different scoring formulas depending on whether the first option or some other is correct. Yet this is precisely the situation that usually arises when the preceding construction is performed. In the example just given, take $k = 3$. Then $g_1(r) = (3/2)r^2 - r^3$ and $g_2(r) = 3r - 3r^2 + r^3$. Therefore, if a subject responds with the distribution (1/2, 1/2), he receives a score of 1/4 if the first option is correct, but 7/8 if the second option is correct.

In order to avoid this imbalance, we investigate symmetric reproducing scoring systems (SRSS's). An SRSS is a reproducing scoring system $\{f_1, \cdots, f_n\}$ such that

$$f_{\pi 1}(r_1, \cdots, r_n) = f_1(r_{\pi 1}, \cdots, r_{\pi n})$$

for every permutation π of $\{1, 2, \cdots, n\}$. In the two-option case, this reduces simply to the requirement that $f_1(r, 1 - r) = f_2(1 - r, r)$ for all r, or equivalently $g_1(r) = g_2(r) = g(r)$. In the example given earlier, by taking $k = 2$, we obtain the SRSS associated with $g(r) = 2r - r^2$.

We continue to consider, for the moment, the case in which just two options exist. Let h be an arbitrary (integrable) nonnegative function. Consider an "option 1 gambling house." This house contains a gambler corresponding to each x in the unit interval. Gambler x is willing to accept an infinitesimal bet of $h(x)\,dx$ on option 1 at odds of $1 : x$. That is, a bet of one dollar with him will return $1/x$ dollars if option 1 is correct. If a subject has subjective probability p on this option, his expectation from a bet with gambler x is

$$\left[p\left(\frac{1}{x} - 1\right) + (1 - p)(-1)\right]h(x)\,dx = \left[\frac{p}{x} - 1\right] h(x)\,dx,$$

and he will prefer to bet only with those gamblers for which $p > x$. If this subject then goes to a similar "option 2 gambling house," he will prefer to bet only with those gamblers for which $1 - p > x$. His payoff from these wagers, if option 1 is correct, will be

$$g(p) = \int_0^p \left(\frac{1}{x} - 1\right) h(x)\,dx + \int_0^{1-p} (-1)h(x)\,dx. \qquad (**)$$

Since the subject maximizes his expectation by revealing (through his bets) his subjective probabilities, this function g corresponds to an SRSS. (#26)

Note that this construction generalizes directly to the n-option case, for which it follows that the scoring system defined by

$$f_i(r_1, \cdots, r_n) = \int_0^{r_i} \frac{h(x)}{x} dx - \sum_k \int_0^{r_k} h(x)\,dx, \qquad i = 1, 2, \cdots, n$$

is an SRSS. Since we are free to vary all scoring functions by an additive constant, the lower limit of integration can actually be set arbitrarily.

Returning to the two-option case, assume that an SRSS $\{f_1, f_2\}$ is given, and let $g(r) = f_1(r, 1 - r) = f_2(1 - r, r)$. Define $h(r) = rg'(r) = (1 - r)g'(1 - r)$. Since $h(r) = h(1 - r)$, we have

$$rg'(r) = h(r) + r[h(1 - r) - h(r)],$$

and therefore

$$g(r) = \int \left[\frac{h(r)}{r} + h(1 - r) - h(r) \right] dr.$$

This may be rewritten as

$$g(r) = \int_0^r \frac{h(x)}{x} dx - \int_0^r h(x)\,dx - \int_0^{1-r} h(x)\,dx + C,$$

which corresponds precisely with (**). Hence, in the two-option case, we have a (constructive) characterization of all symmetric reproducing scoring systems. The "gambling house" construction will yield every two-option SRSS. Whether this is true in the n-option case is not certain. (#27a, b)

Consider the previous example one last time. The SRSS corresponding to $k = 2$ arises (apart from an additive constant) upon taking $h(x) = x$ in (**). More generally, taking $h(x) = 2x$ in the n-option case yields the SRSS

$$f_i(r_1, \cdots, r_n) = -1 + 2r_i - \sum_k r_k^2, \qquad i = 1, 2, \cdots, n.$$

If e^i is the probability vector with $(e^i)_i = 1$, we have

$$f_i(r) = -|e^i - r|^2.$$

Therefore, this "quadratic" scoring system charges a subject the squared deviation of his reported distribution from the "true" distribution (certainty on the correct option). (#28a, b)

One might object to the quadratic scoring system, since the awarded score depends not only on the probability assigned to the correct option, but also on the distribution over the various incorrect options. Let us try to find a reproducing scoring system for which this is not the case.

In such a system, each $f_i(r) = g_i(r_i)$, a function depending only on r_i. For the moment, consider only those subjective distributions p which are positive in all components. For $\{f_i\}$ to be reproducing, we must have all "marginal gains" $(\partial/\partial r_i)E(p, r)|_{r=p}$ equal. (#29) In this particular case, this means that all $p_i g_i'(p_i)$ must be equal. Assume $n \geq 3$. (When $n = 2$, we are dealing with distributions (p_1, p_2) for which the question of the dependence of $f_1(p_1, p_2)$ on p_1, but not on $p_2 = 1 - p_1$, is without meaning.) Considering various redistributions of the probability allocated to sets of $n - 1$ options, we eventually find that a constant K exists such that, for every i and for all $0 < p_i < 1$,

$$p_i g_i'(p_i) = K. \qquad (\#30)$$

Therefore, constants C_i exist such that

$$g_i(p_i) = K \log p_i + C_i, \qquad i = 1, 2, \cdots, n.$$

Because the logarithmic function is discontinuous at zero, we are forced to conclude that no RSS of the desired type exists. However, if we accept the use of a "score" of $-\infty$ in those cases in which a subject assigns probability zero to the correct option, then the "logarithmic" scoring system typified by $f_i(r) = \log r_i$ (for all i) will suffice. In information-theoretic terms, this scoring system charges a subject for the amount of information he (subjectively) gains when he is informed of the correct option. His expected score is $\Sigma\, p_i \log p_i$, or minus the entropy of his subjective distribution. Note that (apart from an additive constant) this scoring system arises from the "gambling house" construction upon taking $h(x) = 1$.

A variety of reproducing scoring systems have been experimentally investigated, with mixed results. Both Rippey (1968) and Hambleton *et al.* (1970) reported an erratic relationship between the reliability of tests on which traditional one-choice responses were elicited, and the reliability of the same tests when probabilistic responses were encouraged. Hambleton *et al.* also compared the validities of test scores obtained from the two types of responses. They found that the scores obtained from probabilistic responses were slightly more valid, but not significantly so. Furthermore, this small gain in validity was at least partially offset by the increased time they observed was necessary for subjects to complete the nontraditional type of test.

These results are somewhat surprising. Since reproducing scoring systems are designed primarily for purposes of response elicitation rather than measurement, one might expect the validity of scores computed by such systems to be low. However, since more explicit information about subject beliefs is obtained from probabilistic responses than from traditional, less complex responses, one would hope to obtain scores of increased reliability. A difficulty seems to be that subjects are generally not very facile at dealing with probabilities. Either they do not have a consistent probabilistic view of the world, or they cannot accurately report their subjective probabilities.

Reproducing scoring systems are applicable to areas other than multiple-choice testing. For example, they may be used to try to get experts to assess the relative likelihood of different future events (where the event which actually occurs corresponds to the "correct option"). Indeed, much of the early work on probabilistic responses was reported in the meteorological literature, relative to the evaluation of weather-forecasting techniques.

The interest in probabilistic response systems has prompted much recent investigation of how individuals deal with their subjective probabilities. When the psychological mechanisms are better understood, perhaps the time for widespread use of tests involving probabilistic responses will be at hand.

5. Comments on the Use of the Chapter as an Educational Module

This first section of the chapter briefly surveys the area of multiple-choice testing and is meant to establish a context for the material in the subsequent sections. It also provides the instructor with an opportunity to involve the class in the kind of general "brain-storming" session not frequently encountered in technical courses. Prior to the presentation of this material, the following questions might be introduced for discussion.

What is the goal of test administration? How can the quantities we seek to measure be precisely defined, and how might they be accurately computed?

What types of multiple-choice tests are there? Can tests be classified by the relation between items and options?

On a given type of multiple-choice test, what kinds of responses might reasonably be elicited?

What are the general purposes of an examination scoring system? (The twofold purpose is to elicit particular subject responses, and to estimate the subjects' criterion scores.) Are these purposes compatible?

How might a particular scheme (consisting of test, system of administration, and scoring system) be evaluated in terms of the original goal of estimating criterion score? What measures of "goodness," "efficiency," or "effectiveness" might be applied?

For students able to devote a sizable amount of time of this topic, the discussion at the end of Section 2 should suggest an interesting project. Obtain the examination papers from a large course in which a relatively long multiple-choice test was given. Divide the examination into two portions of comparable difficulty, one long and the other short. Compute $s(r)$ for the entire group of subjects on the short portion, for several different values of r. Finally, using the subjects' raw scores on the long portion as criterion scores, determine the validity of the scoring system for each value of r. One would be pleased to find that validity increases as a function of r from $r = 0$ to some critical value $r = r_{\max}$, and then decreases for all greater r.

Exercises

Section 2, Option-Weighting. This section presents several classical results on the weighting of options and culminates with the presentation of a new method for deriving option weights for "right–wrong" examinations. A number of details have been omitted from the various arguments. Problems involving these details are presented here.

1. Show that this expression is minimized when $w_{jk} = (\Sigma_i r_{ijk} s_i)/(\Sigma_i r_{ijk}))$.
2. Other measures of "goodness of fit" yield different sets of weights. Show that

$$\sum_i \left| \sum_j w_{jk} r_{ijk} - s_i \right|$$

is minimized when each w_{jk} is equal to the median score of all subjects choosing option j in answer to item k.

3. Verify the recursive equations for $s^{(n)}$ and $w^{(n)}$.

4. Verify that A is doubly stochastic.

5. (a) Make the interpretation of the entries of A more precise.
   (b) Use your result to verify that $A_{i\bar{i}} \neq 0$ iff i and $\bar{i}$ are directly related.

6. Show that some power of A is strictly positive, given that all pairs of subjects are related. (Hint: Observe that all diagonal entries of A are nonzero. Use this fact to show that, once an entry becomes positive in some power of A, it remains positive in every subsequent power. Finally, show that each entry must eventually become positive.) A nonnegative matrix with the property that some power is strictly positive is generally referred to as "primitive."

7. Assume that all entries of A are positive. Prove that $\lim A^n = J/p$. (Hint: Compare the difference between the largest and smallest entries in any row of A^n with the difference between the largest and smallest in the same row of A^{n+1}. Observe that every power of A is doubly stochastic.) This is a very special case of the Perron–Frobenius theory concerning powers of nonnegative matrices.

8. Show that B is stochastic (all row-sums are one), but not doubly stochastic.

9. (a) Explicitly evaluate the $(\boldsymbol{jk}, \boldsymbol{uv})$th entry of $D^{-1}R^T(J/p)(R/q)$.
   (b) Verify the claim concerning the weighting of the terms in the average.

10. Prove that the series converges over the indicated interval (recall problem (7)), and verify the equality of the second and third expressions.

11. (a) Show that the limit exists.
    (b) Determine $s(1)$.

12. Can a subject answer all of the items of a test correctly, yet receive a lower score than some other subject? (Hint: Consider a test on which one subject answers all items correctly, several answer all but the last correctly, and several answer only the last correctly. Now, examine $s(r)$ for r close to 1.)

Section 3, Response-Determined Scoring. Several types of item responses are considered in this section. Various methods of scoring these responses are developed, and optimal subject strategies corresponding to each scoring method are determined.

13. Verify that the strategy "always respond" is optimal.

14. What value of w encourages responses only when $p_1 \geq 2/3$?

15. Verify that this solution satisfies the original relations for all $m \leq k$.

16. (a), (b) Verify these equations.

17. Consider two subjects. Assume that the first has the subjective distribution $(1/k, \cdots, 1/k, 0, \cdots, 0)$ over the n options, and the second has distribution $(1/m, \cdots, 1/m, 0, \cdots, 0)$, where $m < k$. Show that the second (more knowledgeable) subject loses more (subjectively) by guessing at some $m - 1$ options (rather than indicating all of the first m) than the first loses by guessing at some $k - 1$ options.

18. Compare the expectations of two subjects, one with distribution $(1/2, 1/2, 0, \cdots, 0)$ and the other with distribution $(1/2, 1/2(n - 1), \cdots, 1/2(n - 1))$, under this scoring system.

19. Determine a scoring system, similar to that of Coombs *et al.*, for which an optimal response strategy is to indicate only those options of subjective probability greater than P (where $0 \le P \le 1$).

20. A subject with certain belief that the correct option lies in the interval $[0, 1/p]$, but with no further discrimination among the options in this interval, has the subjective density function

$$q(x) = \begin{cases} P, & \text{if } 0 \le x \le 1/P \\ 0, & \text{otherwise.} \end{cases}$$

Determine a scoring system which encourages a subject to indicate those options x for which $p(x) \ge P$. Observe the similarity between this problem and Exercise 19.

21. Show that the maximum of $E(q_1, \cdots, q_n)$ is p_1.

22. An interesting approach to response scoring, which turns out to be a method of item-weighting, has been discussed by Chernoff (1962).
    (a) Assume that a fraction q of the subjects know the correct answer to an n-option item, while the remaining fraction $1 - q$ have no knowledge, and guess randomly. Define the "true score" of a knowledgeable subject to be $+1$, and of an unknowledgeable subject to be 0. What scores s_r and s_w should be awarded for correct and incorrect responses, respectively, in order to minimize the expected squared deviation of item score from true score?
    (Answer:

$$q(s_r - 1)^2 + \frac{1-q}{n}(s_r - 0)^2 + (1-q)\frac{n-1}{n}(s_w - 0)^2$$

is minimized when

$$s_r = q \Big/ \left(q + \frac{1-q}{n}\right) \text{ and } s_w = 0.)$$

    (b) How might q be estimated from the responses of the subjects? (Answer: Let k_r be the number of correct responses from a group of k subjects. Then the expected value of k_r is $\bar{k}_r = qk + (1-q)(k/n)$. Hence an estimate for q is $\tilde{q} = (nk_r - k)/(nk - k)$, yielding empirical scores of $s_r = (nk_r - k)/(nk_r - k_r)$ and $s_w = 0$.)

Section 4, Reproducing Scoring Systems. In this section, scoring systems are developed in order to elicit subjects' true subjective probabilities.

23. What is an "appropriate" definition of the phrase "f_i is nondecreasing in r_i," if the domain of f_i is the set (simplex) of probability n-vectors? (An answer to this conceptual question might be that f_i is nondecreasing in r_i iff

$$w(t) = f_i\left(tr_i; \left\{\frac{1 - tr_i}{1 - r_i} r_k\right\}_{k \ne i}\right)$$

is nondecreasing for all r.)

24. When is the inequality strict?

25. Take $n = 2$ and $p = (2/3, 1/3)$. What is an optimal response r, if $f_i(r) = r_i^{1/2}$?

26. Verify that $pg'(p) = (1 - p)g'(1 - p)$.

27. (a) Show (directly) that $f_i(r) = r_i/(\Sigma r_k^2)^{1/2}$ $(i = 1, 2, \cdots, n)$ defines an SRSS. This is called the "spherical" scoring system.
    (b) Show that the two-option spherical scoring system is generated by taking $h(x) = (1 - x)/(x^2 + (1 - x)^2)^{3/2}$ in the "gambling house" construction. It is not known whether the n-option spherical scoring system arises from the "gambling house" construction.

28. (a) Is this system monotonic, in the sense of problem (1)?
    (b) For a fixed r_1, what distribution over the remaining options maximizes $f_1(r)$?

29. Show that the equality of "marginal gains" is indeed necessary for f_i to be reproducing. (Hint: This can be done directly, or as an exercise in the use of Lagrange multipliers.)

30. Verify this statement, by showing that for all p_i and p_k, $p_i g_i'(p_i) = p_k g_k'(p_k)$. (Hint: Show that both of these expressions equal $p_j g_j'(p_j)$, for $i \neq j \neq k$ and for p_j sufficiently small.)

References

An excellent general introduction to testing theory is Gulliksen's *Theory of Mental Tests* (Wiley, 1950). Other texts will be found under the same library classification. Two excellent critical surveys of the multiple-choice testing literature are

J. C. Stanley and M. D. Wang, "Weighting test items and test item options: An overview of the analytical and empirical literature," *Educational and Psychological Measurement*, vol. 30, pp. 21–35, 1970.

M. D. Wang and J. C. Stanley, "Differential weighting: A review of methods and empirical studies," *Review of Educational Research*, vol. 40, pp. 663–705, 1970.

The references in Section 2 are to the following articles.

L. Guttman, "An outline of the statistical theory of prediction," in *The Prediction of Personal Adjustment*, Social Science Research Council, 1941, especially pp. 284 and 337–342.

C. H. Lawshe and D. H. Harris, "The method of reciprocal averages in weighting personnel data," *Educational and Psychological Measurement*, vol. 18, pp. 331–336, 1958.

S. A. Shiba, "A method for scoring multicategory items," *Japanese Psychological* vol. 7, pp. 75–79, 1965.

The papers referenced in Section 3 are the following.

C. S. Bernhardson, "Comparison of the three-decision and conventional multiple-choice tests," *Psychological Reports*, vol. 20, pp. 695–698, 1967.

H. Chernoff, "The scoring of multiple-choice questionnaires," *Annals of Mathematical Statistics*, vol. 33, pp. 375–393, 1962.

C. H. Coombs, J. E. Milholland, and F. B. Womer, "The assessment of partial knowledge" *Educational and Psychological Measurements*, vol. 16, pp. 13–37, 1956.

C. F. Willey, "The three-decision multiple-choice test," *Psychological Reports*, vol. 7, pp. 475–477, 1960.

An extensive treatment of response-determined scoring system is

B. de Finetti, "Methods for discriminating levels of partial knowledge concerning a test item," *British J. Mathematical and Statistical Psychology*, vol. 18, pp. 87–123, 1965.

Much of the material in Section 4 is drawn from the following.

E. H. Shuford, A. Albert, and H. E. Massengill, "Admissible probability measurement procedures," *Psychometrika*, vol. 31, pp. 125–145, 1966.

Our other source is the excellent survey paper:

T. A. Brown, "Probabilistic forecasts and reproducing scoring systems," RM-6299, The Rand Corporation, Santa Monica, CA, June 1970.

ACKNOWLEDGMENT: The author gratefully acknowledges the value of a conversation with Thomas F. Donlon, of the Educational Testing Service, concerning the multiple-choice testing literature.

Notes for the Instructor

Remarks. The introductory section is followed by three sections which are completely independent. The first of these sections, on option weighting, depends on a knowledge of matrix algebra. The next section, on response-determined scoring, uses only elementary probability theory, with the exception of one optional topic involving continuous probability distributions. The last section, on reproducing scoring systems designed for the elicitation of subjective probabilities, uses both elementary probability theory and some second-year calculus.

Prerequisites. Matrix algebra, elementary probability theory, some second year calculus.

Time. One to six lectures.

CHAPTER 13

Computing Fixed Points, with Applications to Economic Equilibrium Models

Michael J. Todd*

1. Introduction and Notation

The basic problem considered is how to compute fixed points of a given continuous function defined on the standard unit n-simplex S^n, using algorithms based on those of Scarf and Hansen [12], Kuhn [6], Eaves [5], and Kuhn and Mackinnon [8]. Basic references on this problem are Scarf and Hansen [12] and Todd [14].

S^n is defined as follows. R^{n+1} denotes the set of $(n + 1)$-dimensional real (row) vectors with coordinates indexed 0 through n. That is, $R^{n+1} = \{x = (x_0, \cdots x_n) | x_j \text{ is a real number}, 0 \leq j \leq n\}$. A vector x in R^{n+1} is *nonnegative* (we write $x \geq 0$) if each coordinate is nonnegative. (We write $x \geq y$ or $y \leq x$ if the vector $x - y$ is nonnegative. Note that, if $x \geq y$ is false, $y \geq x$ is not necessarily true.) R_+^{n+1} is the set of nonnegative vectors of R^{n+1}. Then S^n is the set of all vectors in R_+^{n+1} whose coordinate sum is one,

$$S^n = \left\{ x = (x_0, \cdots, x_n) \geq 0 \,\middle|\, \sum_{j=0}^{n} x_j = 1 \right\}.$$

Figure 13.1 illustrates S^2, a subset of R^3. We often draw S^2 merely as an equilateral triangle.

EXERCISE

1. Give two vectors x and y in R^2 such that neither $x \geq y$ nor $y \geq x$ is true.

* School of Operations Research and Industrial Engineering, Cornell University, Ithaca, NY 14853.

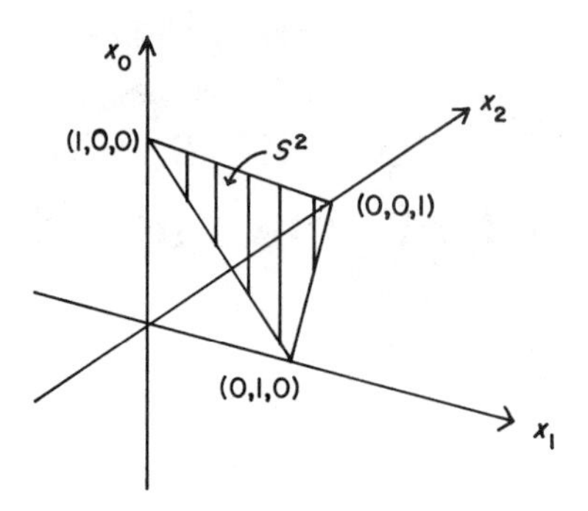

Figure 13.1

If x and y are in R^{n+1}, their inner product $\Sigma_{j=0}^{n} x_j y_j$ is denoted $x \cdot y$. The norm of x, written $\|x\|$, is $\sqrt{x \cdot x}$; note that $\|x\| \geq 0$, with equality if and only if $x = 0$. Hence $\|x\| > 0$ whenever $x \in S^n$. If C is a subset of R^{n+1}, the diameter of C, denoted diam C, is the supremum of distances between its points: diam $C = \sup\{\|x - y\| \,|\, x \text{ and } y \text{ lie in } C\}$.

EXERCISE

2. Suppose a, b and c are in R^{n+1}, $a \geq 0$, and $b \geq c$. Then prove $a \cdot b \geq a \cdot c$.

A set C in a Euclidean space is called convex if, whenever x and y are in C, so is the vector $\lambda x + (1 - \lambda)y$ for $0 < \lambda < 1$; that is, the line segment joining x and y lies entirely within C. The vector $\lambda x + (1 - \lambda)y$ is called a *convex combination* of x and y; more generally, given vectors $x^0, x^1, \cdots, x^j$, the vector $\Sigma_{i=0}^{j} \lambda_i x^i$ is a *convex combination* of the vectors if λ_i is nonnegative for $0 \leq i \leq j$ and $\Sigma_{i=0}^{j} \lambda_i = 1$.

EXERCISE

3. If C is convex, any convex combination of vectors of C lies in C.

Denote by v^j, $0 \leq j \leq n$, the jth unit vector in R^{n+1}; that is, the jth coordinate of v^j is one and all other coordinates are zero.

EXERCISE

4. S^n is convex. S^n is the set of all convex combinations of $v^0, v^1, \cdots, v^n$. Each point of S^n is a unique convex combination of $v^0, v^1, \cdots, v^n$.

We will show that every continuous function f: $S^n \to S^n$ has a fixed point x^*; that is, a point $x^* \in S^n$ such that $f(x^*) = x^*$. The function is *continuous* if, for every $x \in S^n$ and every $\varepsilon > 0$, there is a $\delta > 0$ with $\|f(x) - f(y)\| \leq \varepsilon$

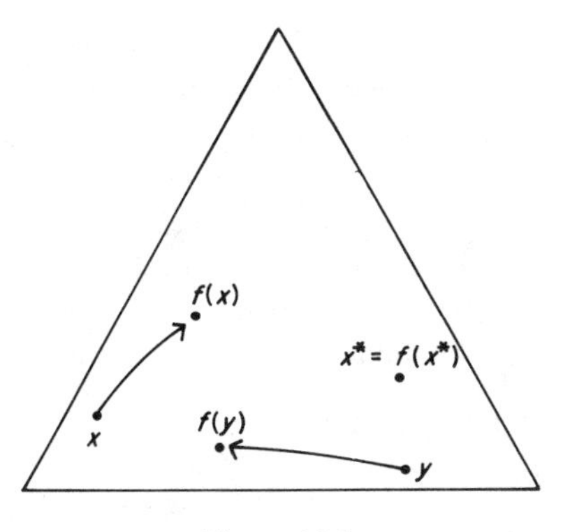

Figure 13.2

whenever $\|x - y\| \leq \delta$. More generally, a continuous function $f: C \to C$ has a fixed point whenever C is a convex, closed, bounded set in R^n. See Figure 13.2.

The organization of this paper is as follows. Section 2 describes an application of the computation of fixed points to an economic problem. Section 3 proves the existence of a fixed point by means of a combinatorial lemma. Section 4 describes how S^n can be cut up into small simplices, and in Section 5 we show how an algorithm can compute approximate fixed points. Section 6 gives some extensions of the algorithm, and Section 7 sketches generalizations to the problems of computing (complex) roots of a polynomial and equilibria in economies with production.

There are several exercises throughout the text. Almost all have solutions worked out in more or less detail at the end of the chapter. Depending on the level of the students, these can be assigned as exercises with no solutions given, assigned as exercises with solutions given and some checking remaining to be done; or presented as full-fledged examples.

2. Why Compute Fixed Points?

We will describe an application to a simple pure trade (or exchange) economy. No production is present in such an economy—the economic agents (or consumers) merely trade among themselves their initial endowments until (in some sense) everyone is happy. General references on economic models are Scarf [12] (the first chapter is particularly good), Arrow and Hahn [1], and Debreu [4].

We suppose that $n + 1$ goods exist. Besides the normal goods such as potatoes, cars, houses, etc., we have goods such as labor, savings, life insurance, and so on. All goods are assumed divisible, so that a vector in R_+^{n+1}, $(x_0, x_1, \cdots, x_n)$, corresponds to an amount x_j of the jth good, $0 \leq j \leq n$, and can be called a *bundle* of goods. There are m consumers, and each consumer may not be happy with his bundle of goods; one consumer may have a lot of labor which he would like to translate into savings, potatoes, and cars, while another has a lot of savings and would like to purchase labor to mow his lawn, clean his house, and so on. We can express these preferences by associating with each consumer a preference relation on vectors in R_+^{n+1}.

Hence consumer i, as well as an initial bundle w^i in R_+^{n+1}, has a preferable relation $\succcurlyeq_i$, where $x \succcurlyeq_i y$ means that consumer i either prefers or is indifferent to x compared to y. The relation satisfies, for all x, y, z in R_+^{n+1},

(i) $x \succcurlyeq_i y$ or $y \succcurlyeq_i x$ (completeness);
(ii) $x \succcurlyeq_i x$ (reflexivity); and
(iii) $x \succcurlyeq_i y$ and $y \succcurlyeq_i z \Rightarrow x \succcurlyeq_i z$ (transitivity).

How can the initial bundles of the consumer be reallocated to satisfy the consumers? Many answers can be found (depending in part on what is meant by "satisfy"), but one reasonable one (especially if the number of traders is large) is by means of a price mechanism. A price vector p is a nonzero vector in R_+^{n+1}; if p_j is positive, p_i/p_j units of good j will buy one unit of good i. Note that no "money" is used—the prices are solely for barter. If the prices p are given, consumer i can sell his initial bundle w^i; his "wealth" will then be

$$p \cdot w^i = \sum_{j=0}^{n} p_j w_j^i.$$

With this wealth he can "buy" any bundle in his *budget set* $\{x \in R_+^{n+1} | p \cdot x \le p \cdot w^i\}$. Note that such an x can be obtained without money, solely by barter; also that the "$\le$" contains the hidden assumption that goods can be thrown away. Among all possible bundles in his budget set, the consumer will pick one, say x^*, such that $x^* \succcurlyeq_i x$ for all x in his budget set. We assume that, in fact, only one such x^* exists, and we call it the *demand vector* of consumer i at prices p and denote it $d^i(p)$.

Note that, for any positive λ, $p \cdot x \le p \cdot w^i$ if and only if $(\lambda p) \cdot x \le (\lambda p) \cdot w^i$. Hence the budget set at prices λp is the same as at prices p, and so $d^i(\lambda p) = d^i(p)$. This is natural, for in the case of pure trade all that matters is *relative* prices and $\lambda p_i/\lambda p_j = p_i/p_j$. Thus given any price vector p, we can associate with it as equivalent price vector $p' = (1/\Sigma_{j=0}^n p_j)p$ whose coordinates sum to one. Henceforth, we restrict the set of price vectors to have coordinate sum one; in other words, S^n is the set of price vectors. We now assume that each d^i is a continuous function from S^n into R_+^{n+1}.

Exercise

5. Preference relations are often defined by some utility function u from R_+^{n+1} into R. We write $x \succcurlyeq y$ if $u(x) \ge u(y)$. One particular type of utility function we will use often has the form $u(x) = x_0^{a_0} x_1^{a_1} \cdots x_n^{a_n}$, where $a = (a_0, a_1, \cdots, a_n)$ is a vector of positive integers.

(a) Prove that, given a price vector p in S^n with each coordinate positive, the demand vector $d(p)$ corresponding to the function u above is such that a proportion $a_j/(\Sigma_{k=0}^n a_k)$ of the consumer's wealth is spent on good j.

Hint: The consumer maximizes $\Pi_{j=0}^n x_j^{a_j}$ subject to $x \geq 0$ and $\Sigma_{j=0}^n p_j x_j \leq I$, where $I = \Sigma_{j=0}^n p_j w_j$ is his wealth. This is equivalent to maximizing $\Pi_{j=0}^n (p_j x_j/a_j)^{a_j}$ subject to $p_j x_j/a_j \geq 0$ for $0 \leq j \leq n$ and $\Sigma_{j=0}^n (p_j x_j/a_j) a_j \leq I$. Now use the arithmetic-geometric mean inequality: for any positive real numbers $b_0, \cdots, b_n$, we have $(b_0 + \cdots + b_n)/(n+1) \geq \sqrt[n+1]{b_0, b_1 \cdots b_n}$, with equality if and only if $b_0 = b_1 = \cdots = b_n$.

(b) Suppose $a = (2, 4, 3)$ and $w = (3, 1, 1)$. Calculate $d(2/5, 2/5, 1/5)$.

Before we leave this example we note a difficulty. If any price is zero, the consumer demands an infinite amount of this good and $d(p)$ is undefined. We therefore limit the consumption of each good to some large number M. Given a price vector p, $d(p)$ is calculated as follows: Can you motivate what follows here?

STEP 0. Set $J = \{j | 0 \leq j \leq n, p_j > 0\}$. Set $d_k(p) = M$ if $k \notin J$. (J is the set of indices of goods whose demand has not yet been calculated).

STEP 1. Calculate the "disposable income" $I = \Sigma_{j=0}^n p_j w_j - \Sigma_{k=0, k \notin J}^n p_k M$. ($I$ is the income left after buying an amount M of each very cheap good.) For each $j \in J$, calculate $x_j = a_j I / p_j \Sigma_{h \in J} a_h$. If all $x_j \leq M$, set $d_j(p) = x_j$ for all $j \notin J$ and STOP.

STEP 2. Otherwise, for each $j \in J$ such that $x_j \geq M$, set $d_j(p) = M$ and remove j from the set J. (Such a good j is very cheap; without the limit M the consumer would purchase more than M. Demand is set to M and the remaining "disposable" income is apportioned.) Return to Step 1.

(c) If $M = 2$, use the procedure above to calculate $d(2/5, 2/5, 1/5)$ with $a = (2, 4, 3)$ and $w = (3, 1, 1)$.

For each consumer i, we have a demand vector $d^i(p)$ at prices p. The aggregate demand vector is $d(p) = \Sigma_{i=1}^m d^i(p)$. To see whether these demands can be met we compare this total demand to the total supply $w = \Sigma_{i=1}^m w^i$. If $d(p) \leq w$, then the consumers can all achieve their desired demand vectors.

By setting $g(p) = d(p) - w$, the *excess demand*, we see that we want $g(p) \leq 0$. Clearly, we will not have $g(p) \leq 0$ for arbitrary prices p. If good j is very cheap, the total demand $d_j(p)$ may be much greater than w_j, and $g_j(p) > 0$. The problem is to find a vector p^* in S^n such that $g(p^*) \leq 0$; then every consumer is satisfied. Such a vector is called an *equilibrium price vector*.

EXERCISE

6. Suppose there are three goods and two consumers. Each consumer has a utility function as given in exercise 5. For consumer 1, the "a" vector is $a^1 = (2, 4, 3)$ and $M = 100$; his initial endowment is $w^1 = (3, 1, 1)$. For consumer 2, $a^2 = (6, 4, 1)$. $M = 100$; and $w^2 = (1, 3, 3)$. Verify that $p^* = (2/5, 2/5, 1/5)$ is an equilibrium price vector. What is the final distribution of goods?

To say anything about whether an equilibrium price vector exists, we must make a further assumption. We suppose that each consumer spends all his wealth to get his demand vector. In symbols, $p \cdot d^i(p) = p \cdot w^i$ for all i and all p in S^n. Roughly, this says that each consumer does not prefer less of any good and that he can never be saturated with respect to every good. Summing over i we get $p \cdot d(p) = p \cdot w$, or in terms of the excess demand vector, $p \cdot g(p) = 0$. This important equation is known as Walras' Law; it says that the value of excess demand is zero. Note that the law holds for all price vectors, not just those in equilibrium.

We now transform the question of whether an equilibrium price vector exists into the question of whether a continuous function from S^n into itself has a fixed point. To define the function f, we consider a price vector p that is not in equilibrium. This means that the excess demand $g(p)$ has some positive coordinates and, by Walras' Law, some negative coordinates, too. That is, some goods are in excess demand and others are in excess supply. How does a price vector change under normal market pressures? The prices of goods in excess demand increase; the prices of goods in excess supply decrease. It therefore seems natural to change p to the vector $p + g(p)$.

Note that here we are adding "apples and oranges" or, more precisely, prices and quantities. We could insert a positive scale factor multiplying $g(p)$ with the dimensions of price/quantity to resolve the paradox. However, we will leave this rule as it stands—it highlights the fact that we are interested in the direction of change and not its magnitude. Two problems remain. First, some prices may now be negative if the excess supply was large. Second, the prices may no longer sum to one. To take care of the first difficulty, we raise any negative price to zero. Hence replace $p + g(p)$ by $(p + g(p))^+$, where for any vector h with coordinates h_j, h^+ is the vector with coordinates $\max\{0, h_j\}$. Now we normalize the result so that its coordinates sum to one. Hence define, for any p in S^n,

$$f(p) = \frac{(p + g(p))^+}{\sum_{j=0}^{n} (p + g(p))_j^+}.$$

Theorem 1. *If g is a continuous function from S^n to R^{n+1} satisfying Walras' Law $p \cdot g(p) = 0$ for all p in S^n, then f is a continuous function from S^n into itself. The fixed points of f are equilibrium price vectors, i.e., vectors p^* with $g(p^*) \leq 0$.*

PROOF. Note first that, if $f(p)$ is defined, it is a nonnegative vector in R^{n+1} whose coordinates sum to one, so that $f(p)$ lies in S^n. We now show that it is defined and continuous. Let $\alpha(p)$ denote the sum $\Sigma_{j=0}^{n} (p + g(p))_j^+$ for any p in S^n. The function f is well-defined at p if $\alpha(p)$ is positive. But $\alpha(p)$ is a sum of nonnegative terms, so that if $\alpha(p)$ is zero each term is zero. Assume $\alpha(p)$ is zero; then $(p + g(p))_j^+ = 0$ for $0 \leq j \leq n$ and so $p + g(p)$

≤ 0. However, using Walras' Law, we get $0 < p \cdot p = p \cdot p + p \cdot g(p) \leq p \cdot 0 = 0$, a contradiction.

We conclude that $\alpha(p)$ is positive for all p is S^n. Hence $f(p)$ is defined and f is a continuous function from S^n to itself.

Now suppose p^* is a fixed point of f, $f(p^*) = p^*$. Then $p^* = (p^* + g(p^*))^+/\alpha(p^*)$ and for $0 \leq j \leq n$

$$\alpha(p^*)p_j^* = (p^* + g(p^*))_j^+.$$

If p_j^* is zero, $(p^* + g(p^*))_j^+ = g(p^*)_j^+$ is zero, so $g_j(p^*) \leq 0$. (1)

If p_j^* is positive, $(p^* + g(p^*))_j^+$ is positive, so $\alpha(p^*)p_j^* = p_j^* + g_j(p^*)$, and thus

$$g_j(p^*) = (\alpha(p^*) - 1)p_j^*, \qquad \text{if } p_j^* > 0. \tag{2}$$

We now prove that $\alpha(p^*) = 1$. Indeed, using Walras' Law,

$$\begin{aligned} 0 = p^* \cdot g(p^*) &= \sum_{j=0}^{n} p_j^* g_j(p^*) = \sum_{j:p_j^*>0} p_j^* g_j(p^*) \\ &= \sum_{j:p_j^*>0} p_j^*(\alpha(p^*) - 1)p_j^* = (\alpha(p^*) - 1)\sum_{j=0}^{n} p_j^{*2} \\ &= (\alpha(p^*) - 1)p^* \cdot p^*. \end{aligned}$$

However, $p^* \cdot p^*$ is positive, since $p^* \neq 0$. Hence $\alpha(p^*) = 1$. By (1) and (2), $g(p^*) \leq 0$, and p^* is an equilibrium price vector. □

The result that f does have a fixed point (and so the economy has an equilibrium) is the famous fixed-point theorem of Brouwer. We will prove this theorem in the next two sections.

We now show that the allocation of goods at an equilibrium price vector does preclude any further incentives to trade.[1] Let $I \subseteq \{1, 2, \cdots, m\}$ be any set of traders. Each consumer i, $i \in I$, currently has $d^i(p^*)$; the set in total has $\Sigma_{i\in I} d^i(p^*)$. Suppose there is an incentive to trade; that is, a set $\{x^i\}$ of commodity bundles exists with $\Sigma_{i\in I} x^i = \Sigma_{i\in I} d^i(p^*)$. Further, for all $i \in I$, $x^i \succcurlyeq_i d^i(p^*)$, and for at least one $i \in I$, $d^i(p^*) \not\succcurlyeq_i x^i$. In other words, all consumers in I are no worse off, and at least one is strictly better off. Since $d^i(p^*)$ is the unique maximal vector in consumer i's budget set, we conclude that either $x^i = d^i(p^*)$ or $p^* \cdot x^i > p^* \cdot d^i(p^*)$ for all $i \in I$, and the latter holds for some i. Hence $p^* \cdot \Sigma_{i\in I} x^i = \Sigma_{i\in I} p^* \cdot x^i > \Sigma_{i\in I} p^* \cdot d^i(p^*) = p^* \cdot \Sigma_{i\in I} d^i(p^*)$, a contradiction to $\Sigma_{i\in I} x^i = \Sigma_{i\in I} d^i(p^*)$. Therefore, there is no incentive for further trade.

Exercise

7. Show in a similar way that there is no incentive for a group of consumers to trade their initial bundles w^i among themselves.

[1] The rest of this section can be omitted without loss of continuity.

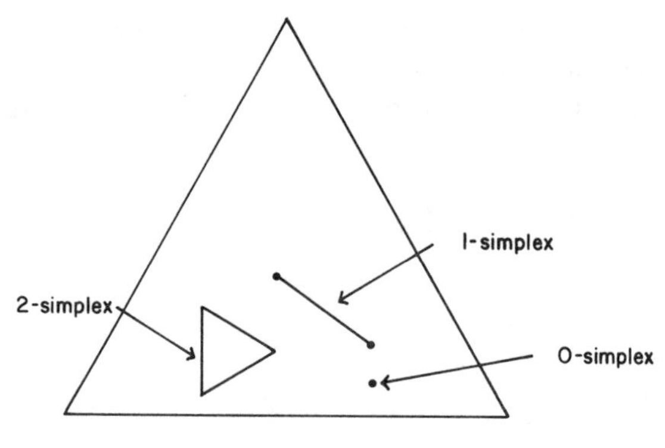

Figure 13.3

3. A Combinatorial Lemma

We prove Brouwer's theorem by using a combinatorial lemma of Sperner. Other simple proofs of this lemma can be found in Burger [2] and (for the case $n = 2$) in Lyusternik [9]. The lemma is concerned with "cutting up" S^n into smaller n-simplices. We need some preliminary definitions.

A j-simplex σ in S^n is the set of all convex combinations of $j + 1$ points of S^n, called its vertices, so that any point of σ can be expressed *uniquely* as a convex combination of the vertices. Thus a 0-simplex is (a set consisting of) a point, a 1-simplex is a line segment, and a 2-simplex is a triangle. See Figure 13.3. If the vertices of a j-simplex σ are $y^0, y^1, \cdots, y^j$ we write $\sigma = [y^0, \cdots, y^j]$.

Exercise

8. Prove that S^n is an n-simplex in itself. In S^2, show that the vertices (1/5, 2/5, 2/5), (2/5, 1/5, 2/5) and (2/5, 2/5, 1/5) determine a 2-simplex, while (1, 0, 0), (0, 1, 0) and (1/2, 1/2, 0) do not.

A k-simplex τ is a *k-face* of the simplex σ if all vertices of τ are vertices of σ. A 2-simplex has three 1-faces, its sides, and three 0-faces, its vertices. If $\sigma = [y^0, \cdots, y^j]$, the $(j - 1)$-face of σ whose vertices are all those of σ except y^i is called the face of σ opposite y^i. For example, the face of S^n opposite v^i is $\{x \in S^n | x_i = 0\}$ for $0 \leq i \leq n$.

We now cut up S^n into smaller n-simplices. A *triangulation* of S^n is a finite collection of n-simplices so that

(i) every point of S^n is in one of the smaller n-simplices; and
(ii) if two such n-simplices intersect, their intersection is a common face of the two simplices.

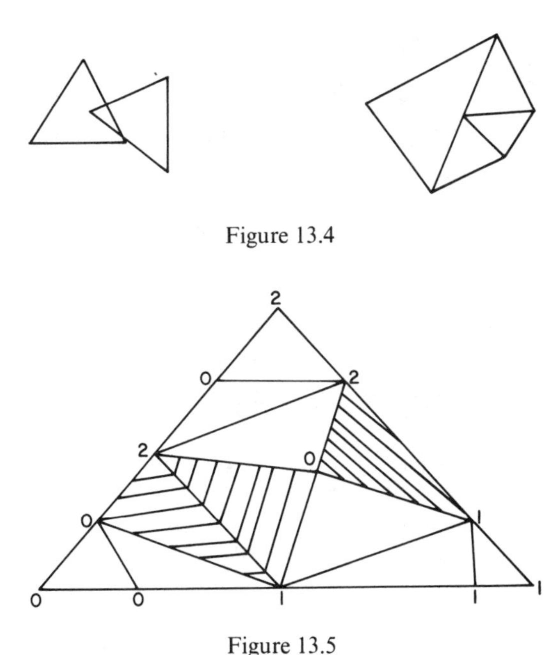

Figure 13.4

Figure 13.5

The second requirement rules out configurations such as those shown in Figure 13.4. Figure 13.5 shows a triangulation of S^2.

Before we state Sperner's Lemma, we need one more preliminary concept. A triangulation T of S^n is *labelled* if there is an integer between 0 and n associated with each vertex of an n-simplex of T. We denote the set of all such vertices by T^0. So a labeling is a function from T^0 to the set $\{0, 1, \cdots, n\}$. Figure 13.5 shows a labelled triangulation. The labeling is *proper* if the label of a vertex y is i only if y_i is positive. Hence the label of v^i must be i, and the label of a vertex on the edge joining v^i and v^j must be i or j, etc. The labeling in Figure 13.5 is proper. Finally, an n-simplex of T is *completely labeled* if its vertices carry all the labels 0 through n. Figure 13.5 has three completely labeled 2-simplices.

Sperner's Lemma. *Every properly labeled triangulation of S^n contains a completely labeled n-simplex.*

We now seem to be a long way from proving Brouwer's theorem, that every continuous function f from S^n to itself has a fixed point. Before proving the lemma, let us show how it implies Brouwer's theorem. The idea is to consider a sequence of triangulations, cutting S^n into n-simplices of smaller and smaller diameters. Each triangulation is given a proper labeling, with the labeling rule dependent on the particular function f. In each triangulation we find a completely labeled n-simplex. Then any limit point of these n-simplices will be a fixed point of the function f.

Let $\varepsilon > 0$ be given, and choose δ, $0 < \delta \leq \varepsilon$, so that whenever x and y in S^n are such that $\|x - y\| \leq \delta$, $\|f(x) - f(y)\| \leq \varepsilon$. (The fact that we can choose δ independently of x relies on f being *uniformly* continuous, but every continuous function on S^n is uniformly continuous.) We now let T be a triangulation of S^n so that all n-simplices of T have diameter at most δ. We say T has mesh at most δ.

Next we label the triangulation T. If $y \in T^0$, we label y with the smallest integer in $\{j | f_j(y) \leq y_j \text{ and } y_j \text{ is positive}\}$.

EXERCISE

9. Show that this set is always nonempty, so that we do have a labelling.

Clearly this labeling is proper, since the label of every vertex corresponds to a positive coordinate of that vertex.

Proposition 1. *If x lies in a completely labeled n-simplex of T, $|f_j(x) - x_j| \leq 2n\varepsilon$ for $0 \leq j \leq n$.*

PROOF. Let the simplex be $\sigma = [y^0, \cdots, y^n]$, where y^j has the label j, $0 \leq j \leq n$. Then $f_j(y^j) \leq y_j^j$ for $0 \leq j \leq n$. Now consider $f_j(x) - x_j$. Since $\|x - y^j\| \leq \delta$, $\|f(x) - f(y^j)\| \leq \varepsilon$ and $f_j(x)$ is within ε of $f_j(y^j)$. Also, x_j is within δ of y_j^j. Hence

$$f_j(x) - x_j \leq (f_j(y^j) + \varepsilon) - (y_j^j - \delta) \leq (f_j(y^j) - y_j^j) + \varepsilon + \delta \leq \varepsilon + \delta \leq 2\varepsilon.$$

This is true for $0 \leq j \leq n$, but also

$$\sum_{j=0}^{n} f_j(x) = 1 = \sum_{j=0}^{n} x_j$$

because x and $f(x)$ lie in S^n. Hence

$$f_j(x) - x_j = - \sum_{i=0, i \neq j}^{n} (f_i(x) - x_i) \geq - \sum_{i=0, i \neq j}^{n} 2\varepsilon = -2n\varepsilon.$$

Combining the inequalities, we have $|f_j(x) - x_j| \leq 2n\varepsilon$ for $0 \leq j \leq n$, as desired. □

Now let $\varepsilon_k = 1/k$, $k = 1, 2, 3, \cdots$. We can find for each k a corresponding δ_k and then, using the proposition, a point x^k with $|f_j(x^k) - x_j^k| \leq 2n\varepsilon_k = 2n/k$ for $0 \leq j \leq n$. (Note that we have used Sperner's lemma to guarantee the existence of x^k.) We now have an infinite sequence x^k in S^n. By a famous theorem (since S^n is closed and bounded), such a sequence has a convergent subsequence, that is, an infinite set of integers K exists such that x^k tends to a limit x^* in S^n as k tends to infinity in the set K. Now we use the continuity of f and find that $|f_j(x^*) - x_j^*| \leq 0$ for $0 \leq j \leq n$, but this means that $f(x^*) = x^*$, and f has a fixed point.

We have now shown that Brouwer's theorem follows from Sperner's lemma, with one proviso. We have implicitly assumed that we can find triangulations of S^n of arbitrarily small mesh. In the next section we will show how to construct such triangulations. Note that the proposition states that we need only examine a finite number of n-simplices to find an approximate fixed point. By this we mean a point close to its image; such a point may not be close to a fixed point.

EXERCISE

10. Draw any triangulation of S^2. Label the triangulation according to the mapping $f: S^2 \to S^2$ defined by $f(x_0, x_1, x_2) = (x_1, x_2, x_0)$. Find a completely labeled 2-simplex of your triangulation. Pick any x in this 2-simplex and compare x and $f(x)$. Verify proposition 3.1.

We now prove Sperner's lemma, following the method of Cohen [3].

PROOF OF SPERNER'S LEMMA. In fact, we prove the stronger assertion, that every properly labeled triangulation contains an odd number of completely labeled n-simplices, by induction on n.

For $n = 0$, S^0 is the single point 1 in R^1, and just one triangulation of S^0 exists with just one simplex [1]. The vertex 1 is labeled 0, and [1] is a completely labeled 0-simplex.

If this seems an artificial basis for the induction, the case $n = 1$ is just as easy. S^1 is the line segment joining (1, 0) and (0, 1) in R^2. A triangulation is a division of this segment into several segments, and a labeling then assigns the integers 0 or 1 to each subdivision point. The labeling is proper if (1, 0) is labeled 0 and (0, 1) is labeled 1. Figure 13.6 shows a proper labeling of a triangulation of S^1. Now proceed from (1, 0) to (0, 1). The labels encountered are a string of zeros, then a string of ones, ... then a string of zeros, and finally a string of ones. Clearly, we find an odd number of transitions from a zero to a one or a one to a zero. Hence we have an odd number of completely labeled 1-simplices.

We now prove the inductive step. We suppose the lemma true for $n - 1$, and consider a properly labeled triangulation T of S^n. Let $\tilde{S}^{n-1}$ be the face

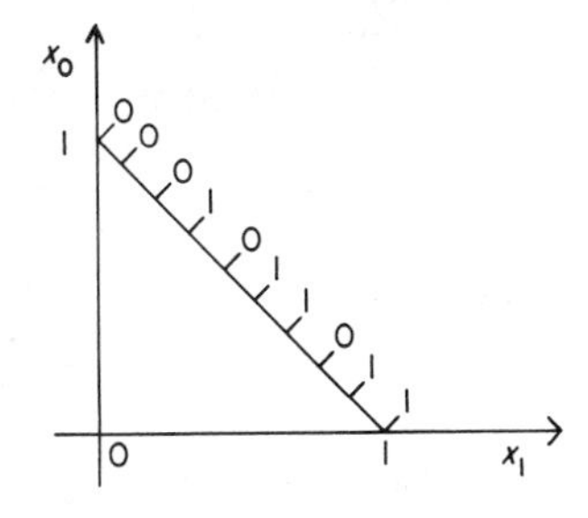

Figure 13.6

of S^n opposite v^n, $\tilde{S}^{n-1} = \{x \in S^n | x_n = 0\}$. The $(n-1)$-faces of simplices of T that lie in $\tilde{S}^{n-1}$ form a triangulation of $\tilde{S}^{n-1}$, say $\tilde{T}$. Moreover we can identify $\tilde{S}^{n-1}$ with S^{n-1} (the only difference is the final zero coordinate in $\tilde{S}^{n-1}$) and $\tilde{T}$ with a triangulation of S^{n-1}. The labeling of T induces a proper labeling of $\tilde{T}$; note that the label n does not appear on $\tilde{S}^{n-1}$ and the remaining labels satisfy the requirements. Hence by the induction hypothesis there is an odd number of $(n-1)$-simplices in $\tilde{T}$ whose vertices have all the labels 0 through $n-1$.

We now link these simplices with completely labeled n-simplices of T by forming a *graph* G. The nodes of G are $(n-1)$-simplices of $\tilde{T}$ or n-simplices of T, whose vertices have all the labels 0 through $n-1$. Refer to Figure 13.7. Two nodes of G are adjacent if one is a face of the other or if both are n-simplices and they have a common $(n-1)$-face with all the labels 0 through $n-1$.

On examining Figure 13.7 we see that the graph G as a very special form: it splits into disjoint paths and circuits. To show that this is true in general we find the number of nodes adjacent to each node of G. There are three cases.

(a) The node is an $(n-1)$-simplex of $\tilde{T}$. Then the $(n-1)$-simplex is a face of just one n-simplex of T, and this n-simplex is a node of G. Thus in this case the node is adjacent to exactly one node of G.

(b) The node is a completely labeled n-simplex of T. Then each of its $n+1$ vertices has a different label, and the simplex has exactly one $(n-1)$-face with all the labels 0 through $n-1$. Either this face is in $\tilde{T}$ (and is a node of G) or leads to another n-simplex that is a node of G. Thus in this case also the node is adjacent to exactly one node of G.

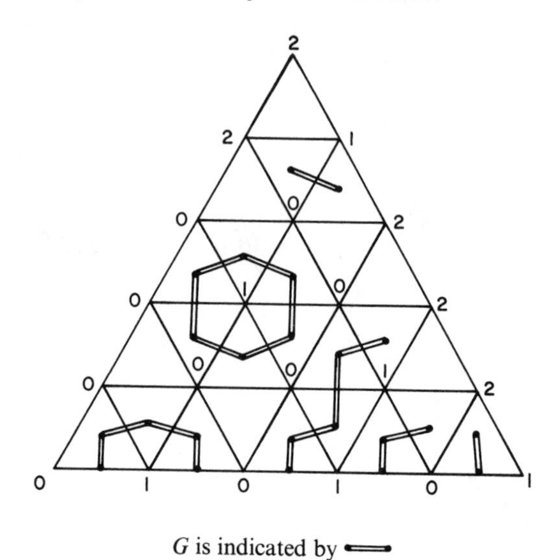

Figure 13.7

(c) The node is an n-simplex of T with all the labels 0 through $n - 1$, but not the label n. Since the $n + 1$ vertices have exactly n different labels, precisely one label is duplicated. That is, two vertices have the same label, and the two $(n - 1)$-faces opposite these vertices have all the labels 0 through $n - 1$. Each of these $(n - 1)$-faces is either a node of G or leads to a node of G as in case (b). In this case, the node is adjacent to exactly two nodes of G.

Now a graph in which each node is adjacent to one or two nodes is clearly a union of paths and circuits. The number of endpoints, or nodes adjacent to just one node, must be even. So there is an even number of nodes of types (a) and (b). By induction hypothesis there is an odd number of type (a). Hence there is an odd number of type (b), and this is precisely the claim of the inductive step. Hence the lemma is finally proved. □

4. Triangulations

In order to complete the proof of Brouwer's theorem from Sperner's lemma, we must demonstrate that there are triangulations of S^n of arbitrarily fine mesh. Also, our algorithms will find approximate fixed points of functions on S^n by locating completely labeled simplices. Such simplices are found by tracing paths similar to those used in the proof of Sperner's lemma. We therefore need triangulations that are simple to manipulate. The following was used by Kuhn [6].

Let m be a positive integer. We construct a simple triangulation of S^n with mesh at most $\sqrt{n + 1}/m$. Let Q be the $n \times (n + 1)$ matrix

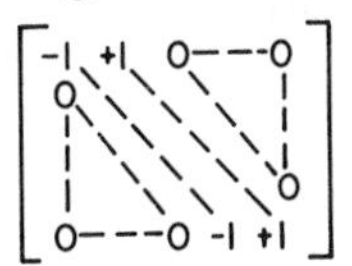

and let q^j, $1 \leq j \leq n$ denote the jth row of Q, a vector in R^{n+1}. Note that adding q^1, then $q^2, q^3, \cdots$, up to q^n successively to the vertex v^0 of S^n takes one through $v^1, v^2, \cdots$, up to v^n. We will create small n-simplices by adding only $(1/m)$ times these vectors, and also adding these vectors in possibly a different order.

Let the set of vertices $T^0(n, m)$ by $\{y \in S^n | my_j \text{ is integer for } 0 \leq j \leq n\}$. Choose y^0 in $T^0(n, m)$, and let π be any permutation of the integers $\{1, 2, \cdots, n\}$; that is, $(\pi(1), \pi(2), \cdots, \pi(n))$ is just a reordering of $(1, 2, \cdots, n)$. Now define the vertices $y^1, y^2, \cdots, y^n$ in turn by setting $y^i = y^{i-1} + (1/m)q^{\pi(i)}$. In other words, y^i is obtained from y^{i-1} by moving a "distance" $1/m$ in the "direction" that goes from $v^{\pi(i)-1}$ to $v^{\pi(i)}$. If π is the identity permutation ($\pi(i) = i$ for $1 \leq i \leq n$), then the n-simplex $[y^0, y^1, \cdots, y^n]$ will be similar to (and $(1/m)$th the linear size of) S^n. In any case, we write $[y^0, \cdots, y^n]$ as $k(y^0, \pi)$. Some such simplices may not lie in S^n. However, those that do form a triangulation $T(n, m)$ of S^n. Figure 13.8 shows $T(2, 5)$.

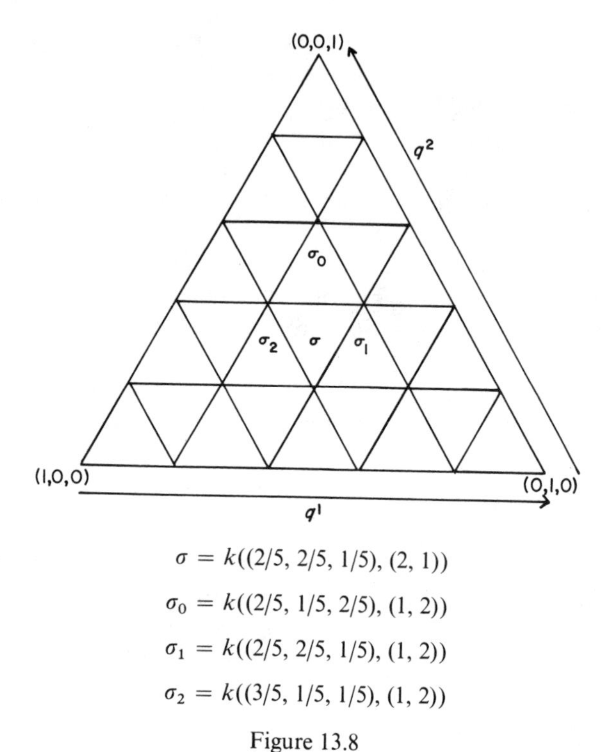

$$\sigma = k((2/5, 2/5, 1/5), (2, 1))$$

$$\sigma_0 = k((2/5, 1/5, 2/5), (1, 2))$$

$$\sigma_1 = k((2/5, 2/5, 1/5), (1, 2))$$

$$\sigma_2 = k((3/5, 1/5, 1/5), (1, 2))$$

Figure 13.8

EXERCISE

11. (a) Let $n = m = 2$. Show that when $y^0 = v^0$ and $\pi = (\pi(1), \pi(2)) = (2, 1)$, $k(y^0, \pi)$ does not lie in S^2. Draw $T(2, 2)$ and identify each 2-simplex.
    (b) Draw $T(2, 6)$ and identify each 2-simplex.

After doing Exercise 11, you may feel reasonably familiar with $T(n, m)$. However, from your experience with the case $n = 2$, you may think that $T(n, m)$ consists of regular n-simplices. The situation is not that simple.

EXERCISE

12. Find each 3-simplex of $T(3, 2)$. Calculate the diameter of each one. (Note that the diameter of a simplex is the maximum distance between two of its vertices.) How many of the 3-simplices of $T(3, 2)$ are regular? (A simplex is regular if the distance between each pair of vertices is the same).

For our algorithms, we will need a slightly extended triangulation. Let $\hat{S}^n = \{x \in R^{n+1} | \Sigma_{j=0}^n x_j = 1, x_j \geq 0 \text{ for } 0 \leq j \leq n, x_n \geq -1/m\}$. $\hat{S}^n$ can be thought of as S^n with another "layer" beyond the face of S^n opposite v^n.

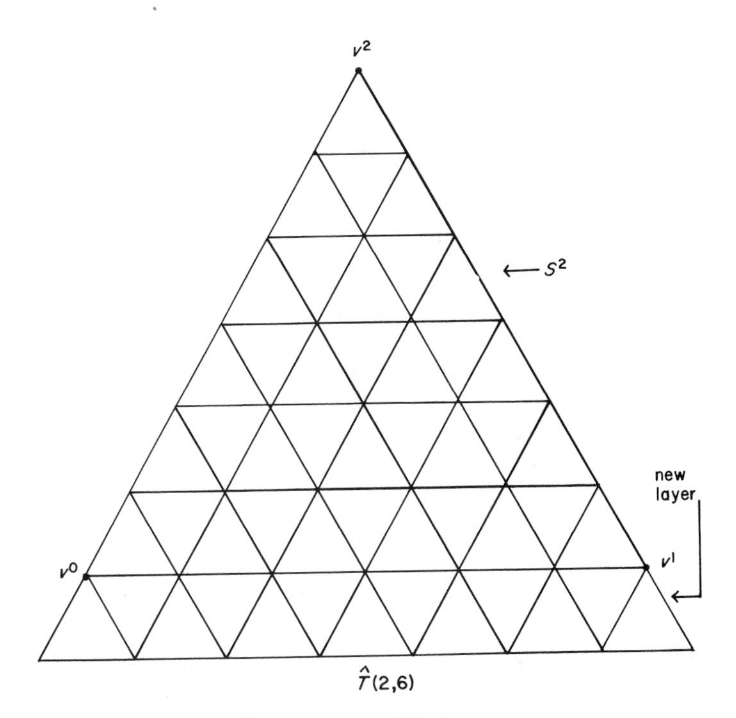

Figure 13.9

See Figure 13.8. Now let $\hat{T}^0(n, m)$ be $\{y \in \hat{S}^n | my_j \text{ is integer for } 0 \leq j \leq n\}$. Define $k(y^0, \pi)$, for $y^0 \in \hat{T}^0(n, m)$ and π a permutation of $\{1, 2, \cdots, n\}$ as before. Let $\hat{T}(n, m)$ be the collection of such $k(y^0, \pi)$ that lie in $\hat{S}^n$. Figure 13.9 shows $\hat{T}(2, 6)$.

The algorithm we describe in the next section traces a path of adjacent n-simplices in $\hat{T}(n, m)$. (Two n-simplices are adjacent if they have a common $(n - 1)$-face.) The efficiency of the algorithm depends on being able to generate easily an n-simplex adjacent to the current one. This procedure is very simple in $\hat{T}(n, m)$. Let $\sigma = k(y^0, \pi) = [y^0, y^1, \cdots, y^n]$ be an n-simplex of $\hat{T}(n, m)$, and choose i between 0 and n. Now suppose the face of σ opposite y^i does not lie in the boundary of $\hat{S}^n$. We want to find the n-simplex τ of $\hat{T}(n, m)$ that shares this $(n - 1)$-face. The simplex τ can be described as $k(z^0, \rho)$, with $z^0 \in \hat{T}^0(n, m)$ and ρ a permutation of $\{1, 2, \cdots, n\}$. From this description, every vertex of τ (in particular, the new vertex not in σ) can be obtained easily. Table 1 shows how we obtain z^0 and ρ from y^0, π and i. We also note which z^j is the new vertex. Figure 13.8 illustrates the use of this table. If we write $\sigma = k((2/5, 2/5, 1/5), (2, 1))$ as $[y^0, y^1, y^2]$ with $y^0 = (2/5, 2/5, 1/5)$, $y^1 = (2/5, 1/5, 2/5)$, $y^2 = (1/5, 2/5, 2/5)$, then σ_i is obtaining by dropping y^i for $0 \leq i \leq 2$. For example, if $i = 2$, we should have $z^0 = y^0 - (1/5)(-1, +1, 0) = (3/5, 1/5, 1/5)$ and $\rho = (\pi(2), \pi(1)) = (1, 2)$; in fact, $\sigma_2 = k(z^0, \rho)$ as expected.

Table 1

	z^0	ρ	New Vertex
$i = 0$	$y^0 + (1/m)q^{\pi(1)}$	$(\pi(2), \pi(3), \cdots, \pi(n), \pi(1))$	z^n
$0 < i < n$	y^0	$(\pi(1), \cdots, \pi(i + 1), \pi(i), \cdots, \pi(n))$	z^i
$i = n$	$y^0 - (1/m)q^{\pi(n)}$	$(\pi(n), \pi(1), \pi(2), \cdots, \pi(n - 1))$	z^0

EXERCISE

13. Draw $\hat{T}(2, 3)$. For each simplex σ in $\hat{T}(2, 3)$ and each vertex y of σ so that the face of σ opposite y is not in the boundary of $\hat{S}^2$, use the table above to find the other simple τ of $\hat{T}(2, 3)$ containing this face. Next, to get used to the rules without having a picture to help you, consider $\hat{T}(3, 2)$. What simplex arises when you drop the vertex y^0 from $k((1, 0, 0, 0), (1, 2, 3))$? When you drop y^2 or y^3 from $k((1/2, 1/2, 0, 0), (2, 3, 1))$?

5. Two Algorithms

We now have all the ingredients for finding an approximate fixed point of a given continuous function f from S^n to itself. We can divide S^n into small n-simplices using $T(n, m)$, then label the triangulation $T(n, m)$, and finally locate a completely labeled n-simplex. Proposition 1 in Section 3 then gives an approximate fixed point.

The only problem is the exorbitant amount of work involved. There are $(n + m)!/m!n!$ vertices in $T(n, m)$. (Why?) Having labeled all of these, do we have to search all the n-simplices of $T(n, m)$? There are m^n of these. For even relatively small values of n and m, these numbers are exorbitant. We need a way to search only a small fraction of the n-simplices, labeling vertices only when necessary.

The proof of Sperner's lemma gives an idea of how this can be done. If we knew all completely labeled $(n - 1)$-simplices in $\tilde{S}^{n-1}$, we could search the paths in G starting from each of these. At least one has to end with a completely labeled n-simplex in S^n. However, we still have to find all completely labeled $(n - 1)$-simplices in $\tilde{S}^{n-1}$.

The difficulty can be resolved in two ways. First, we can find completely labeled $(n - 1)$-simplices by starting with completely labeled $(n - 2)$-simplices in $\tilde{S}^{n-2} \equiv \{x \in S^n | x_{n-1} = x_n = 0\}$, and so on. If we denote by G^n the graph we described in the proof of Sperner's lemma, we can similarly define graphs $G^{n-1}, G^{n-2}, \cdots, G^1$. That is, G^j is composed of disjoint circuits and paths. The endpoints of the paths are either completely labeled $(j - 1)$-simplices in $\tilde{S}^{j-1} = \{x \in S^n | x_j = x_{j+1} = \cdots = x_n = 0\}$ (i.e., simplices with the labels $0, 1, \cdots, j - 1$) or completely labeled j-simplices in $\tilde{S}^j$. Figure 13.10 shows G^1 and G^2 in the labeled triangulation $T(2, 5)$.

Next note that just one completely labeled 0-simplex exists in $\tilde{S}^0$, that is, v^0 itself. We now join up all the graphs $G^1, \cdots, G^n$. The resulting graph

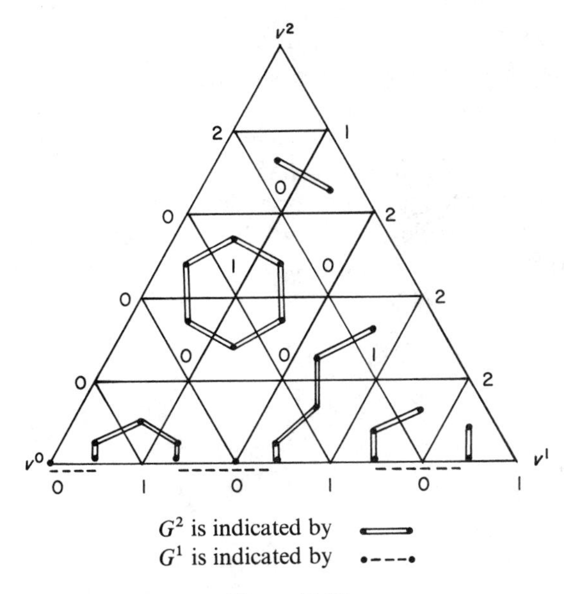

Figure 13.10

consists of disjoint circuits, paths joining completely labeled n-simplices and one long path from v^0 to a completely labeled n-simplex. Note that the dimensions of the simplices encountered can decrease as well as increase—see Figure 13.10 again. We call this algorithm the variable dimension algorithm. It is very similar to the original algorithm of Scarf [11]; see also Shapley [13].

The main complication with this algorithm is caused by the dimension varying. If we could guarantee the existence of just one completely labeled $(n - 1)$-simplex and if we knew where it was, we could start there and search only n-simplices. But how can we make this guarantee? The answer is (as some readers may have guessed from the last section) that we add an "artificial layer" beneath $\tilde{S}^{n-1}$ to create a new face $\tilde{\tilde{S}}^{n-1}$.

This algorithm, due to Kuhn [6], is dependent on the triangulation used: we employ $\hat{T}(n, m)$ with $m = hn$ a multiple of n. Although the labels of vertices of $\hat{T}(n, m)$ in S^n are determined by the particular function f, the new vertices (with nth coordinate $-1/m$) can be labeled as we please. We choose the following rule. If $y \in \hat{T}^0(n, m)$ has $y_n = -1/m$, the label of y is the minimum integer in $\{j | y_j = \max_{0 \le k \le n} y_k\}$.

Exercise

14. With the labelling of $\hat{T}(n, m)$ above, show that y has the label i only if y_i is positive for $0 \le i \le n$, and the label n only if y_n is greater than $-1/m$. Hence the labeling of $\hat{T}(n, m)$ is proper.

    Next show that if a simplex of $\hat{T}(n, m)$ has any vertex with last coordinate positive, then every vertex of this simplex has last coordinate nonnegative. Conclude that any completely labeled n-simplex of $\hat{T}(n, m)$ lies in S^n, where all labels are computed from the function f.

Now the new face of $\hat{S}^n$ opposite v_n is $\tilde{\hat{S}}^{n-1} = \{x \in \hat{S}^n | x_n = -1/m\}$. Every vertex in this face is labeled by our special rule. Consider the $(n-1)$-simplex τ_0 in this face whose vertices are $y^0, \cdots, y^{n-1}$ with $y^i = (h/m, \cdots, h/m, (h+1)/m, h/m, \cdots, h/m, -1/m)$. The entry $(h+1)/m$ appears in the ith position, $0 \leq i \leq n-1$. Clearly, y^i has the label i, so that $[y^0, \cdots, y^{n-1}]$ has all the labels 0 through $n-1$. In addition, this $(n-1)$-simplex is an $(n-1)$-face of the n-simplex $\sigma_0 = k(y^0, (1, 2, \cdots, n))$ of $\hat{T}(n, m)$. The final vertex of σ_0 is $y^n = (h/m, \cdots, h/m, 0)$. We can show further that τ_0 is the *only* $(n-1)$-simplex of $\hat{T}(n, m)$ lying in $\tilde{\hat{S}}^{n-1}$ with all the labels 0 through $n-1$.

Exercise

15. Find every 2-simplex that is a face of a 3-simplex of $\hat{T}(3, 3)$ in $\tilde{\hat{S}}^2$. Show that only $[(2/3, 1/3, 1/3, -1/3), (1/3, 2/3, 1/3, -1/3), (1/3, 1/3, 2/3, -1/3)]$ has all the labels 0 through 2.

Now our algorithm is clear. Trace the path of $\hat{G}$ leading from τ_0 to a completely labeled n-simplex of $\hat{T}(n, m)$. We illustrate the algorithm by finding an approximate fixed point of the function $f: S^2 \to S^2$ given by $f(x_0, x_1, x_2) = (x_1, x_2, x_0)$. We use the triangulation $\hat{T}(2, 6)$. The reader

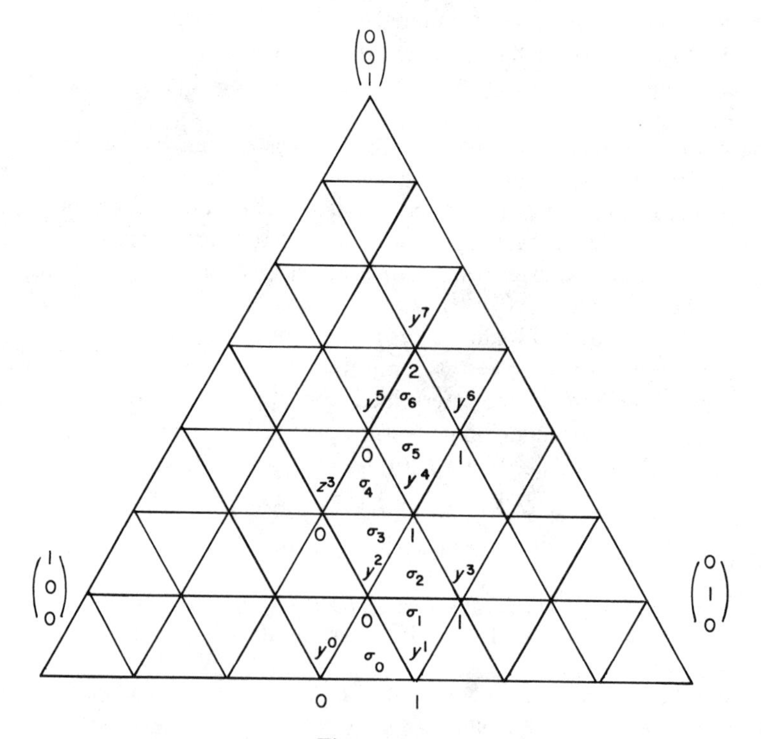

Figure 13.11

can follow the path of the algorithm in Figure 13.11. However, note that the formal steps do not need the picture to guide us.

We start with the simplex $\sigma_0 = [y^0, y^1, y^2] = k(y^0, (1, 2))$ where $y^0 = (4/6, 3/6, -1/6)$, $y^1 = (3/6, 4/6, -1/6)$, and $y^2 = (3/6, 3/6, 0)$. The labels of y^0 and of y^1 are 0 and 1, respectively. Our first task is to find the label of y^2. We calculate $f(y^2) = (3/6, 0, 3/6)$ so that y^2 has the label 0 $(f_0(y^2) \le y_0^2 > 0)$.

Next we find the simplex σ_1 that contains y^1 and y^2 but not y^0. (We drop the vertex y^0 because its label is duplicated by y^2.) Using Table 1, the new simplex is $\sigma_1 = k(y^1, (2, 1))$ with vertices y^1, y^2 and $y^3 = y^2 + (1/6)q^1 = (2/6, 4/6, 0)$.

Now we calculate the label of y^3. $f(y^3) = (4/6, 0, 2/6)$ so this label is 1. We therefore drop the vertex y^1 from σ_1. Using the table, the new simplex is $\sigma_2 = k(y^2, (1, 2))$ with vertices y^2, y^3, and $y^4 = y^3 + (1/6)q^2 = (2/6, 3/6, 1/6)$. The calculations from here on are collected in the following table.

Current Simplex	New Vertex	f(New Vertex)	New Label	Vertex to be Dropped
$\sigma_2 = k(y^2, (1, 2))$	$y^4 = (2/6, 3/6, 1/6)$	$(3/6, 1/6, 2/6)$	1	y^3
$\sigma_3 = k(y^2, (2, 1))$	$z^3 = (3/6, 2/6, 1/6)$	$(2/6, 1/6, 3/6)$	0	y^2
$\sigma_4 = k(z^3, (1, 2))$	$y^5 = (2/6, 2/6, 2/6)$	$(2/6, 2/6, 2/6)$*	0	z^3
$\sigma_5 = k(y^4, (2, 1))$	$y^6 = (1/6, 3/6, 2/6)$	$(3/6, 2/6, 1/6)$	1	y^4
$\sigma_6 = k(y^5, (1, 2))$	$y^7 = (1/6, 2/6, 3/6)$	$(2/6, 3/6, 1/6)$	2	STOP

The algorithm now terminates with $\sigma_6 = [y^5, y^6, y^7]$, since these vertices have the labels 0, 1, and 2, respectively. As an approximate fixed point we can choose any x in σ_6. Perhaps, most natural is to choose $x = (1/3)y^5 + (1/3)y^6 + (1/3)y^7 = (2/9, 7/18, 7/18)$, which is a distance $1/3\sqrt{2}$ from its image $(7/18, 7/18, 2/9)$. Of course, anyone with an ounce of common sense would have stopped at stage * when he found that $f(y^5) = y^5$—we continued the algorithm for illustrative reasons.

Note that only seven of the 36 triangles in $T(2, 6)$ were examined. More importantly, only seven of the 28 vertices of $T(2, 6)$ were examined; since we only evaluated the function f at such a vertex when it was encountered, our systematic search minimized the work necessary.

Exercises

16. Use this second algorithm to find an approximate equilibrium price vector in the economic model in exercise 6, using the triangulation $\hat{T}(2, 4)$. Use the simpler labeling rule for vertices in S^2: label p with the smallest integer j for which $p_j > 0$ and $g_j(p) \le 0$.

17. Write and test a computer program to compute approximate equilibrium price vectors in economic models as described in exercise 5.

6. Extensions of the Algorithms

The algorithms of Section 5 share a certain drawback. One must make an *a priori* selection of the parameter m. If m is large, to obtain an accurate answer, we must take a large number of steps. This is, to be expected, of course, except that most of these steps are taken a long way from the solution. If m is small, the algorithm is fast, but the final answer is often a poor approximation. Again, this would not be bad, except that we cannot use this poor approximation applying the algorithm again with a larger m. All starts must be made from the boundary of S^n, either at a vertex or on an $(n-1)$-face.

This difficulty can be circumvented in two ways. The first is due to Merrill [10] and Kuhn and MacKinnon [8], and the second to Eaves [5]. We describe the first in some detail and merely illustrate the second for $n = 1$.

The first method uses the algorithm of Section 5 as a base and applies it with a sequence of increasing m's. To avoid computational inefficiency, it uses the information obtained with small m's to obtain a good starting point for larger m's. This statement seems to contradict the first paragraph of the section. To motivate the algorithm and resolve the paradox, we reexamine the last algorithm of Section 5. Recall that we started with the simplex $[y^0, \cdots, y^n]$ with $y^i = (h/m, \cdots, h/m, (h+1)/m, \cdots, h/m, -1/m)$ for $0 \le i \le n-1$, and $y^n = (h/m, \cdots, h/m, 0)$. In a sense, we chose y^n as our best guess for a fixed point, and the simplex was then determined.

However, if we had information that $\bar{y}^n = (a_0/m, \cdots, a_n/m, 0)$ was close to a fixed point, we could instead have started with $k(\bar{y}^0, (1, 2, \cdots, n))$ with $\bar{y}^0 = (a_0+1)/m, a_1/m, \cdots, a_n/m, -1/m)$. In order to ensure that this simplex had an $(n-1)$-face in $\hat{S}^{n-1}$ with all the labels 0 through $n-1$,

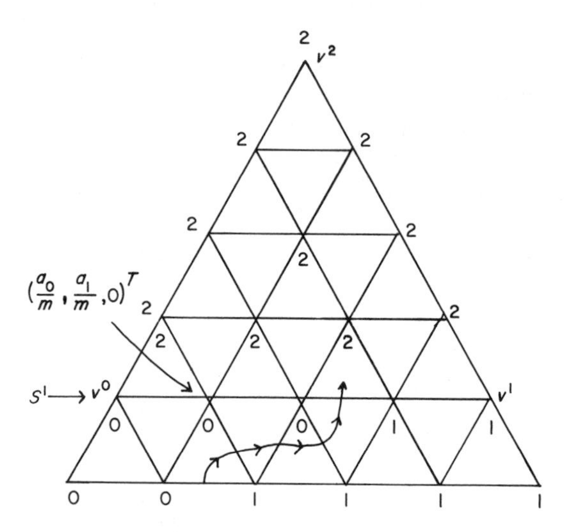

Figure 13.12

we would have had to change our labeling rule: if $y \in \hat{S}^n$ has $y_n = -1/m$, label y with the smallest integer j so that $y_j - (a_j/m) = \max_{0 \le k \le n-1}(y_k - (a_k/m))$.

Hence we can now move our starting point close to any point in S^n with last coordinate zero. The next step is to eliminate this last restriction. We do this by making the last coordinate a dummy. Hence the dimension is increased by 1. Figure 13.12 shows how S^1 is embedded in $\hat{S}^2$.

For a problem in S^n, we now take the triangulation $\hat{T}(n+1, m)$. Let $(a_0/n, \cdots, a_n/m)$ be an approximation to a fixed point of $f: S^n \to S^n$. Label $\hat{T}(n+1, m)$ as follows. If $y_{n+1} = -1/m$, y has as label the smallest integer $j(0 \le j \le n)$ so that $y_j - (a_j/m) = \max_{0 \le k \le n}(y_k - (a_k/m))$. If $y_{n+1} = 0$, y has the label of $y' = (y_0, \cdots, y_n)$ using the function f. If $y_{n+1} > 0$, y has the label $n+1$. Figure 13.12 is labeled in this way with $f(x_0, x_1) = (1/3, 2/3)$.

Now we merely apply the algorithm of the last section to $\hat{T}(n+1, m)$. The algorithm terminates when it finds a vertex with the label $n+1$. This happens as soon as a vertex with final ($(n+1)$st) coordinate positive is encountered. At this stage, the other $n+1$ vertices of the $(n+1)$-simplex all have final coordinates zero, by Exercise 14. They therefore form an n-simplex in S^n that is completely labeled. Further, all its labels arise from the function f of interest, and the n-simplex gives an approximate fixed point. The path marked in Figure 13.12 illustrates the algorithm.

What is the advantage of this algorithm? Given the approximate fixed point just found, we may restart the algorithm with $m' > m$, choosing $(a'_0/m', \cdots, a'_n/m')$ close to the point we have found. Figure 13.13 shows the results of such a strategy when $n = 1$.

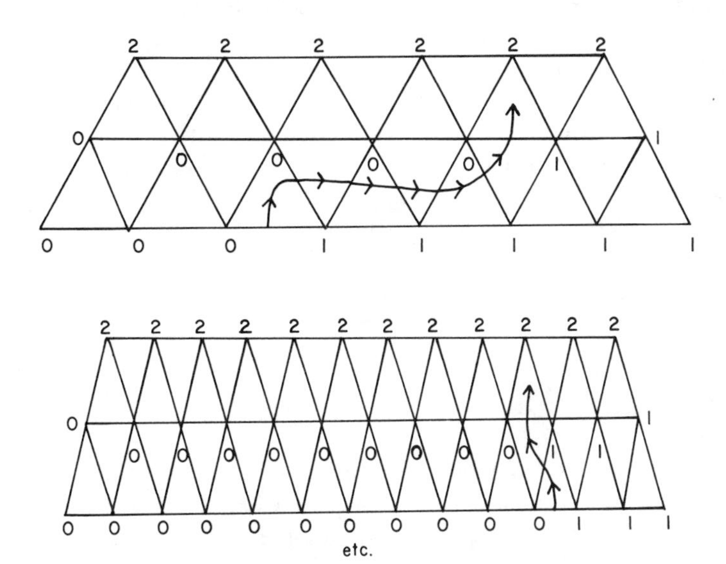

Figure 13.13

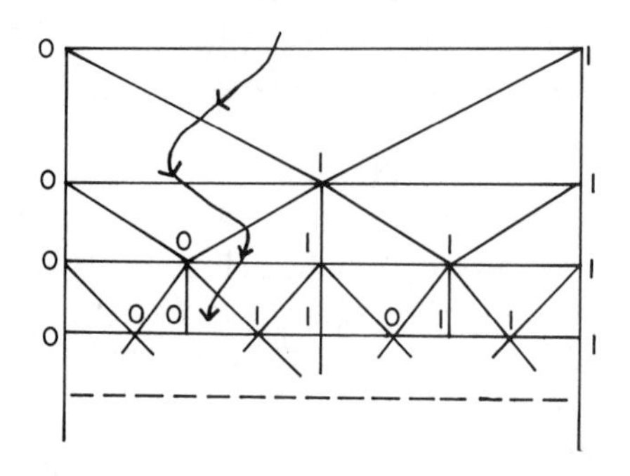

Figure 13.14

EXERCISE

18. Taking the final simplex found in exercise 16 to give a starting point, find an approximate equilibrium price vector in exercise 6 using $\hat{T}(3, 8)$.

Finally, we illustrate the second method, which requires no restarting only in the case $n = 1$. Several copies of S^n are placed one above the other, each triangulated, with the meshes getting smaller and smaller. Then those triangulations are all joined up (to form a triangulated set of dimension $n + 1$ and a sequence of $(n + 1)$-simplices followed, giving "better and better" approximate fixed points. See Figure 13.14. Each horizontal line segment is a copy of S^1. Again, all simplices met have all the labels 0 through n; in this algorithm, the label $n + 1$ is never found. Note the resemblance of Figure 13.14 to what happens in the bisection algorithm to find the zero of a continuous function g defined on the interval $[0, 1]$ with $g(0)$ and $g(1)$ having opposite signs. The bisection algorithm examines $g(1/2)$. If it has the same sign as $g(0)$, $1/2$ replaces 0; otherwise, $1/2$ replaces 1. In fact, the two algorithms are exactly equivalent. Eaves' algorithm is therefore a generalization of the bisection algorithm to n dimensions.

We cannot leave our description of Eaves' algorithm without noting that the picture for $n = 1$ is misleading, in that it implies that the simplices get continually smaller. For larger n, it can happen that the simplices get larger for a while. However, given any positive ε a simplex of diameter at most ε is generated in a finite number of steps.

7. Generalizations

In this section we describe briefly applications of some generalized algorithms to two problems; computing roots of a complex polynomial and computing economic equilibria with production is permitted. The first of these applications requires only some simple notions from complex analysis, while the second is more demanding and needs some results from linear programming.

You are probably familiar with algorithms for approximating real roots of a polynomial, for instance Newton's method. However, polynomials exist that have no real roots, and also polynomials with complex coefficients. How can we approximate the roots (in general, complex) of these polynomials? One way is to use an algorithm very similar to those of Sections 5 and 6. The paper by Kuhn [7] is easy to read, so we will not prove the rather messy bounds required, but the ideas involved are very simple.

We are given a monic polynomial $f(z) = z^n + a_1 z^{n-1} + \cdots + a_n$. Note that here superscripts mean powers. The variable z and the coefficients $a_1, \cdots, a_n$ are complex numbers. We want to find a root of f, i.e., a complex number z^* with $f(z^*) = 0$. Consider two complex planes, the z plane and the w plane, where $w = f(z)$. Instead of finding z^* with $w^* = f(z^*) = 0$, we will find there points z_0, z_1, z_2 forming a small triangle in the z plane such that $w_i = f(z_i)$, $i = 0, 1, 2$, all lie close to 0 in the w plane. We label z as 0 if $f(z) = 0$ or $\arg f(z)$ lies in the range $(-\Pi/3, \Pi/3]$; 1 if $\arg f(z)$ lies in the range. $(\Pi/3, \Pi]$; and 2 if it lies in the range $(-\Pi, -\Pi/3]$. Here, by convention, $-\Pi < \arg f(z) \leq \Pi$.

Clearly, if z_0, z_1, and z_2 are close and z_i has the label i for $i = 0, 1, 2$, then $f(z_i)$ is "small." In fact, one can easily show that if z_i has the label i and $|f(z_i) - f(z_j)| \leq \varepsilon$ for all i, j, in $\{0, 1, 2\}$, then $|f(z_k)| \leq 2\varepsilon/\sqrt{3}$.

We therefore cut up the z plane with small triangles. In fact, we need only consider the square consisting of $z = u + iv$ with $|u| \leq R$, $|v| \leq R$. We divide this square into $4m^2$ small squares of side R/m, and each square by a diagonal into two triangles—see Figure 13.15.

The algorithm starts from an edge on the boundary of the square with endpoints labeled 0, 1 (in that order, proceeding counterclockwise around the square). We continue exactly as in Section 5 until either we find a label 2, and hence a completely labeled triangle, or until we exit from the square. However, since the path of triangles always keeps a "0" on the left and a

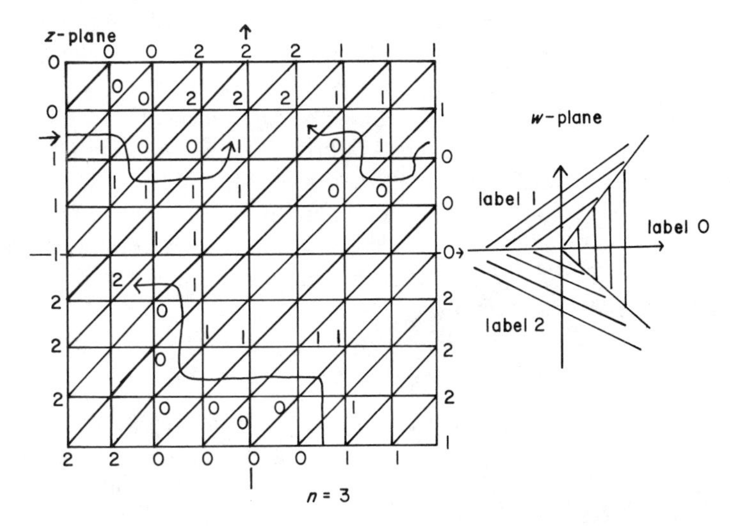

Figure 13.15

"1" on the right, it can exit only if there is an edge on the boundary of the square labeled 1, 0 (in *that* order).

To ensure an algorithm that works, we need to guarantee that there *is* an edge labeled 0, 1 and there is *no* edge labeled 1, 0. For this we must choose m and R suitably. Now if R is large $f(z) \approx z^n$ for z on the boundary, and as z traverses the boundary, $w = f(z)$ circles the origin in the w-plane n times. If m is sufficiently large, the labels encountered during such a traverse of the boundary go from 0 to 1 n times, and never from 1 to 0. In fact, one can prove that there is a (0, 1)-edge and no (1, 0)-edge if $R > 48 \max|a_k| + 4$ and $m > 1/2(n + 2)$. This argument says more: for large R and m, there will be n starting edges leading to n different completely labeled triangles. These approximate the complete set of roots to the polynomial with the correct multiplicities.

EXERCISE

19. Compute three approximate roots of $z^3 - 1$, using $R = 2$ and $m = 2$.

The algorithm just described has also been extended using the techniques of Section 6 to allow better and better approximations. Also, a new start mechanism that does not require R and m to be estimated and a start on the boundary has been developed by Kuhn.

We now consider generalizations of our algorithms that use linear programming steps at each iteration. Besides being more efficient, these algoithms allow us to tackle more general problems. For example, we can find equilibria in economic models where production possibilities are present.

One unsatisfactory feature of the algorithms of Section 5 and 6 is that after evaluating the function f at the current vertex y, most of this information is discarded and replaced by an integer between 0 and n. To avoid throwing away information, we now label each vertex of the triangulation with a vector rather than an integer. In the case of a vertex y whose integer label depended on the function f, we use the vector label $\ell(y) = f(y) - y$. If y is a fixed point, its vector label will be 0. Instead of trying to find a simplex whose vertices carry all labels 0 through n, we find a simplex (called *complete*) whose vector labels are all close to 0. More precisely, the simplex is *complete* if the vector 0 is a convex combination of the vector labels of its vertices. Note that if $\sigma = [y^0, \cdots, y^n]$ is an n-simplex and for some vector w, $w \cdot \ell(y^j) > 0$ for $0 \leq j \leq n$, then σ cannot be complete, for $w \cdot \ell > 0$ for any convex combination of $\ell(y)$, $0 \leq j \leq n$.

EXERCISE

20. Draw $T(2, 4)$ in S^2. Using the function $f\colon S^2 \to S^2$ defined by $f(x_0, x_1, x_2) = (x_1, x_2, x_0)$, find the label of each vertex. Show that there is just one complete 2-simplex of $T(2, 4)$. (*Hint:* use the remark above. It is convenient to indicate the labels of the vertices by drawing an arrow from each vertex y to $f(y)$. The arrow then represents $\ell(y) = f(y) - y$.)

We now describe one algorithm (basically that of Merrill [10]) using vector labels. Given $f\colon S^n \to S^n$, consider the triangulation $\hat{T}(n+1, m)$. Associate vector labels in R^{n+1} with the vertices of $\hat{T}(n+1, m)$ as follows. If $y_{n+1} = 0$, $\ell(y) = f(y') - y'$, where $y' = (y_0, y_1, \cdots, y_n)$. If $y_{n+1} = -1/m$, $\ell(y) = ((m+1)/m)a - y'$, where $a \in S^n$ is the best guess to a fixed point of f. If $y_{n+1} > 0$, $\ell(y)$ is arbitrary. The situation is illustrated in Figure 13.16 for $n = 1$.

The point $((m+1)/m)a_0, \cdots, ((m+1)/m)a_n, -1/m)$ lies in the boundary of $\hat{S}^{n+1}$, so it lies in some n-face $[y^0, \cdots, y^n]$ of a simplex of $\hat{T}(n+1, m)$. Hence $((m+1)/m)a = \Sigma_{j=0}^n \lambda_j y^j$ for some nonnegative λ_j summing to one. By our labeling rule, it follows that

$$\sum_{j=0}^{n} \lambda_j \left[\left(\frac{m+1}{m} \right) a - y^j \right] = \sum_{j=0}^{n} \lambda_j \ell(y^j) = 0,$$

so $\tau_0 = [y^0, \cdots, y^n]$ is complete.

Now τ_0 is an n-face of some $(n+1)$-simplex σ_0. We next find another complete n-face of σ_0 by a linear programming step. We have a basic feasible solution to the linear system.

$$\begin{bmatrix} 1 & \cdots\cdots\cdots & 1 \\ \ell^T(y^0) & \cdots\cdots\cdots & \ell^T(y^n) \end{bmatrix} \begin{bmatrix} \lambda_0 \\ \vdots \\ \lambda_n \end{bmatrix} = \begin{pmatrix} 1 \\ 0 \end{pmatrix}, \qquad \lambda_0, \cdots, \lambda_n \geq 0;$$

here $\ell^T(y^0)$, for example, is the column vector that is the transpose of the row vector $\ell(y^0)$. We now introduce the column

$$\begin{pmatrix} 1 \\ \ell^T(y^{n+1}) \end{pmatrix},$$

where y^{n+1} is the vertex of σ_0 not in τ_0. Barring nondegeneracy, exactly one column leaves the basis, and we have found another complete n-face of σ_0. We then find the $(n+1)$-simplex σ_1 of $\hat{T}(n+1, m)$ sharing this face with σ_0. By arguments analogous to those used for integer labeling, we generate a path of distinct complete n-simplices. When we first encounter a vertex with last coordinate positive, the current complete n-simplex lies completely in $\{x \in \hat{S}^{n+1} | x_{n+1} = 0\}$ and is therefore a complete n-simplex labeled according to f. If the simplex is $[z^0, \cdots, z^n]$ and we have $\Sigma_{j=0}^n \mu_j \ell(z^j) = 0$ with μ_j nonnegative and summing to one, our approximate fixed point of f is given by the first $n+1$ coordinates of $\Sigma_{j=0}^n \mu_j z^j$. We can then increase m and repeat the process.

The argument that the algorithm does not cycle relies crucially on the fact that each $(n+1)$-simplex of $\hat{T}(n+1, m)$ that has one complete n-face, has exactly one other complete n-face. This fact requires avoiding degeneracy in the linear systems. Making use of lexicographic rules to circumvent the problems of degeneracy is usual.

Instead of spelling out the algorithm in more detail and performing an example in detail, we merely indicate how the algorithm proceeds with the

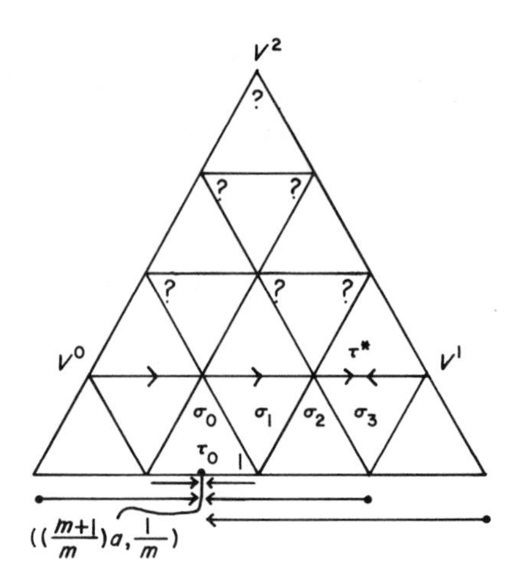

Figure 13.16

labels of Figure 13.16. The 2-simplices generated are $\sigma_0, \sigma_1, \sigma_2$, and σ_3. The 1-simplex τ^* corresponds to a complete 1-simplex in S^1, $[(1/3, 2/3), (0, 1)]$.

As promised we next describe the application to an economic model with production. The demand side of the model is exactly as before. The production side is modeled by a finite number of activities, represented by vectors $b^1, b^2, \cdots, b^k$ in R^{n+1}. The vector $(-2, -1, 4)$, for example, means that it is possible to use two units of good 0 and one unit of good 1 to produce four units of good 2. Each activity can be used at any nonnegative level and produces proportionate amounts of output with proportionate amount of input. For example, four units of good 0 and two units of good 1 can be used to produce eight units of good 2. We are therefore assuming *constant returns to scale*. More realistic assumptions can be made but with a sacrifice of simplicity. See Scarf [11], Arrow and Hahn [1], and Debreu [4].

The set of net outputs that can be achieved with these activities is $B = \{\sum_{r=1}^{k} z_r b^r | z_r \geq 0 \text{ all } r\}$. We assume that the only nonnegative vector in B is zero—otherwise an infinite amount of some goods can be produced from nothing. We also assume that the only way to produce zero is to choose each $z_r = 0$. These assumptions are equivalent to assuming the existence of a vector $q \in R^{n+1}$ with $q > 0$ and $q \cdot b^r < 0, 1 \leq r \leq k$. By choosing the units of each good appropriately, we can assume that $q = (1, \cdots, 1) \equiv v$. Now by scaling each activity if necessary, we assume that $v \cdot b^r \leq -1$ for $1 \leq r \leq k$.

Given a price vector $p \in S^n$, what happens? The demand side of the economy has already been considered: an excess demand $g(p) \in R^{n+1}$ can be found. On the production side, producer r calculates the profit $p \cdot b^r$ generated by using activity r at unit level. If the profit is negative, he will not use the activity at all. If the profit is positive, he will use the activity at as high a level as he can. With our assumption of constant returns to scale, he wants to use it at an infinite level. Only if the profit is zero can be produce at a finite positive level and still maximize his profit. (You may ask, Why should he bother? The answer is that "managerial skill" is a good consumed by the activity, so that he is rewarded for his labor. Similarly, "capital" is also consumed by the activity. A profit of zero does not mean the activity is worthless, merely that after paying for managerial skill and a fair return on capital, nothing is left over.)

In an equilibrium, each consumer attains his demand vector, and each producer maximizes his profit; finally, all demand is satisfied. The equilibrium is specified by a pair (p^*, z^*) with $p^* \in S^n$ a price vector and $z^* = (z_1^*, \cdots, z_k^*)$ a vector of activity levels. The demand must be satisfied, so we require $g(p^*) \leq \Sigma_{r=1}^k z_r^* b^r$; the excess demand is no bigger than the net production. To avoid infinite production no activity can make a positive profit. So we require $p^* \cdot b^r \leq 0$ for $1 \leq r \leq k$. Now from Walras' Law, we get

$$0 = p^* \cdot g(p^*) \leq p^* \cdot \sum_{r=1}^{k} z_r^* b^r = \sum_{r=1}^{k} z_r^*(p^* \cdot b^r) \leq 0.$$

Therefore equality exists throughout, and so $p^* \cdot b^r = 0$ if $z_r^* > 0$; that is, an activity is only used if the profit is zero. Therefore, z^* is consistent with profit maximization.

Definition. (p^*, z^*) is an equilibrium if $g(p^*) \leq \Sigma_{r=1}^k z_r^* b^r$ and $p^* \cdot b^r \leq 0$ for $1 \leq r \leq k$.

EXERCISE

21. Consider the economy below. There are three goods and two consumers. The consumer's demands are as described in exercise 5, with consumer 1 having "a" vector (1, 6, 9), "w" vector (3, 3 1/2, 2) and $M = 100$; consumer 2 has "a" vector (1, 4, 3), "w" vector (1, 1/2, 2) and $M = 100$. There are two productive activities; $b^1 = (-4, -1, 2)$ and $b^2 = (-3, 3, -1)$. Show that $p^* = (1/6, 1/3, 1/2)$, $z^* = (1/5, 2/5)$ is an equilibrium.

To compute an approximation to an equilibrium for such an economy, we will find a complete n-simplex in S^n using special labeling rules. The rules are motivated by setting $\ell(y) = f(y) - y$ where $f(p)$ is a meaningful modification of prices p. However, in this case f is not a continuous function. We will discuss this further after presenting the economic model.

Given a price vector p, what is a reasonable modification $f(p)$? If any p_j is zero, we may want to increase that coordinate of p. We therefore choose $f(p) = 1/2(p + v^j)$ where j is the least integer with $p_j = 0$. The factor 1/2 insures that $f(p) \in S^n$. Next, suppose each coordinate of p is positive, but for some activity r, $p \cdot b^r > 0$. We want to modify p to decrease the profit. An easy way is to increase the prices of the inputs and decrease those of the output. Hence we change p to $p - b^r$. Since the coordinates may not sum to one, we normalize by dividing by $v \cdot (p - b^r) = 1 - v \cdot b^r$ which is ≥ 2 by assumption. For such a p, let $f(p) = (p - b^r)/(1 - v \cdot b^r)$. Note that some coordinates of $f(p)$ may be negative, so that $f(p)$ may not lie in S^n.

The final case is when all coordinates of p are positive and $p \cdot b^r \leq 0$ for $1 \leq r \leq k$. In this case we want to decrease the excess demand $g(p)$. As in the pure trade case we can change p to $p + g(p)$. Normalizing this, we would like to set $f(p) = (p + g(p))/(1 + v \cdot g(p))$, but the denominator may be zero. We therefore scale $g(p)$, and let $f(p) = (p + \rho g(p))//1 + \rho v \cdot g(p))$, where ρ is any positive number such that the denominator is positive. Indeed, we can choose ρ independent of p. Recall that $g(p) \geq -w$ since aggregate demand $d(p)$ is nonnegative. Thus $v \cdot g(p) \geq -v \cdot w$, and any ρ smaller than $1/v \cdot w$ suffices. Again note that $f(p)$ may not lie in S^n.

We now use the algorithm sketched above to find a complete n-simplex in S^n, with $\ell(y) = f(y) - y$ for all y. There is a question as to whether such an algorithm will work, since $f(p)$ is sometimes not in S^n. The answer is that whenever p is in the boundary of S^n, $f(p)$ is in S^n; hence the algorithm cannot escape.

Now we show that such a complete n-simplex gives an approximate equilibrium. We have prices $p^0, \cdots, p^n$, all close, with $\Sigma_{j=0}^n \kappa_j \ell(p^j) = 0$ for some nonnegative κ_j summing to one. Equivalently,

$$\sum_{j=0}^{n} \kappa_j f(p^j) = \sum_{j=0}^{n} \kappa_j p^j.$$

If the mesh of the triangulation is small, all p^j's are close. In the limit, they converge to some p^*. We will replace each p^j by p^* to obtain an approximate result.

Now $f(p^j)$ can have one of three forms. First it may be $\frac{1}{2}(p^j + v^i)$ for some i if $p_i^j = 0$, or p_i^* is close to 0. Secondly, it may be $(p^j - b^r)/(1 - v \cdot b^r)$ if $p^j \cdot b^r > 0$, or $p^* \cdot b^r$ is not very negative. Lastly, it may be $(p^j + \rho g(p^j))/(1 + \rho v \cdot g(p^j))$ if $p^j \cdot b^r \leq 0$ for all $1 \leq r \leq k$, or if $p^* \cdot b^r$ is not very positive for all r. Collecting similar terms together, the following holds approximately:

$$\sum_{i=0}^{n} \lambda_i \left(\frac{p^* + v^i}{2}\right) + \sum_{r=1}^{k} \mu_r \left(\frac{p^* - b^r}{1 - v \cdot b^r}\right) + \nu \left(\frac{p^* + \rho g(p^*)}{1 + \rho v \cdot g(p^*)}\right) = p^*; \quad (*)$$

here

$$\begin{aligned} &\lambda_i \geq 0 \text{ with } \lambda_i = 0 \text{ unless } p_i^* = 0; \\ &\mu_r \geq 0 \text{ with } \mu_r = 0 \text{ unless } p^* \cdot b^r \geq 0; \qquad (**) \\ &\nu \geq 0 \text{ with } \nu = 0 \text{ unless } p^* \cdot b^r \leq 0 \text{ for } 1 \leq r \leq k; \end{aligned}$$

and

$$\sum_{i=0}^{r} \lambda_i + \sum_{r=1}^{k} \mu_r + \nu = 1.$$

Multiplying the equation $(*)$ by p^*, we obtain

$$\sum_{i=0}^{n} \frac{\lambda_i}{2}(p^* \cdot p^* + p^* \cdot v^i) + \sum_{r=1}^{k} \left(\frac{\mu_r}{1 - v \cdot b^r}\right)(p^* \cdot p^* - p^* \cdot b^r)$$
$$+ \left(\frac{\nu}{1 + \rho v \cdot g(p^*)}\right)(p^* \cdot p^* + \rho p^* \cdot g(p^*)) = p^* \cdot p^*.$$

Using Walras' Law and the conditions $(**)$, we get

$$\left(\frac{1}{2}\sum_{i=0}^{n} \lambda_i + \sum_{r=1}^{k} \left(\frac{\mu_r}{1 - v \cdot b^r}\right) + \frac{\nu}{1 + \rho v \cdot g(p^*)}\right) p^* \cdot p^* \geq p^* \cdot p^*. \qquad (^\dagger)$$

If $\nu = 0$, then since $\mu_r/(1 - v \cdot b^r) \leq \mu_r/2$, we find

$$\frac{1}{2}\sum_{i=0}^{n} \lambda_i + \frac{1}{2}\sum_{r=1}^{k} \mu_r \geq 1,$$

a contradiction. Hence ν is positive, and from $(**)$, $p^* \cdot b^r \leq 0$ for $1 \leq r \leq k$. Therefore, $\mu_r = 0$ unless $p^* \cdot b^r = 0$, and $\mu_r(p^* \cdot b^r) = 0$ all r. It follows that we have equality in $(^\dagger)$ and so the p^* terms in $(*)$ cancel, leaving

$$\sum_{i=0}^{n} \left(\frac{\lambda_i}{2}\right) v^i + \sum_{r=1}^{k} \left(\frac{\mu_r}{1 - v \cdot b^r}\right)(-b^r) + \left(\frac{\nu\rho}{1 + \rho v \cdot g(p^*)}\right) g(p^*) = 0.$$

If we denote the coefficient of $g(p^*)$ by $\bar{\nu}$, we can divide by $\bar{\nu}$ and obtain $g(p^*) \leq \sum_{r=1}^{k} z_r^* b^r$ with $z_r^* = \mu_r/\bar{\nu}(1 - v \cdot b^r)$, but this implies that p^* is an equilibrium price vector together with the activity levels z^*!

The existence of such an equilibrium follows from a generalization of Brouwer's theorem called Kakutani's fixed-point theorem. It deals with point-to-set mappings F defined on S^n, such that the image of a point x in S^n is a nonempty convex *set* $F(x)$ in S^n. Such a mapping is *upper semi-continuous* if for every $x \in S^n$ and $\varepsilon > 0$, there is a positive δ such that whenever $\|x - y\| \leq \delta$ and $y \in S^n$, each f_y in $F(y)$ is within a distance ε of some f_x in $F(x)$. Kakutani's fixed point theorem then asserts that such a mapping F has a fixed point, that is, for some $x^* \in S^n$, $x^* \in F(x^*)$.

Solutions to Selected Exercises

5. (b) We have $w = (3, 1, 1)$, $a = (2, 4, 3)$. When the price vector is $p = (2/5, 2/5, 1/5)$, the consumer's wealth is $(2/5) \times 3 + (2/5) \times 1 + (1/5) \times 1 = 9/5$. He spends $(2/9) \times (9/5) = 2/5$ on good 0, $(4/9) \times (9/5) = 4/5$ on good 1, and $(3/9) \times (9/5) = 3/5$ on good 2. His demand is therefore $(2/5)/(2/5) = 1$ of good 0, $(4/5)/(2/5) = 2$ of good 1, and $(3/5)/(1/5) = 3$ of good 2, so $d(p) = (1, 2, 3)$.

6. Consumer 1 gets (1, 2, 3), consumer 2 (3, 2, 1).

8. Suppose some point x can be expressed as a convex combination of (1/5, 2/5, 2/5), (2/5, 1/5, 2/5) and (2/5, 2/5, 1/5). Then, using matrix notation, we have

$$x = (\lambda_0, \lambda_1, \lambda_2)\begin{bmatrix} 1/5 & 2/5 & 2/5 \\ 2/5 & 1/5 & 2/5 \\ 2/5 & 2/5 & 1/5 \end{bmatrix}.$$

with the λ's nonnegative and summing to one. However, the matrix

$$\begin{bmatrix} 1/5 & 2/5 & 2/5 \\ 2/5 & 1/5 & 2/5 \\ 2/5 & 2/5 & 1/5 \end{bmatrix}$$

is nonsingular (its determinant is 1/25) so that $(\lambda_0, \lambda_1, \lambda_2)$ is unique.

9. If the set is empty, then $f_j(y) > y_j$ whenever $y_j > 0$, but then

$$1 = \sum_{j=0}^{n} f_j(y) \geq \sum_{j=0, y_j>0}^{n} f_j(y) > \sum_{j=0, y_j>0}^{n} y_j = \sum_{j=0}^{n} y_j = 1,$$

a contradiction.

10. As an example, see Figure 13.17. $x = (1/2, 1/4, 1/4)$, $f(x) = (1/4, 1/4, 1/2)$; $|f_j(x) - x_j| = 1/4$ for $j = 0, 2$, 0 for $j = 1$. Here $\varepsilon = \sqrt{2}/4$ and $\delta = \sqrt{2}/4$; $2n\varepsilon = 2 \times 2 \times \sqrt{2}/4 = \sqrt{2}$. Certainly, $1/4 \leq \sqrt{2}$ and $0 \leq \sqrt{2}$.

11. (a), (b) see Figure 13.18(a), (b). α denotes the permutation (1, 2), β (2, 1). Each 2-simplex (triangle) also contains an arrow pointing to its 0th vertex y^0. The coordinates of each point are easily read off; for example, the heavy dot is (1/2, 1/6, 1/3).

12. Simplices are
$k((1, 0, 0, 0), (1, 2, 3))$, diameter $\sqrt{2}/2$; regular
$k((1/2, 1/2, 0, 0), (1, 2, 3))$; diameter $\sqrt{2}/2$; regular
$k((1/2, 0, 1/2, 0), (1, 2, 3))$; diameter $\sqrt{2}/2$; regular
$k((1/2, 0, 0, 1/2), (1, 2, 3))$; diameter $\sqrt{2}/2$; regular
$k((1/2, 1/2, 0, 0), (2, 1, 3))$; diameter 1; irregular
$k((1/2, 1/2, 0, 0), (2, 3, 1))$; diameter 1; irregular
$k((1/2, 0, 1/2, 0), (3, 1, 2))$; diameter 1; irregular
$k((1/2, 0, 1/2, 0), (1, 3, 2))$; diameter 1; irregular.

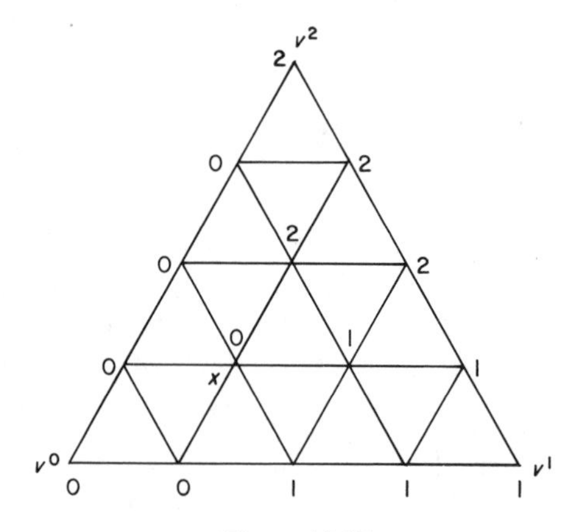

Figure 13.17

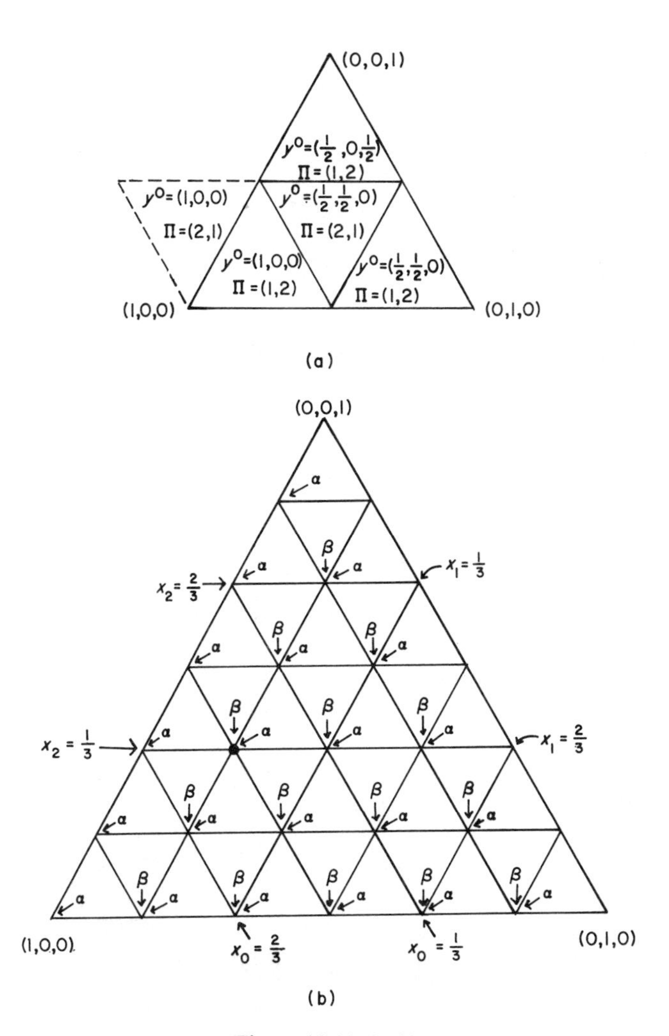

Figure 13.18 (a, b).

13. $k((1, 0, 0, 0), (1, 2, 3)) = [(1, 0, 0, 0), (1/2, 1/2, 0, 0), (1/2, 0, 1/2, 0), (1/2, 0, 0, 1/2)] = [y^0, y^1, y^2, y^3]$. Dropping y^0 the table gives $z^0 = y^0 + (1/m)(-1, 1, 0, 0) = (1, 0, 0, 0) + (1/2)(-1, 1, 0, 0) = (1/2, 1/2, 0, 0)$; also $\rho = (\pi(2), \pi(3), \pi(1)) = (2, 3, 1)$. We therefore get $k((1/2, 1/2, 0, 0), (2, 3, 1)) = [(1/2, 1/2, 0, 0), (1/2, 0, 1/2, 0), (1/2, 0, 0, 1/2), (0, 1/2, 0, 1/2)]$. The vertex $(0, 1/2, 0, 1/2)$ is the new one.

Now $y^0 = (1/2, 1/2, 0, 0)$, $\pi = (2, 3, 1)$, and we wish to drop $y^2 = (1/2, 0, 0, 1/2)$. From the table, $z^0 = y^0 = (1/2, 1/2, 0, 0)$ and $\rho = (\pi(1), \pi(3), \pi(2)) = (2, 1, 3)$. The new simplex is $k((1/2, 1/2, 0, 0), (2, 1, 3)) = [(1/2, 1/2, 0, 0), (1/2, 0, 1/2, 0), (0, 1/2, 1/2, 0), (0, 1/2, 0, 1/2)]$.

Similarly, dropping $y^3 = (0, 1/2, 0, 1/2)$ we find $z^0 = y^0 - (1/m)q^{\pi(n)} = (1, 0, 0, 0)$ and $\rho = (\pi(3), \pi(1), \pi(2)) = (1, 2, 3)$. We therefore have moved back to the simplex from which we started.

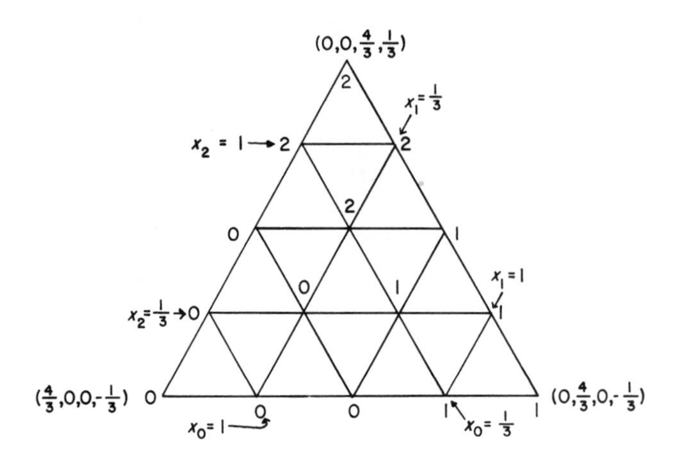

Figure 13.19

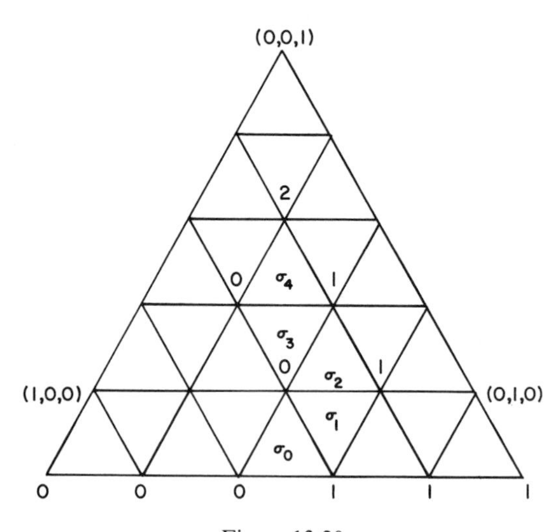

Figure 13.20

15. The simplices are shown in Figure 13.19. All vertices have $x_3 = -1/3$.

16. See Figure 13.20. We start with $\sigma_0 = k((3/4, 2/4, -1/4), (1, 2, 3))$ with vertices $y^0 = (3/4, 2/4, -1/4)$, $y^1 = (2/4, 3/4, -1/4)$ and $y^2 = (2/4, 2/4, 0)$. y^0 and y^1 have the labels 0 and 1, respectively. We must calculate the label of y^2.

At the price $(1/2, 1/2, 0)$, consumer 1's wealth is $(1/2) \times 3 + (1/2) \times 1 + 0 \times 1 = 2$ and consumer 2's $(1/2) \times 1 + (1/2) \times 3 + 0 \times 3 = 2$. Consumer 1 wants to spend $(2/9) \times 2 = 4/9$ on good 0, $(4/9) \times 2 = 8/9$ on good 1, and $(3/9) \times 2 = 6/9$ on good 2. Since good 2 is free, he cannot, so he gets $M = 100$ of good 2. He still has wealth 2, so he spends $(2/6) \times 2 = 4/6$ on good 0, $(4/6) \times 2 = 8/6$ on good 1. His demand is therefore $d^1(y^2) = (4/3, 8/3, 100)$. Similarly, $d^2(y^2) = (12/5, 8/5, 100)$, so the aggregate demand is $d(y^2) = (56/15, 64/15, 200)$ and the excess demand is

$g(y^2) = (-4/15, 4/15, 196)$. (Note that $y^2 \cdot g(y^2) = 0$ as required by Walras' Law.) Since $g_0(y^2) \leq 0$, y^2 is given the label 0.

The vertex y^0 leaves σ_0, and from the table the new simplex is $\sigma_1 = k(z^0, \rho)$ with $\rho = (2, 1)$ and vertices $z^0 = y^1 = (2/4, 3/4, -1/4)$, $z^1 = y^2 = (2/4, 2/4, 0)$ and $z^2 = (1/4, 3/4, 0)$. We must evaluate the label of z^2. Proceeding as above, we find $g(z^2) = (4, -4/3, 196)$ so that z^2 has the label 1, and z^0 leaves σ_1.

The new simplex is $\sigma_2 = k(t^0, (1, 2))$ with vertices $t^0 = z^1 = (2/4, 2/4, 0)$, $t^1 = z^2 = (1/4, 3/4, 0)$ and $t^2 = (1/4, 2/4, 1/4)$. We calculate $g(t^2) = (92/33, -28/33, -12/11)$ and the label of t^2 is 1 and t^1 leaves σ_2.

The new simplex is $\sigma_3 = k(t^0, (2, 1))$ with vertices t^0, $w^1 = (2/4, 1/4, 1/4)$ and t^2. We calculate $g(w^1) = (-92/99, 244/99, -20/33)$. Hence the label of w^1 is 0 and t^0 leaves σ_3.

The new simplex is $\sigma_4 = k(x^0, (1, 2))$ with vertices $x^0 = w^1 = (2/4, 1/4, 1/4)$, $x^1 = t^2 = (1/4, 2/4, 1/4)$ and $x^2 = (1/4, 1/4, 2/4)$. We calculate $g(x^2) = (92/33, 76/33, -28/11)$ and so x^2 has the label 2. Thus σ_4 is a completely labeled simplex and we stop.

Since the excess demand is not very small at any vertex of σ_4, we might try as our approximation $(1/3)x^0 + (1/3)x^1 + (1/3)x^2 = (1/3, 1/3, 1/3)$. In fact, $g(1/3, 1/3, 1/3) = (92/99, 76/99, -56/33)$, which while not very small is a reasonable approximation.

18. The center of the simplex found in Exercise 16 is (1/3, 1/3, 1/3). We approximate this by $(a_0/8, a_1/8, a_2/8) = (2/8, 3/8, 3/8)$. The initial simplex in $\hat{T}(3, 8)$ is $k((3/8, 3/8, 3/8, -1/8), (1, 2, 3))$ and the iterations proceed as follows, starting with $[y^0, y^1, y^2, y^3]$ with $y^0 = [(3/8, 3/8, 3/8, -1/8)$, $y^1 = (2/8, 4/8, 3/8, -1/8)$, $y^2 = (2/8, 3/8, 4/8, -1/8)$, $y^3 = (2/8, 3/8, 3/8, 0)]$.

Current Simplex	Labels	New Vertex	New Label	Dropped Vertex
$k(y^0, (1, 2, 3)) = [y^0, y^1, y^2, y^3]$	y^0 y^1 y^2 0 1 2	$y^3 = (2/8, 3/8, 3/8, 0)$	2	y^2
$k(y^0, (1, 3, 2)) = [y^0, y^1, z^2, y^3]$	y^0 y^1 y^3 0 1 2	$z^2 = (2/8, 4/8, 2/8, 0)$	1	y^1
$k(y^0, (3, 1, 2)) = [y^0, z^1, z^2, y^3]$	y^0 z^2 y^3 0 1 2	$z^1 = (3/8, 3/8, 2/8, 0)$	2	y^3
$k(w^0, (2, 3, 1)) = [w^0, w^1, w^2, w^3]$ $= [w^0, y^0, z^1, z^2]$, $w^0 = (3/8, 4/8, 2/8, -1/8)$	w^1 w^2 w^3 0 2 1	$w^0 = (3/8, 4/8, 2/8, -1/8)$	0	w^1
$k(w^0, (3, 2, 1)) = [w^0, t^1, w^2, w^3]$	w^0 w^2 w^3 0 2 1	$t^1 = (3/8, 4/8, 1/8, 0)$	1	w^3
$k(x^0, (1, 3, 2)) = [x^0, x^1, x^2, x^3]$ $= [x^0, w^0, t^1, w^2]$, $x^0 = (4/8, 3/8, 2/8, -1/8)$	x^1 x^2 x^3 0 1 2	$x^0 = (4/8, 3/8, 2/8, -1/8)$	0	x^1
$k(x^0, (3, 1, 2)) = [x^0, u^1, x^2, x^3]$	x^0 x^2 x^3 0 1 2	$u^1 = (4/8, 3/8, 1/8, 0)$	0	x^0
$k(u^1, (1, 2, 3)) = [u^1, x^2, x^3, q]$	u^1 x^2 x^3 0 1 2	$q = (3/8, 3/8, 1/8, 1/8)$	?	STOP

The 2-simplex $[u^1, x^2, x^3] = [(4/8, 3/8, 1/8, 0), (3/8, 4/8, 1/8, 0), (3/8, 3/8, 2/8, 0)]$ is completely labeled. An approximate equilibrium price vector is the center of this simplex, $(5/12, 5/12, 1/6) \simeq (0.417, 0.417, 0.17)$. The true equilibrium is $(0.4, 0.4, 0.2)$.

19. See Figure 13.21. The labels are computed as follows. Take, for example, the heavy dot, $z = -1 + 2i$. Then $f(z) = (-1 + 2i)^3 - 1 = -1 + 6i + 12 - 8i - 1 =$

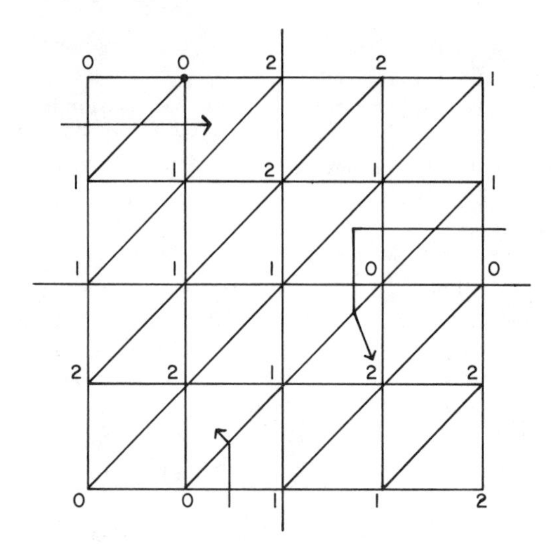

Figure 13.21

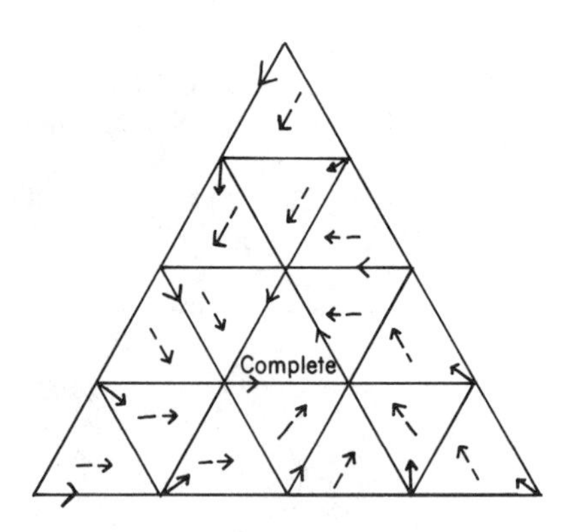

Figure 13.22

$10 - 2i$; thus $-\Pi/3 < \arg f(z) \leq \Pi/3$ and z is labeled 0. The three approximate roots (taking the centroids of the three triangles) are $(2/3) - (2/3)i$ (approximating 1) $(-2/3) + (5/3)i$ (approximating $w = (-1/2) + (\sqrt{3}/2)i$) and $-(2/3) - (4/3)i$ (approximating $w^2 = -(1/2) - (\sqrt{3}/2)i$).

20. See Figure 13.22. (Each arrow is about 1/8 its true length.) The appropriate w for each "incomplete" simplex is shown by a dotted arrow in the simplex.

References

[1] K. J. Arrow, and F. H. Hahn, *General Competitive Analysis*. San Francisco, Holden-Day, 1971.

[2] E. Burger, *Introduction to the Theory of Games*, John E. Freund, Tr. Englewood Cliffs, NJ: Prentice-Hall, 1963.

[3] D. I. A. Cohen, "On the Sperner lemma," *J. Comb, Theory*, vol. 2, pp. 585–587, 1967.
[4] G. Debreu, *Theory of Value*. New York: Wiley, 1959.
[5] B. C. Eaves, "Homotopies for computation of fixed points," *Mathematical Programming*, vol. 3, pp. 1–22, 1972.
[6] H. W. Kuhn, "Simplicial approximation of fixed points," *Proc. Nat. Acad. Sci., U.S.A.*, vol. 61, pp. 1238–1242, 1968.
[7] ——, "A new proof of the fundamental theorem of algebra," *Mathematical Programming Studies*, vol. 1, pp. 148–158, 1974.
[8] H. W. Kuhn and J. G. MacKinnon, "The sandwich method for finding fixed points," *J. Optimization Theory and Applications*, vol. 17, pp. 189–204, 1975.
[9] L. A. Lyusternik, *Convex Figures and Polyhedra*, T. J. Smith, Tr. New York: Dover, 1963.
[10] O. H. Merrill, "Applications and extensions of an algorithm that computes fixed points of certain non-empty, convex, upper semi-continuous point to set mappings," TR 71–7, Dept. of Industrial Engineering, Univ. of Michigan, Ann Arbor, 1971.
[11] H. Scarf, "The approximation of fixed points of a continuous mapping," *SIAM J. Appl. Math.*, vol. 15, pp. 1328–1343, 1967.
[12] H. E. Scarf and T. Hansen, *Computation of Economic Equilibria*. New Haven, CT: Yale Univ. Press, 1973, 249 pp.
[13] L. S. Shapley, "On balanced games without side payments," *Mathematical Programming Study*, vol. 1, pp. 175–189, 1974.
[14] M. J. Todd, "The computation of fixed points and applications," *Lecture Notes in Economics and Mathematical Systems* 124, M. Beckman and H. P. Kunzi, Ed. Berlin: Springer-Verlag, 1976.

Notes for the Instructor

Objectives. The aim of the chapter is to introduce the reader to the recent algorithms for computing fixed points and to show their importance in economic equilibrium models. This chapter is suitable for an upperclass course on mathematical models in the social sciences or economics (for mathematics majors), or for a graduate course in math models for economics graduate students.

Prerequisites. While the formal prerequisites for the main part of the chapter are minimal, quite a lot of maturity is needed. For this reason, students should be upperclassmen. I hope that a fair amount of the material will also be accessible to graduate students in economics with a good mathematical background.

The main techniques used are vector algebra in real Euclidean spaces and combinatorial reasoning. We use some notions of graph theory at a crucial juncture, but the ideas (degree of a vertex, paths and circuits) are very simple. The notion of a matrix is also helpful. The most useful skill for the student is ease of manipulation in vector geometry. The only use of calculus is in showing that a limit of solutions to discrete approximations to a problem gives a solution to the original continuous problem. If one is content with approximate solutions, this step is unnecessary.

In the final section, we give an additional application to computing roots of polynomials, for which some knowledge of complex analysis is necessary. We also use some results from the theory of linear programming in describing more sophisticated algorithms and their application to an economy with production.

Time. Probably at least four weeks. Could also be used as basis for a senior project/thesis.

CHAPTER 14

Production Costing and Reliability Assessment of Electrical Power Generation Systems under Supply and Demand Uncertainty

Jacob Zahavi*

1. Background [1]

The demand for power has constantly increased over the years. To meet this increasing demand, vast amounts of capital are needed to build generation facilities (power plants) as well as transmission and distribution networks. The amount of investment to meet future demand is expected to be even higher because of the introduction of newer and more expensive technology, higher degrees of automation, and a tendency to switch to larger-sized units. The capacity expansion problem in the power field is basically composed of the following major aspects.

(a) the types of generating facilities to be installed (nuclear, conventional oil-fired, gas turbines, etc.);
(b) the characteristics—mainly the size—of the units to be installed;
(c) the time in the future at which the various units should be introduced;
(d) the geographic location of each unit;
(e) the development of the transmission network.

Several factors contribute to making the investment decisions in power system most complex. Among these are the following.

(a) It is fast growing industry. From a humble start in the late 19th century, power has become one of the world's largest industries. This growth is primarily attributed to the rapid increase in demand for power. In many cases, demand for power is growing at a faster rate than the population and the national economy. In fact, up until the energy crisis of 1973, the demand for power followed an exponential growth, with a growth rate of

* Faculty of Management, Tel-Aviv University, Tel-Aviv, Israel.

5–20%. The growth might have slowed down recently but the market is still far from saturation. On the contrary, all indications show that demand for electricity is still growing at a relatively high and continuous rate, even though fluctuations due to uncertainties will always exist.

(b) Unlike many other commodities, electricity is one of the few products that cannot be stored for later use. Indeed, some progress has been made to store energy, usually in pumped storage systems; however, these storage schemes are still very limited and expensive. Hence the bulk of electrical energy has to be generated and distributed to customers concurrently with the demand.

(c) The total instantaneous demand for power, i.e., the sum of the needs of all customers, is highly variable, with changes linked to the time of the day and the season.

(d) It is a highly capital-intensive industry, the largest single investor in plants and equipment of all industries.

(e) The lead time to install a power plant is relatively long, ranging from as long as two years for gas turbines to ten years or more for nuclear units. Hence planning horizons are long, with investment decisions to be made at least four to six years in advance.

(f) The various generating units come in a variety of types, sizes, forced outage rates, operating characteristics, maintenance requirements, emission patterns, etc., resulting in a very large solution space out of which an investment plan is to be chosen.

(g) A strong interaction exists between decision variables at a point in time and long time. Hence present decisions will affect future decisions, and vice versa, future considerations will affect the expansion pattern at present.

(h) Increasing dependence of society on constant electricity supply, thus imposing severe restrictions or the quality of power supply. As a result, power companies are required to meet very high reliability standards.

(i) The existence of a variety of interests and objectives to be satisfied, including cost functions, reliability measures, environmental consideration, security factors, even political goals, etc. Hence investment decisions are to be made in the presence of multiple, often conflicting, interests.

(j) Major uncertainties are introduced because of the random behavior of demand, fluctuation of the installed capacity, changes in prices and other economic conditions, and the like.

(k) Last but not least, power systems are inherently complex, stretching over thousands of miles, reaching every household in the country, thus substantially increasing the magnitude and the dimension of the decision problem.

The above considerations, and others, are therefore the major factors complicating the capacity expansion problem of a power generation system and rendering it a complex and intricate decisionmaking problem.

Obviously, discussing all aspects of power system planning in one short module is not possible. Rather, we will concentrate on analyzing power system performance in the face of uncertainty in both supply and demand for power. We will focus on the generation system only, ignoring the transmission and distribution systems. We start by discussing the two basic components of power systems, namely supply and demand for power and their characteristics.

2. Introduction to Power Generation Facilities

The following are the typical power plants to be found in modern power utilities:

(a) fossil-fueled (oil or coal) steam turbines
(b) nuclear Power stations
(c) gas (combustion) turbines
(d) hydroelectric plants
(e) pumped storage

We will now discuss the characteristics of each of these plants.

2.1 Fossil-Fueled Steam Turbine Plants

Fossil-fueled (oil, coal, lignite, natural gas, etc.) plants are the most common power generation facilities. (Figure 14.1 is a schematical description of such a plant.) The fuels are ignited at the heat source, and the heat discharged by the ensuing chemical reaction is used to convert water to high-pressure high-temperature steam. The pressure of the steam causes the blades of the turbine to turn. Heat energy is thus converted to mechanical energy. The generator, in turn, converts mechanical energy to electricity and passes it through the transmission lines to the customer.

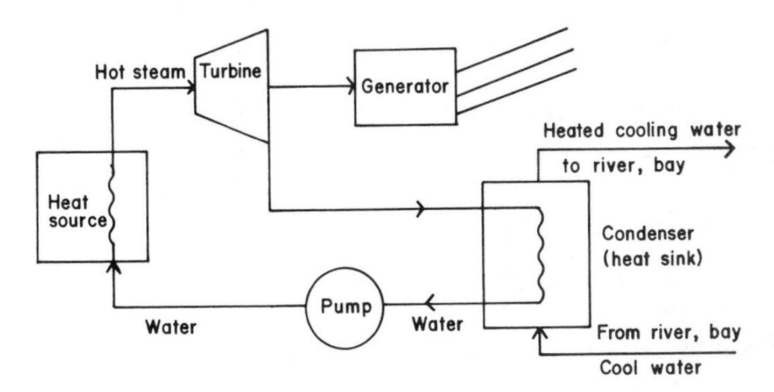

Figure 14.1. Schematic Representation of a Fossil Fueled Steam Turbine.

After passing through the turbine, the steam is cooler than when it went in, because some of the heat was converted to electrical energy. This cooler steam could be used again in the process, by pumping it back to the heat source. However, for technological reasons (mainly back-pressure), this steam has to be cooled before it can be reused. This is done by passing the steam which leaves the turbine through a condenser (heat sink), where it is turned back into water and then recycled.

The cooling process uses water taken from a bay or a river. After this water is used, it is returned to its source, naturally at a higher temperature, because it now carries the excess heat of the steam.

2.2 Nuclear Power Plants

Nuclear power plants produce energy using the same principles as fossil-fueled plants. The only difference is the type of fuel used to provide the heat. All nuclear power plants operating today are based on a fission process, in which the atomic nuclei of very heavy and unstable materials, such as uranium and plutonium, are split, producing nuclei of medium weight, among other fragments, and releasing surplus energy (heat). When an atomic nucleus is fissioned, some of the fragments released are neutrons. Neutrons traveling at appropriate speed may hit other nuclei and cause them to split. This process may be repeated so that a chain of such reactions occurs, each releasing neutrons and discharging heat. When this process is controlled, we have a constant supply of heat which is used in nuclear power plants to generate electricity.

(A more advanced technology based on fusion is now being developed. In fusion, two very light nuclei are forced together to make up a medium-weight nucleus, again releasing a lot of energy. Future generations of nuclear power plants will probably be based on this technology.)

2.3 Gas (Combustion) Turbines

Gas turbines are much more limited means to generate electricity than fossil and nuclear units. Gas turbines use the same principles to generate electricity as modern jets. In jets, the gases that come out of the engine are used to push the plane forward at a high speed. In gas turbines, these hot gases are passed through a turbine, causing it to spin and generate electricity.

2.4 Hydroelectric Stations

Hydroelectric power plants which use the energy of falling water in order to turn turbines and generate electricity. In general, hydroelectric plants are classified as either storage or run-of-river plants.

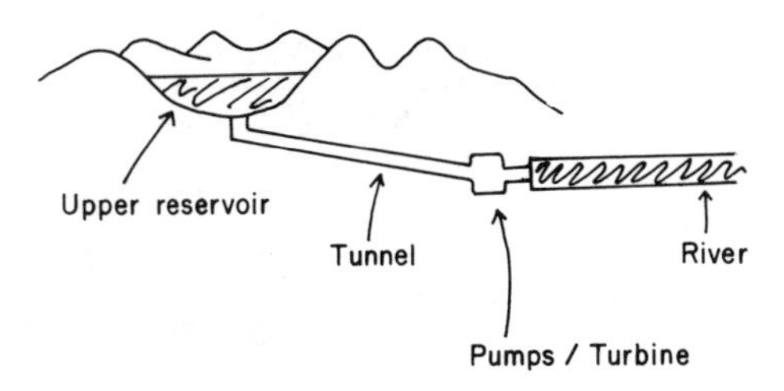

Figure 14.2. Pumped Storage Systems.

Storage Plants. These are plants with large storage capacity. During periods of high flow, water is stored for continued power generation during periods of deficient flows. Since periods of deficient flows may extend to many years, considerable storage capacity may be required. Obviously, storage type hydroelectric plants are capable of delivering only a certain amount of energy, depending upon the quantity of water in the reservior and are therefore referred to as sources of limited energy.

Run-of-River. Having little or no storage capacity, a run-of-river plant produces electricity from the constant flow of water in the river. Certainly, a height difference, or a fall, is required to convert water power into electricity.

2.5 Pumped Storage

Pumped storage systems are a means to store electricity. They are described schematically in Figure 14.2. In such systems we have a lower water source, say a lake or a river, with plenty of water, and an elevated site that can be made a reservior of water. Because of the fluctuating demand for power and based on the economics of the system, it is economically worthwhile to pump water from the lower reservoir to the upper reservoir in off-peak hours (say at night or on weekends), and use this water to produce electricity during in peak demand hours by letting the water from the upper reservoir flow to the lower reservoir.

Certainly, pumped storage systems are also a limited means of generating electricity. The limitations are imposed by the capacity of the elevated reservoir and the amount of off-peak energy that can be use to pump water from the lower to the upper reservoir.

3. Basic Characteristics of Power Generation Means

The various types of power plants behave differently with respect to certain characteristics which play an important role in investment decisions. In view of this variation and the special nature of the demand for power, we find

Table 1. Comparing between Alternative Means of Generation

Characteristic	Oil-Fired Steam	Coal-Fired Steam	Nuclear Plants	Gas Turbines	Hydro Plants	Pumped Storage
Capital (Investment) Cost	moderate	moderate	high	low	high	high
Operating (Variable) Cost	moderate	moderate	low	high	low to zero	moderate
Pollution Emission	moderate to high	high	low (except thermal poll.)	low	zero	moderate
Unit Reliability	high	high	high	low	high	high
Production Capacity	unlimited	unlimited	limited or unlimited	limited	limited or unlimited	limited
Size	up to 1000 MW	up to 1000 MW	up to 1200 MW	10–50 MW	up to 1500 MW	depends upon site
Siting Consideration	Proximity to load & cooling water	Proximity to fuel supply & cooling water	Sparsely pop. area & proximity to cooling water	anywhere	very specific site	very specific site

that the different means of generating electricity actually complement each other, rather than competing. In any well-balanced system we therefore expect to find all means of generation. (See Berrie [1] for further discussion of this point.)

We will devote this section to analyzing various characteristics of power generation. The results are summarized in Table 1, where characteristics are classified either high, moderate, or low relative to the other units of the system. Before we proceed, we will define the two basic terms; power and energy, used in power system analysis and their units of measure.

Power (demand) is the rate of electricity production per unit of time. The smallest unit of power is the watt (W). 1000 watts make up a kilowatt (KW), and 1000 KW are referred to as megawatt (MW).

Energy (consumption) is the output of electricity over a period of time, given by the product of power by time. Thus 1 KW applied over 1 hour is 1 kilowatt-hour (KWh), whereas 1 MW applied over 1 hour is 1 megawatt-hour (MWh).

We now consider the following characteristics of power generation.

Investment (Capital) Costs. The one-time costs required to design, construct, and commence operation of the power plant. Included in the capital cost are maintenance and operating fixed cost, which do not vary with production. We find that capital costs per unit of installed capacity (measured in \$/MW) are the highest for nuclear and hydroelectric stations, are lowest for gas turbines, and moderate for the other types of generation.

Operating (Variable) Costs. Operating costs depend on the amount of energy produced. They consist mainly of fuel cost (oil, coal, nuclear fuel, etc.). These costs can be increased by a certain coefficient to account for transmission losses. Operating costs (\$/MWh) are the highest for gas turbines, the lowest for nuclear and hydroelectric units, and somewhere in between for fossil-fired units.

Pollution Emission. Power plants discharge various pollutants to the atmosphere, including NO, NO_2, NO_3 (grouped together as NO_x), CO_1, CO_2, CO_3 (or CO_x), SO_2, and particulates. Heat discharged to local bodies of water as a result of the cooling process is referred to as thermal pollution. Ignoring radioactive radiation from nuclear power plants, for which the probability is very small, nuclear units and gas turbines are low polluting units, while coal-fired plants are the highest polluting units.

Unit Reliability. The forced outage rate (FOR) of a unit is the ratio of the unit's duration of down-time to the total operation and down-time. It actually describes the failure pattern of the power plant. Typical figures for the FOR are 2–10% for nuclear and fossil-fired plants up to 20–25% for gas turbines.

Production Capacity. Production capacity is unlimited for fossil-fueled, run-of-river, and continuously refueled nuclear plants—provided a constant supply of fuel. It is limited for gas turbines, storage-type hydroelectric, and batch-refueled nuclear units.

Size. Typical sizes run from 40–50 MW for gas turbines, to 1200 MW and up for nuclear and hydroelectric plants.

Siting. Proximity to cooling water is required for fossil-fired and nuclear plants. In addition, nuclear plants require sparsely populated areas (to allow for a quick evacuation of the population in case of emergencies), while coal-fired plants also require proximity to coal supply. Hydroelectric stations and pumped storage systems require very specific sites for their installation. Gas turbines are the only means of generation that can be located practically anywhere.

4. Load Characteristics [1]

The basic characteristics of the demand for power is that it is highly variable. If one observes the demand for power over a period of time, say a year, he is most likely to find a pattern like the one in Figure 14.3. The highest point on the curve is referred to as the peak demand, a quantity of paramount importance in power system analysis. The curve itself is referred to as the load curve. The variation of demand for power can actually be attributed to several factors [1].

(a) Time of day: Demand is usually at its lowest point in the early morning hours. Then it builds up during the day until it reaches a peak usually in the afternoon. Finally, it tapers off as night approaches.
(b) Day of the week: Demand is higher on weekdays than during weekends, mainly because many industries, offices, and processes are closed for the weekend.
(c) Season of the year: Weather conditions cause fluctuations in the demand for power. Winter and summer demands are higher because of space heating and cooling requirements. Generally, where the standard of living is high, power demand peaks in the summer because of the proliferation of air-conditioning equipment. The actual level of demand each season depends on the severity of weather conditions in each area.
(d) Other factors which are responsible for the variability of demand for power are of longer term impact. These include economical conditions, business cycles, unemployment levels, change in standard of living, etc.

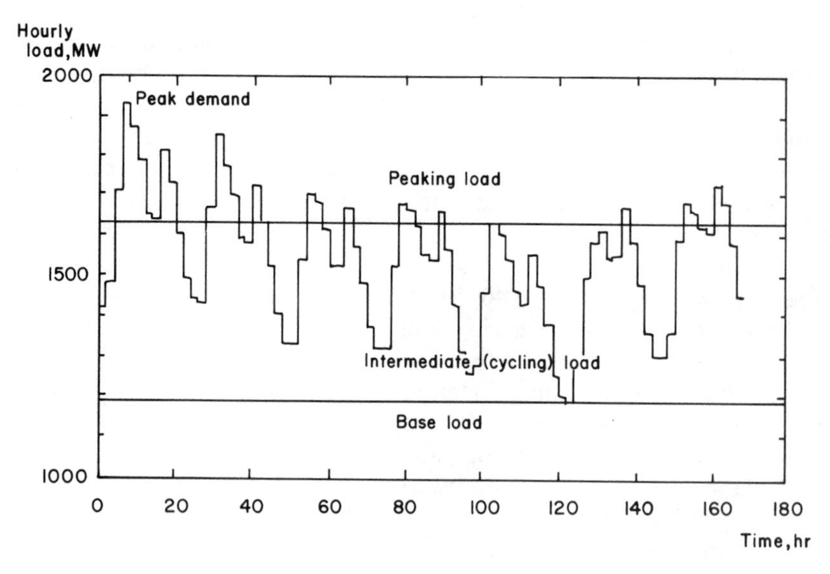

Figure 14.3. The Load Curve.

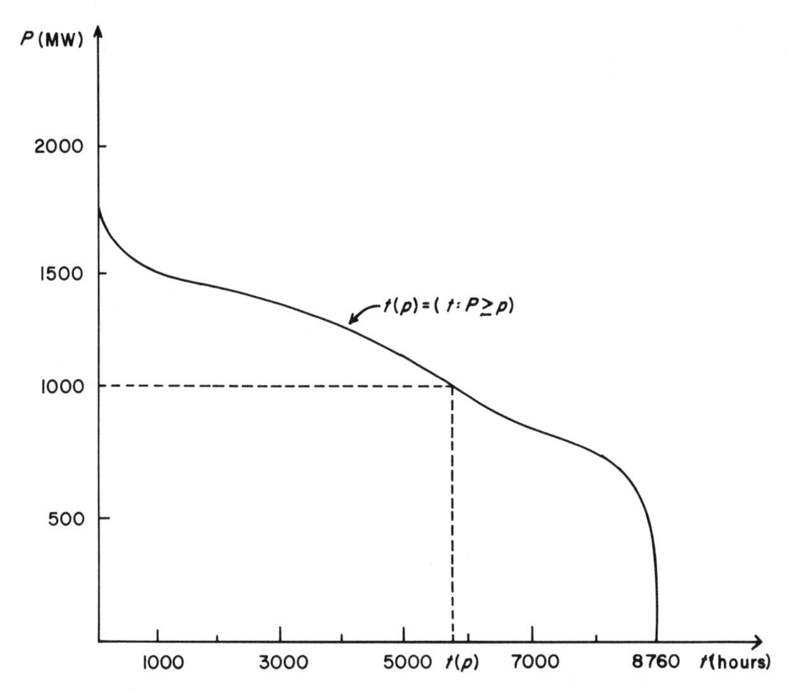

Figure 14.4. The Load Duration Curve.

The load curve in Figure 14.3 is indeed a very inconvenient form for mathematical manipulations. To represent the load data in a more convenient form, we rearrange the demand in a descending order, from the peak demand to the lowest demand. For each demand level we count the number of hours during which this demand level is equaled or exceeded. We then plot the results in two dimensions, with time (hours) on the x axis, and demand (MW) on the y axis. As a result we obtain a curve like the one in Figure 14.4. The curve is not continuous but can be approximated in a continuous manner for all practical purposes. This curve is most fundamental to power system analysis and is referred to as the *load duration curve* (LDC). By definition, its abscissa denotes the number of hours in the given period during which the customers' load equals or exceeds the associated demand on the ordinate. Usually, LDC's are calculated for a year; however, we can derive an LDC for practically any period of time: a day, a week, a month, etc. The LDC offers the following advantages over the load curve:

(a) more compact,
(b) less oscillatory,
(c) more stable.

For these reasons the LDC has become such an important tool for power system analysis.

Mathematically, if we denote by P_D the demand variable, by p_D a particular value of it, and by $t(p_D)$ the value of the LDC at point p_D, we have by definition:

$$t(p_D) = \{t: P_D \geq p_D\}.$$

Let us look on the LDC from a different point of view. Rather than expressing the time axis in terms of number of hours in a given period, we can express the time variable as a fraction, or percentage of time, relative to the total duration of time involved. This is done by simply normalizing the time axis, i.e., dividing it by the number of hours in the period (8760 hours for a year). Then each point on the abscissa can be thought of as the probability that the corresponding load will be equaled or exceeded. By reversing the role of the axes, placing the time variable (the dependent

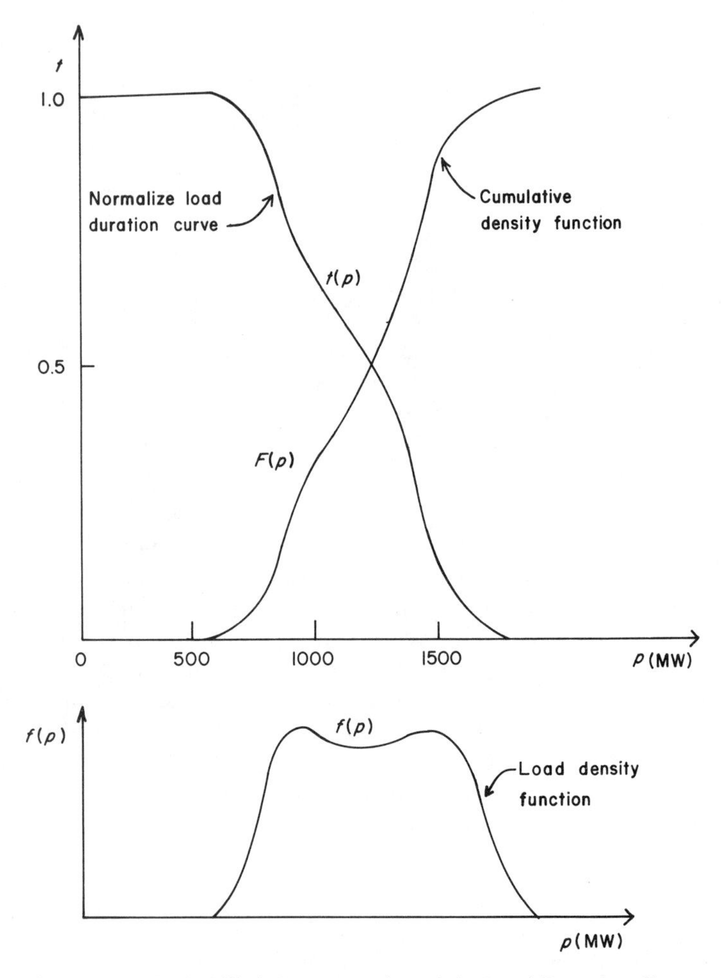

Figure 14.5. Probabilistic Interpretation of the Load Duration Curve.

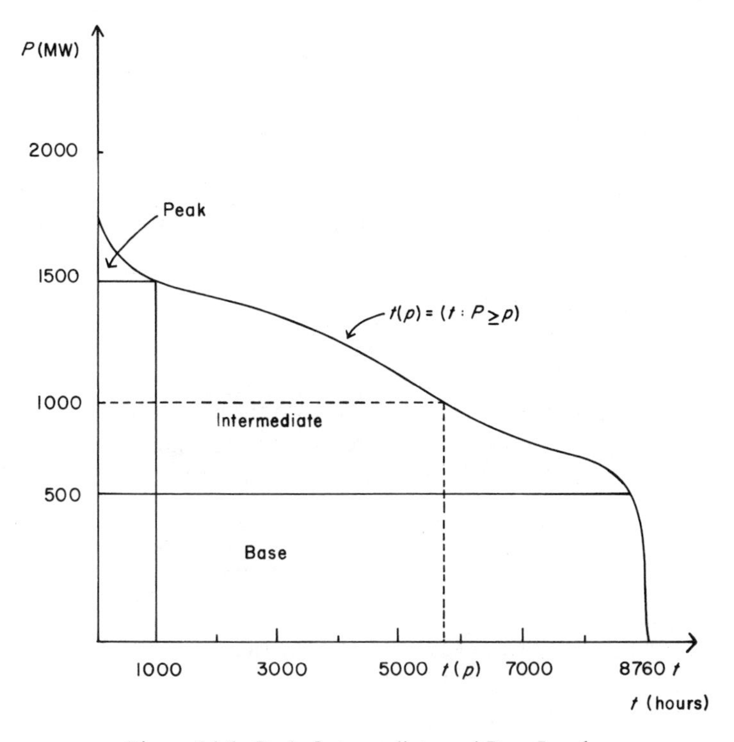

Figure 14.6. Peak, Intermediate and Base Loads.

variable) on the y axis and the demand variable (the independent variable) on the x axis, the LDC can actually be interpreted as the complementary density function of customers' demand. Denoting by Pr the probability operator, we therefore have

$$t(p_D) = \Pr[P_D \geq p_D].$$

Then the cumulative density function (CDF), $F(p_D)$, is given by

$$F(p_D) = 1 - t(p_D),$$

from which the load density function $f(p_D)$ is easily given by differentiation, i.e.,

$$f(p_D) = \frac{d}{dp_D}[1 - t(p_D)] = -t'(p_D),$$

where $t'(p_D)$ is the derivative of the LDC (with the role of the axes reversed) evaluated at point p_D. Figure 14.5 exhibits these curves.

As we will see, in the following, the LDC is used for many purposes. Two immediate applications are [1] as follows:

(a) estimating the energy delivered by the system in the given period—by definition this energy is equal to the area under the LDC, i.e.,

$$E = \int_0^{p_{D\max}} t(p_D)dp_D$$

where $p_{D\max}$ denote the peak demand for power;

(b) distinguishing between the so-called base, intermediate, and peaking capacity.
   - Base capacity is the capacity which is required continuously over the period.
   - Peaking capacity is the capacity which is required during a certain amount of hours in a year (say 1000 hours) or less.
   - Intermediate capacity is any capacity in between base and peaking capacity.

   By definition of the LDC, base capacity is at the base of the LDC, peaking capacity is at the "tail" of the curve, while intermediate capacity is any level in between. Figure 14.6 demonstrates this fact.

EXAMPLE. Consider the following daily peaks for an electric system in a given year:

8000 MW 5% of the time
6000 MW 65% of the time
4000 MW 100% of the time.

Since we only consider three levels of demand, the LDC for this system is given as a step functions as shown in Figure 14.7. By multiplying the time axis by the number of hours in a year (8760), the load data can be expressed in terms of the number of hours, rather than a fraction of time, that each load level is equaled or exceeded.

Given this LDC, the base load is clearly 4000 MW, as it is the level of

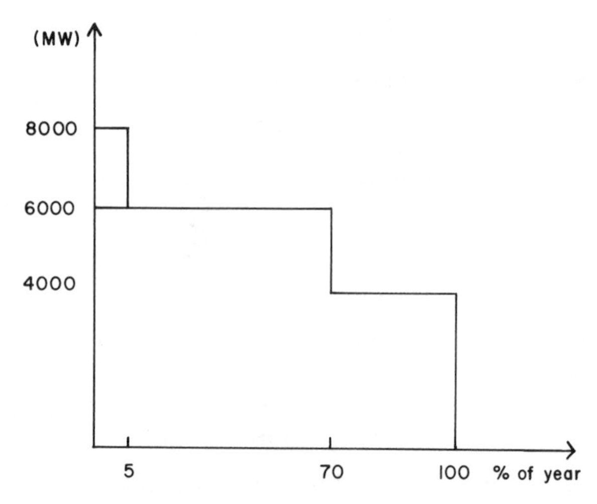

Figure 14.7. LDC for the Example.

demand required 8760 hours a year. Likewise, we can define the peak demand to be 8000 MW and the intermediate demand 6000 MW.

Finally, as mentioned, the energy required by the customers in this system is equal to the area under the LDC. Being a step function, this area is calculated easily as follows:

$$\begin{aligned} 5\% \text{ of the year} &= 438 \text{ hours} \\ 70\% \text{ of the year} &= 6132 \text{ hours} \\ 100\% \text{ of the year} &= 8760 \text{ hours.} \end{aligned}$$

$$\begin{aligned} \text{Peak energy} &= 438 * 2000 = 876{,}000 \text{ MWh} \\ \text{Intermediate energy} &= 6132 * 2000 = 12{,}264{,}000 \text{ MWh} \\ \text{Base energy} &= 8760 * 4000 = 35{,}040{,}000 \text{ MWh} \\ \text{Total} &\quad 48{,}180{,}000 \text{ MWh.} \end{aligned}$$

5. Reliability Analysis

5.0 Introduction

Like any other piece of equipment, generating units tend to be out of service from time to time, resulting in unpredicted forced outages. In addition, generating units have to undergo periodic preventive maintenance in order to keep them in proper running conditions. Both maintenance and forced outages might decrease the installed capacity to a point where the system will not be able to accomodate all the demand for power and a shortage results. The capacity of the system to meet the demand for power at any given point in time is defined as the reliability of the system. Due to the vital nature of electricity and the increasing dependence of society on a reliable and uninterrupted supply of power, reliability has become one of the most important design criteria of power systems. It is a most complex quantity, depending on many factors such as unit size, unit type, forced outage rate, shape of load, number of units in the system, etc.

5.1 Basic Reliability Measures

Reliability evaluation of power generating systems is usually related to either static or spinning requirements [2]. The former is concerned with long-term evaluation of reliability where no attention is given to the operating characteristics of the system at any specific point in time. The latter is concerned with short-run reliability (an hour, a day) for which the operating characteristics of the system at a specific point in time play a major role. Since we are interested in planning aspects, rather than in operation aspects of power systems, we will only be concerned with long-run measures of reliability. In the following, reliability will always mean long-run reliability; the words "long-run" will be omitted for the sake of simplicity.

Four primary measures have been devised to evaluate the reliability performance of power systems [2].

(a) The loss-of-load probability (LOLP): This is by all means the simplest and most common reliability criterion. Basically, it specifies the expected number of hours in a given period, usually a year, during which the system has experienced a shortage of one magnitude or another. The LOLP is usually expressed in terms of "days per year," "hours per day," etc. When the LOLP is divided by the number of hours in the period, we obtain a measure of the probability of shortage. Obviously, the higher the LOLP, the poorer is the reliability performance of the system.

(b) The loss-of-energy probability (LOEP): The main disadvantage of the LOLP is that it only indicates cumulative expected duration of interruptions; yet major differences may exist in the amount of unsatisfied energy even for systems with equal levels of LOLP. Hence a supplementary measure of reliability is the loss-of-energy probability (LOEP), a quantity indicating the expected amount of energy required by customers that was not supplied because of the inability of the system to meet the demand for power.

(c) Nonsupply distribution: A more sophisticated reliability measure, which provides information on the magnitude as well as the expected duration of shortages, is the complementary density function of the unsupplied demand, known in short as the nonsupply distribution. If $S(x)$ denotes this distribution, then any point on the abscissa of the curve will denote

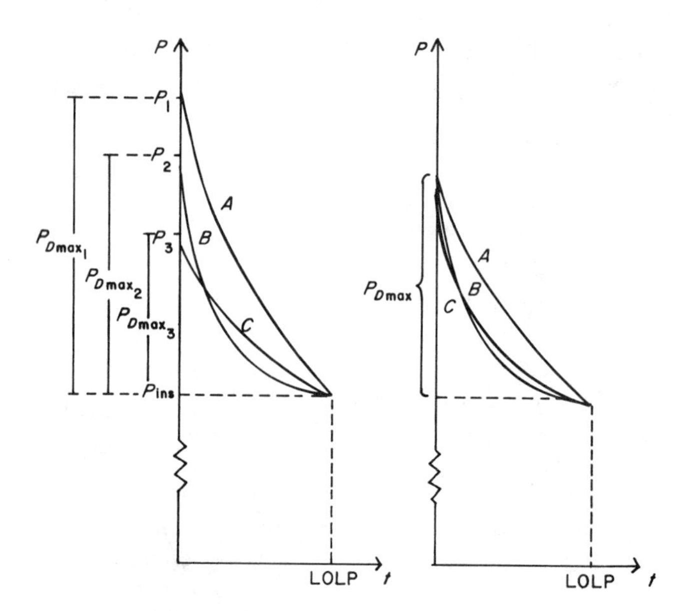

Figure 14.8. Nonsupply Distributions.

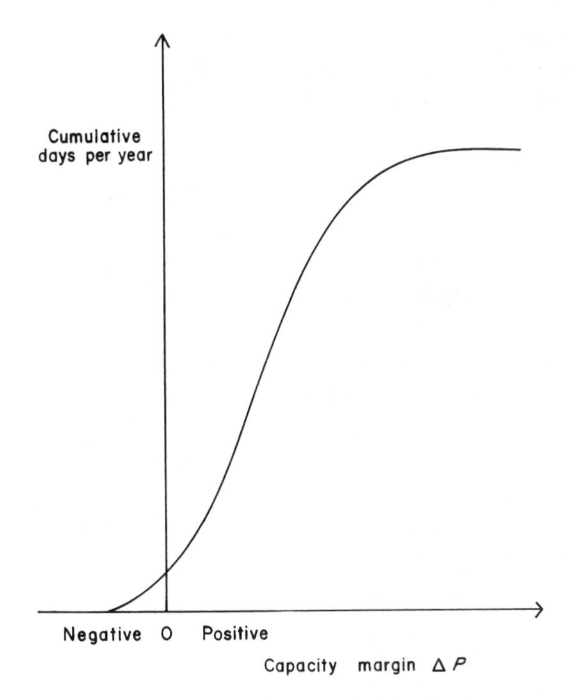

Figure 14.9. Capacity Margin Distribution.

the expected number of hours in the given period, during which the shortage for power equals or exceeds x MW. By definition, the base of this distribution is the LOLP. Figure 14.8 shows typical nonsupply distributions.

(d) Capacity margin distribution: An extended version of the nonsupply distribution is the capacity margin distribution. Each point on the abscissa of the capacity margin distribution will denote the expected number of hours in the given period, usually a year, during which the associated capacity margin, defined as the difference between the installed capacity and the capacity in maintenance or forced outages, is equaled or exceeded. By definition, the tail of the capacity margin distribution, which corresponds to negative capacity margin, is the nonsupply distribution. A typical capacity margin distribution is shown in Figure 14.9. The main application of the capacity margin distribution is for analysis of power pools.

5.2 Calculating the Reliability Measures [2]

Distribution of Outages. To find the LOLP, as well as the other reliability measures, we first have to find the capacity in outage distribution. The calculation is based on the unit reliability of each individual unit in the system. We define the unit reliability as the probability that the unit will be out of service because of a forced outage. For a continuously operated unit, such as base-loaded units, the unit reliability is measured by the forced outage rate (FOR) of the unit, defined as:

$$\text{FOR} = \frac{\text{duration of down-time}}{\text{service time} + \text{down-time}} \times 100.$$

However, except for base-loaded units, all units in the system are loaded to generation only part of the time to meet intermediate and peak demand (for further explanation see Section 6.2). An approximate way to calculate the unit reliability for these units is to multiply the unit's FOR by the fraction of time the unit is called upon to meet customer demand during the period.[1] Taking one year (8760 hours) as a basis, we have

$$q = \text{FOR}\,\frac{\text{duration of service time} + \text{duration of down-time}}{8760}.$$

We will refer to this probability as the outage probability of the unit and will denote it by q.

To derive the capacity in outage distribution we assume that

(1) each generating unit can be in either one of two possible states: working with probability $1 - q$ and thus capable of full power generation, or not working with probability q and incapable of delivering any energy, and
(2) the outage probability of any given unit is independent of its generation level and the demand for power.

Then the generation capacity which is out of service because of forced outages is multinomially distributed with the outage probabilities of the individual units as parameters. The distribution might take on as many as 2^N values, equal to the number of combinations of up and down units in an N-unit system.

The following example illustrates the calculation process.

EXAMPLE. Assume, for simplicity, a system with three units, with the following characteristics.

Unit	*Capacity* (MW)	*Outage Probability*
1	400	0.06
2	300	0.04
3	250	0.02

The capacity in outage distribution will assume $2^3 = 8$ values, the occurrence probability of each of which is described in Table 2.

[1] An exact derivation of the unit reliability measures even for base-loaded units, let alone for intermediate and peaking units, is based on the up- and down-time distribution of the unit and is beyond the scope of this module.

Table 2. Calculating Distribution of Outages

Case j	Unit 400 MW	300 MW	250 MW	Available Capacity (MW)	Probability $P(p_f)$	Capacity in outage P_F (MW)
0	up	up	up	950	$0.94 \cdot 0.96 \cdot 0.98 = 0.8844$	0
1	up	up	down	700	$0.94 \cdot 0.96 \cdot 0.02 = 0.0180$	250
2	up	down	up	650	$0.94 \cdot 0.04 \cdot 0.98 = 0.0368$	300
3	down	up	up	550	$0.06 \cdot 0.96 \cdot 0.98 = 0.0564$	400
4	up	down	down	400	$0.94 \cdot 0.04 \cdot 0.02 = 0.0008$	550
5	down	up	down	300	$0.06 \cdot 0.96 \cdot 0.02 = 0.0012$	650
6	down	down	up	250	$0.06 \cdot 0.04 \cdot 0.98 = 0.0023$	700
7	down	down	down	0	$0.06 \cdot 0.04 \cdot 0.02 = 0.00005$	950

The last two columns give the capacity in outage distribution. Case 0 refers to the case where all the units are up, while case 7 refers to the case where all the units are down. Obviously, the capacity in outage range in the interval $[0, p_{\text{INS}}]$, where p_{INS} is the installed capacity of the system.

Calculating the LOLP. Given the capacity in outage distribution, the LOLP is calculated using the formula

$$\text{LOLP} = \sum_{j=0}^{2^N-1} t(p_{\text{INS}} - p_{Fj})P(p_{Fj})$$

where

p_{Fj}	various capacities in outages, $j = 0, 1, \cdots, 2^N - 1$,
$P(p_{Fj})$	corresponding occurrence probabilities,
$t(x)$	LDC evaluated at point x,
N	number of units in the system,
p_{INS}	installed capacity of the system.

For any capacity in outage, P_{Fj}, the number of shortage hours is given by the LDC evaluated at the point of available capacity $(p_{\text{INS}} - p_{Fj})$, denoted by $t(p_{\text{INS}} - p_{Fj})$. Since each outage level occurs with a probability $P(p_{Fj})$, the expected contribution of this outage level to the LOLP is given by multiplying the duration of shortage by the corresponding occurrence probability. Then, to obtain the total LOLP for the system, we have to sum up these contributions for all possible outage levels.

As an example, we consider the LDC in Figure 14.10 and the sample system described above. We note that the demand variable is expressed as percentage of the peak demand for power. For calculation purposes, we assume the peak demand is 800 MW. The process is demonstrated in Table 3.

The LOLP is therefore 280.0 hours, implying that during 280 hours of that year we can expect that at least one customer will not get all the electricity

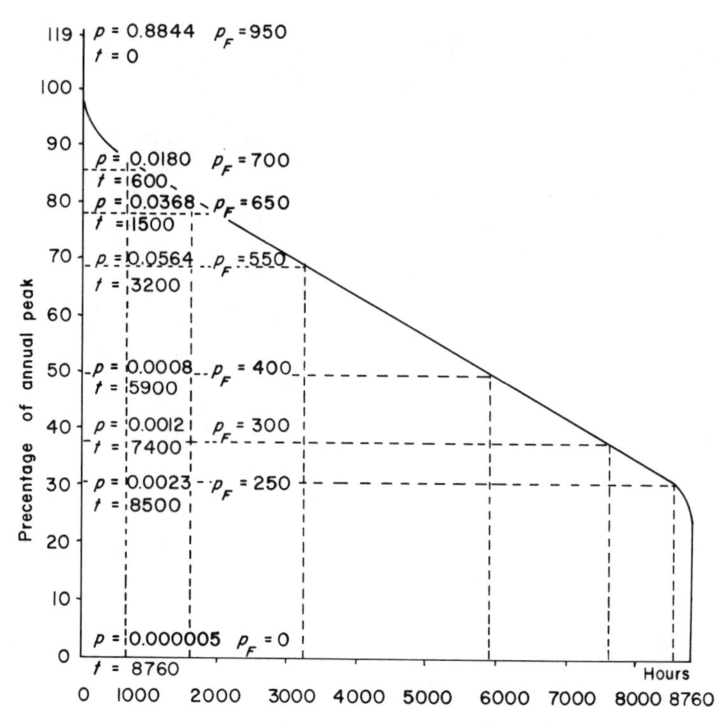

Figure 14.10. Calculating the LOLP.

Table 3. Calculating the LOLP

Case	Capacity in outage p_{Fj}	Available Capacity $p_{\text{INS}} - p_{Fj}$	% of Annual Peak	Probability $P(p_{Fj})$	Nonsupply Time (hours) $t(p_{\text{INS}} - p_{Fj})$	Nonsupply time × Probability $t(p_{\text{INS}} - p_{Fj})P(p_{Fj})$
0	0	950	119	0.8844	0	0
1	250	700	87	0.0180	600	10.8
2	300	650	81	0.0368	1500	55.2
3	400	550	69	0.0564	3200	180.4
4	550	400	50	0.0008	5900	4.7
5	650	300	38	0.0012	7400	8.9
6	700	250	31	0.0023	8500	19.6
7	950	0	0	0.00005	8760	0.4
Sum				1.0000		280.0

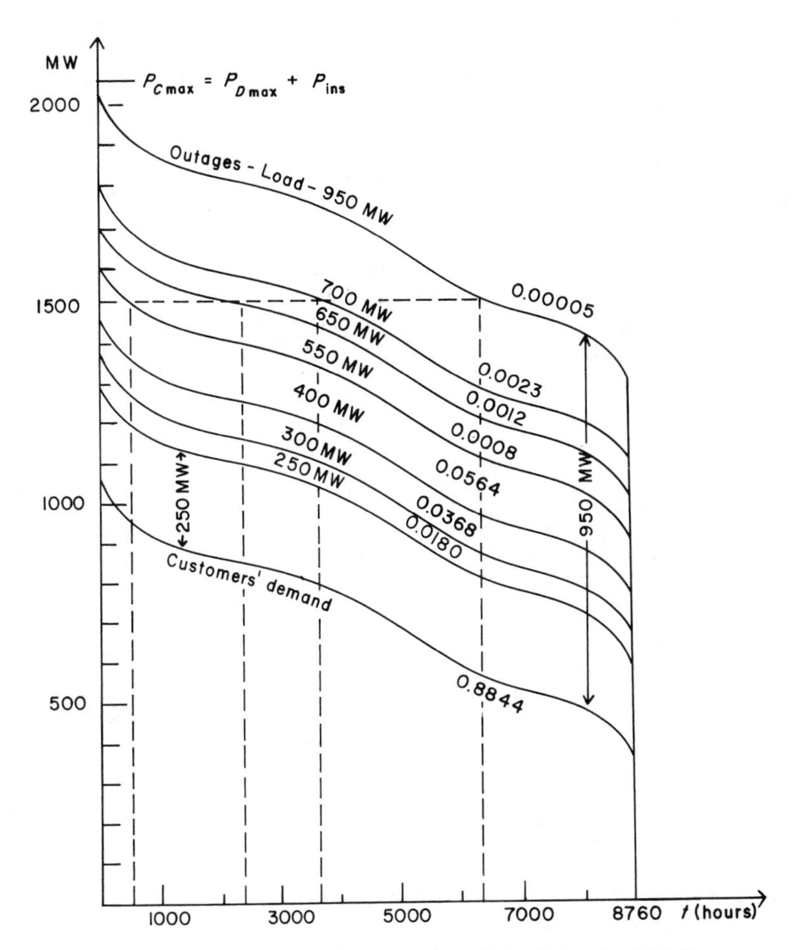

Figure 14.11. Graphical Approach to Calculating the LOLP.

he wants. The proportion of hours of shortage is expected to be 280/8760 = 0.032. This can be used as an estimate of the probability of a shortage at any given time.

Figure 14.11 demonstrates the calculation process. Forced outages are conceived of as an additional type of "demand" (in fact, this is the demand the power company makes on itself for installed capacity) and are therefore superimposed on the LDC, with each outage level raising the LDC in a parallel manner at a distance equal to the corresponding outage level. The occurrence probabilities associated with each outage level are the same as shown in Table 2 and are also listed in Figure 14.11. To calculate the LOLP from the graph, we draw a horizontal line through the installed capacity p_{INS}, which cuts the various curves at times $t(p_{\mathrm{INS}} - P_{Fj})$, $j = 0, 1, \cdots, 2^N - 1$. This line defines all events for which the sum of customers' demand and forced outage exceeds the installed capacity of the system, thus resulting in a shortage of power. The expected duration of shortage is therefore obtained as a summation of the duration of shortage in each event times its occurrence probability, as also suggested by the previous formula.

Calculating the Nonsupply Distribution. The LOLP provides the expected duration of time, or the expected fraction of time during which the shortage for power exceeds 0 MW. To find the nonsupply distribution, we have to find the expected duration of time, or the expected fraction of time, during which the shortage level $S(x)$ exceeds x MW, for various values of x in the range $[0, p_{\mathrm{INS}}]$. We do this by repeating the calculation process described above using the formula:

$$S(x) = \sum_{j=0}^{2^N-1} t(p_{\mathrm{INS}} + x - p_{Fj})P(p_{Fj}) \qquad x \varepsilon [0, p_{\mathrm{INS}}].$$

The capacity in outage probabilities remains as before. Only the values of $t(p_{\mathrm{INS}} - x - p_{Fj})$ change for the various values of x. In the context of Figure 14.11 we "slide" the installed capacity by an amount of x MW and repeat the calculation process for various levels of x. The number of times we repeat the calculation process will actually determine the accuracy of of the resulting nonsupply distribution.

The LOEP. Given the nonsupply distribution, the loss-of-energy probability (LOEP) is calculated very easily by taking the area under the curve.

Capacity Margin Distribution. The capacity margin distribution can be calculated using a similar procedure as described for the nonsupply distribution, except that we have to repeat the calculation for various excesses of the installed capacity over customers' demand.

5.3 Incorporating Maintenance Requirements

Maintenance requirements bear a substantial impact on power system reliability, since they increase the chances of a shortage situation, thus increasing the system's LOLP. It is therefore important to include maintenance requirements in calculating reliability measures. However, we note that while independent of customers' load, forced outage probabilities are strongly dependent upon the maintenance schedule, as each schedule leads to different combinations of up and down units, and hence to different failure probabilities. In order to account for this dependence, we need to partition the period involved to subperiods of constant maintenance and arrange the load in each subperiod into a load duration curve, with maintenance requirements considered as part of the demand for power. We then calculate the LOLP, or the other reliability measures, separately for each subperiod, and combine the results to yield the reliability measures for the whole period.

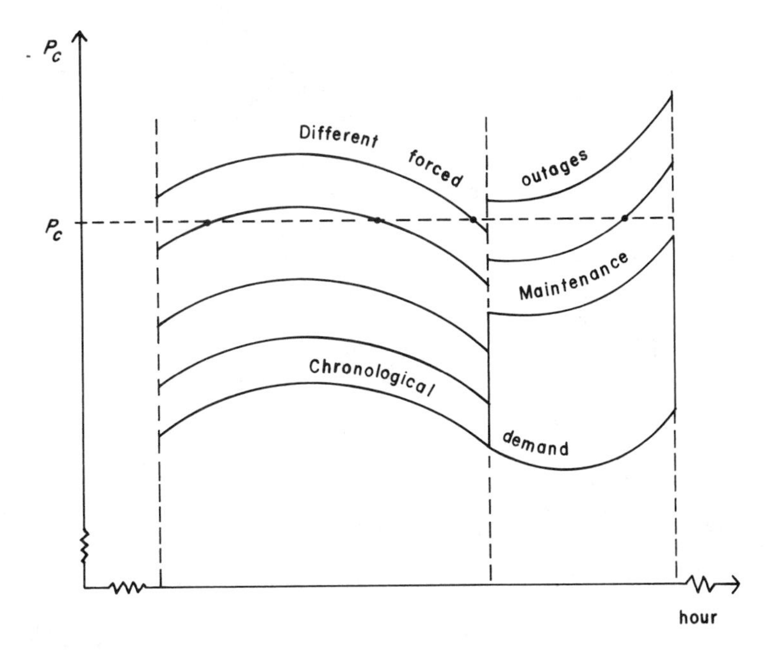

Figure 14.12. Partitioning into Subperiods of Constant Maintenance.

The paritioning into subperiods of constant maintenance is demonstrated in Figure 14.12 along with various outage probabilities.

Appendix A: The LDC with Random Component of Demand [4]

When estimating the LDC for future years, we should take cognizance of demand uncertainties. We do this by interpreting the demand values on the LDC as representing the expected value of demand, or rather the deterministic component of customers' demand. The random deviation of load is then distributed around this expected value in a random manner, which we assume to follow a normal distribution with the expectation given by the corresponding value on the LDC and a known standard deviation. Thus the customers' demand is actually made up of two components: the deterministic and random components. In mathematical terms,

$$P_C = P_D + P_R$$

where

- P_C summation variable (referred to as the combined demand),
- P_D deterministic component of customers' load as given by values on the LDC,
- P_R random component of customers' load.

Since P_R is a random variable, P_C is also a random variable. As in the case of the LDC, it might be more convenient to represent the combined load data using the complementary distribution function of the combined demand. We refer to the resulting curve as the combined load duration curve (CLDC). To find the CLDC we superimpose the random component of customers' load on the deterministic component of customers' load as demonstrated in Figure 14.13. The curve on the (t, P_D) plane, representing the collection of the expected customers' demand for all hours, is actually the LDC. We then cut the surface by a vertical plane going through p_C. The curve which is formed by the intersection of this plane with the density functions of the random component of demand, defines all combinations of random and deterministic components which satisfy:

$$p_C = p_D + p_R.$$

For each level of p_D, a random component p_R therefore exists, with occurrence probability $P(p_R)$, that brings the combined demand to a level p_C. For a given p_D the number of hours in the period (or the fraction of time, if we normalize the time axis) the combined demand equals or exceeds p_C is given by the corresponding $t(p_D)$. Since p_C might result from various combinations of p_D and p_R, the *expected* number of hours (or the expected fraction of time) the combined demand p_C is equaled or exceeded is given by

$$\bar{t}(p_C) = \int_{p_D=0}^{p_{D\max}} t(p_D)P(p_R = p_C - p_D)\,dp_D$$

where

$\bar{t}(p_C)$	CLDC evaluated at point p_C, interpreted as the expected number of hours in the given period during which the combined demand equals or exceeds p_C,
$t(p_D)$	LDC evaluated at p_D,
$P(p_R = p_C - p_D)$	probability of the random deviation of demand that brings the combined demand to a level p_C.

This process is sometimes referred to as *convolution*.

By repeating the above calculation for various levels of p_C, we obtain the LDC including the random component of demand, i.e., the CLDC. By normalizing the time axis, as explained above, the CLDC can actually be interpreted as the complementary density function of the combined demand. This curve will be used as the vehicle to introduce the random deviations of customers' load into studying the performance of the system.

For example, if we want to account for demand uncertainties in calculating reliability measures, we simply use the CLDC instead of the LDC as the basis for calculations. All the formulas defined above remain the same with $\bar{t}(p_C)$ replacing $t(p_D)$.

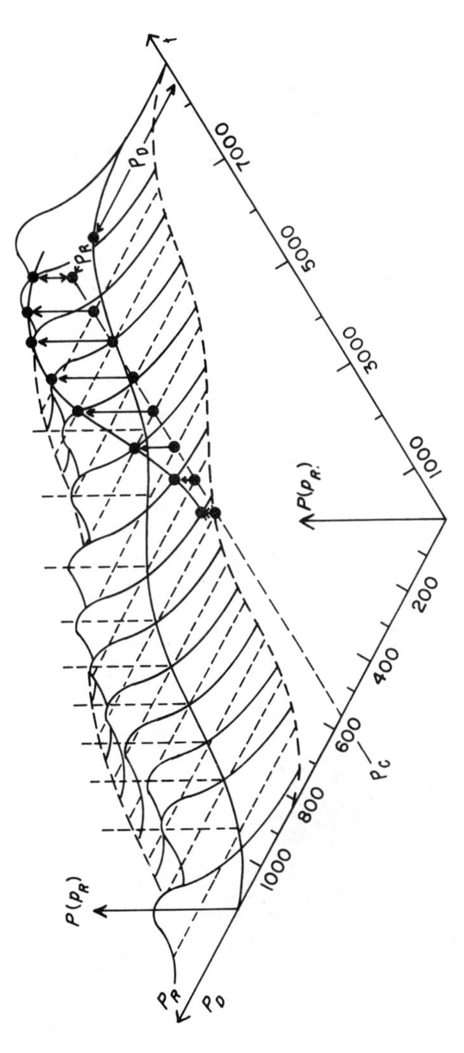

Figure 14.13. Superposition of Random Component of Demand on the Deterministic Component of Demand.

Appendix B: An Extension of the CLDC

We have demonstrated procedures for calculating reliability measures while accounting for random and deterministic components of demand, forced outages, and maintenance requirements. The process can be consolidated by extending the CLDC presented in Appendix A to include all four components.

The combined demand will be extended to include the deterministic component of customers' load, the random component of customers' load, "demand" for outages, and "demand" for maintenance. For better comprehension, the latter two demands can be conceived as the self demand of the power company. The sum of all components, or the combined demand, clearly specifies what portion of the installed capacity is occupied in satisfying customers' demand or engaged in maintenance and forced outages. The CLDC is then defined as the complementary density function of the combined demand, i.e., it specifies the fraction of time that each combined demand (the total requirement for installed capacity) will be equaled to or exceeded.

Letting

P_C combined demand for power,
P_D deterministic component of customers' load,
P_R random component of customers' load,

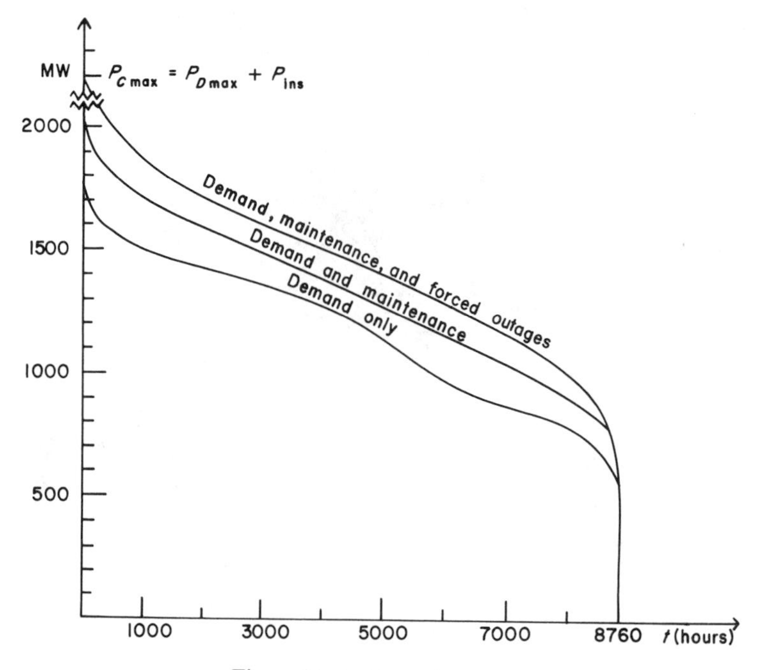

Figure 14.14. Typical CLDC.

P_M requirements for maintenance,
P_F requirements for forced outages,
$\bar{t}(p_C)$ CLDC evaluated at point p_C,

then by definition

$$\bar{t}(p_C) = P[P_C \geq p_C]$$

with P denoting the probability operator, and where P_C, the combined demand, is given by

$$P_C = P_D + P_R + P_M + P_F.$$

By multiplying $\bar{t}(p_C)$ by the number of hours in the period (8760 for a year), the CLDC can be expressed in terms of the expected number of hours in the given period that each combined load is equaled or exceeded. A typical CLDC is shown in Figure 14.14.

Obviously, the CLDC is obtained by superimposing the four components of demand one on top of the other. Graphically, this can be described by adding one component at a time. We first partition the period involved,

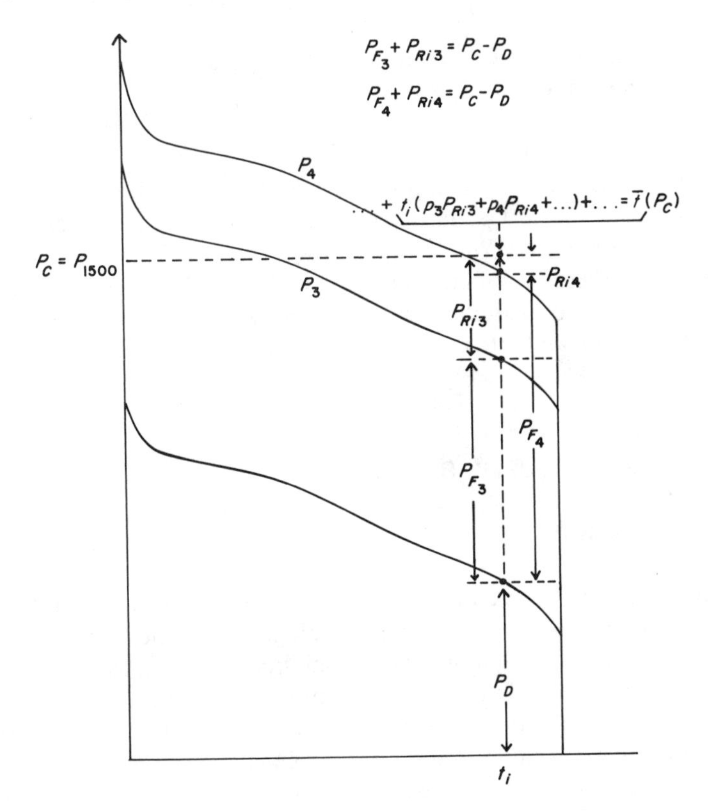

Figure 14.15. Effect of the Random Component of Customers Demand on the Combined LDC.

say a year, to subperiods of constant maintenance to take care of the dependence of outage probabilities on the maintenance schedule, and arrange the load in each subperiod, with maintenance requirements considered as part of the load, into a LDC. We then superimpose the random deviation of customers' load on the LDC as suggested by Figure 14.13. Finally, we superimpose forced outages on the resulting curve. Figure 14.11 actually describes the convolution process for this case. The combined effect of the random component of demand and forced outages is described in Figure 14.15.

Having calculated the CLDC, all reliability measures can actually be calculated as a by-product. By definition of the combined demand, each excess of the installed capacity over the combined demand results in a shortage of energy. We thus have

$$\text{LOLP} = \bar{t}(p_{\text{INS}}).$$

By the same logic, the nonsupply distribution is simply the "tail" of the CLDC above the installed capacity,

$$S(x) = \bar{t}(x + p_{\text{INS}}), \qquad 0 \leq x \leq p_{\text{INS}}.$$

Given $S(x)$, the LOEP is then derived as the area under the curve. By similar argumentation, we also have

$$\tilde{t}(x) = \bar{t}(p_{\text{INS}} - x)$$

where $\tilde{t}(x)$ is the capacity margin distribution evaluated at point x, interpreted as the expected fraction of time during which the capacity margin equals to or exceeds x MW.

The CLDC therefore offers a consolidated and efficient means to calculate the reliability measures of a power system while taking both uncertainty in supply and demand into consideration.

6. Operating Cost Calculations

6.0 Introduction

The prediction of the operating cost of a power utility is one of the most important aspects of power system planning. In capacity expansion decisions, cost data are used, along with reliability measures, as a major criterion for comparing various alternative investment policies. For operating purposes, cost calculations are needed for cash-flow analysis, fuel budgeting, etc. Cost estimations are also important elements in setting up the price of electricity, as well as in various other applications. Some basic models applied to operating cost predictions will be described in this section.

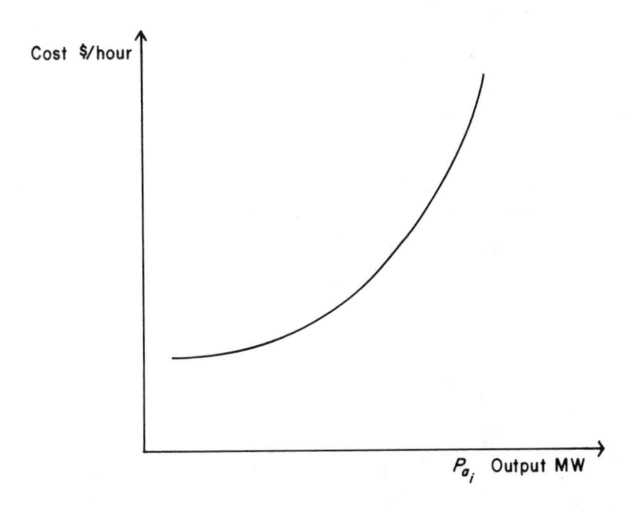

Figure 14.16. A Typical Cost Curve for a Generating Unit.

6.1 Cost Function of Generating Units

An important input to operating cost calculations is the cost function of the individual generating units in the system. The major component of the production expenses of any given unit, ignoring such costs as startup or shutdown costs, is the fuel cost. It is well-known that fuel consumption is a nonlinear function of the unit's output; in particular, it has been empirically established that the cost function of generating units in a power system, with the exception of hydroelectric and pumped storage units, is usually an increasing quadratic function of the form [3]

$$C_i = \alpha_i + \beta_i P_{a_i} + \gamma_i P_{ai}^2$$

where

C_i operating cost ($/hour) of unit i,
P_{a_i} output (MW) of unit i,
$\alpha_i, \beta_i, \gamma_i$ empirical parameters.

The curve is, of course, valid only for values of P_{a_i} that are less than the installed capacity of the unit. A typical cost curve is exhibted in Figure 14.16.

In order to derive the cost function for any unit, it is necessary to measure fuel consumption at various output levels and then fit a quadratic curve to these measurements using, say, the least square method. The procedure is demonstrated in an appendix.

Most fundamental to cost calculation is the marginal cost associated with each output level, defined as the incremental fuel cost incurred when the output of the generator is increased by 1 unit (1 KW). Given the cost function C_i, the marginal cost is given by differentiation, i.e.,

$$C_i = \frac{d}{dp_{a_i}}(\alpha_i + \beta_i P_{a_i} + \gamma_i P_{a_i}^2) = \beta_i + 2\gamma_i P_{a_i}.$$

That is, the marginal cost function is a linear function of the output P_{a_i}.

In particular, since α_i, β_i, and γ_i are nonnegative, the marginal cost for generator i is a nondecreasing function, implying that the marginal cost increases as we keep increasing the generator's output. A typical marginal cost function for a generating unit is described in Figure 14.17. We note that the curve is increasing in a linear pattern up to the installed capacity of the unit $P_{a_i \max}$, at which point it becomes infinite.

6.2 Loading Procedures

The loading procedure is the manner by which the various units in a power system are scheduled to operate to meet the demand for power. As such, the loading procedure determines the number, type, and production level of the units which are loaded to generation at any given point in time. It is therefore an important part of the cost calculation process. Two basic loading procedures are usually used for this purpose.

The Incremental Loading Procedure. This is the loading scheme which is actually used by power utilities to dispatch power. According to this procedure, units are loaded to generation at the point where their marginal cost are equal to one another (with the exception of units which are working at either their upper or lower limit). This common marginal cost then defines the marginal cost for the system. The procedure is demonstrated in Figure 14.18 for a three-unit system, with marginal cost function C_1', C_2', and C_3'. The marginal cost function for the system is obtained by summing up the abscissa of the individual units. The resulting curve is also demonstrated in Figure 14.18.

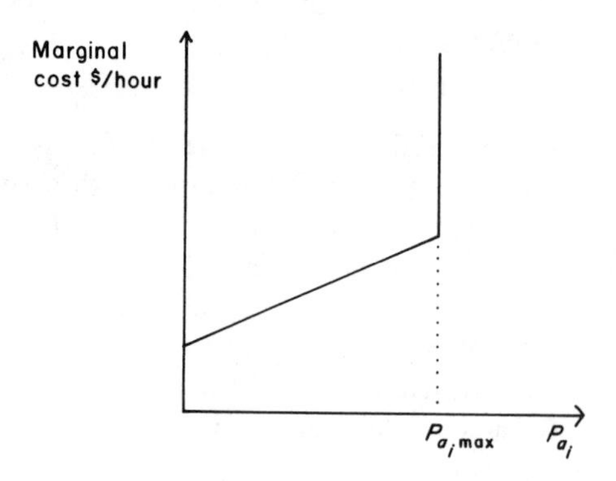

Figure 14.17. Marginal Cost Function of Generating Units.

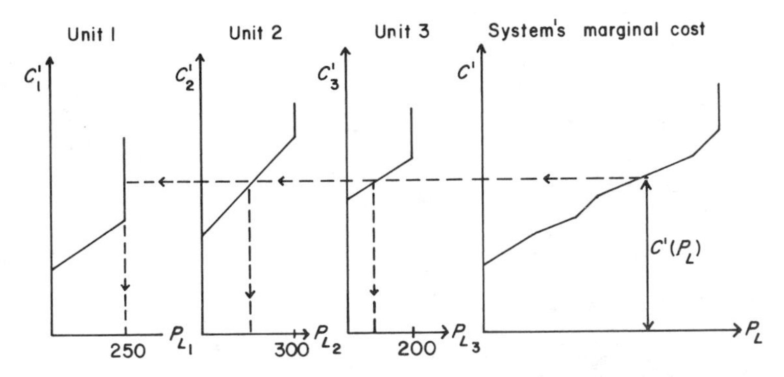

Figure 14.18. The Incremental Loading Procedure.

Given a demand level P_L, the corresponding generation levels at each generator P_{L_i}, $i = 1, 2, 3$, are determined, according to the incremental loading procedure, at the point where the marginal costs of the various generators are equal. To derive these levels graphically, we draw a horizontal line through $C'(p_L)$ (the marginal cost for the system) which cuts the ordinates of the individual cost functions at the points shown in the graph. The generation level of each unit is then read directly off the abscissa. We can prove [3] that the incremental loading procedure yields the minimu-cost procedure to dispatch power. It is therefore also referred to as the economic dispatch Procedure.

The Merit Order Loading. Employing the incremental loading procedure for cost calculation is very difficult, except, perhaps, in a simulation model. However, assuming (a) the operating cost for a generating unit is constant regardless of the unit's output, and (b) no setup and shutdown costs exist to bring the unit on or off line, then the least cost procedure for operating the power system is to load the units to production in increasing order of their variable (mainly fuel) cost. This loading procedure, widely used in power system analysis, is referred to as the merit order loading [1].

The merit order is described in Figure 14.19. We start by ordering the generating units in increasing order of their average production cost (the average is calculated over all possible outputs), which serves as an indicator to the unit's merit order, denoting the most efficient unit (the unit with the least average production cost) as unit 1, the second most efficient unit as unit 2, etc. Under merit order loading, we first load unit 1 to operation at its rated capacity, then unit 2, 3, and so on until demand is satisfied. This procedure is certainly more convenient for cost calculation than the incremental loading procedure. In the following we demonstrate two procedures, deterministic and probabilistic for calculating the production costs of a power system using the merit order loading.

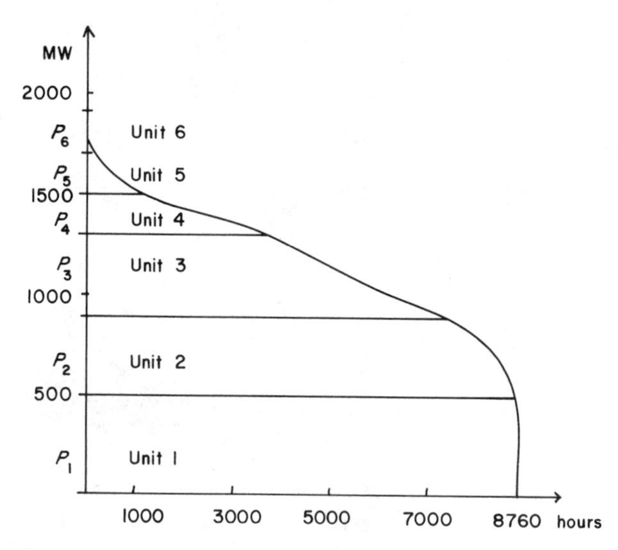

Figure 14.19. The Merit Order Loading.

6.3 The Effective Capacity Approach

The main problem in applying the merit order loading for cost calculation is how to account for maintenance requirements and forced outages. In the effective capacity approach, this is done by reducing the capacities of the units by a few percent to account for the fact that generating units are usually not available 100% of their time. Denoting by a_j the average percentage of time unit j is available for production (i.e., it is not out of service because of maintenance and forced outages) and by p_j the capacity of the unit, then the "effective" capacity, p_j', is given by

$$p_j' = a_j p_j.$$

Units are then loaded to generation at their effective capacities, under the merit order procedure. The area under the LDC formed between the lines of "effective" capacities, represents an approximation to the energy delivered by each unit in the given period. The operating costs are then obtained by multiplying the energy by the corresponding average production cost for the unit and summing up the results for all units. The procedure is demonstrated in Figure 14.20.

While very simple to implement, the effective capacity approach provides only an approximate measure of the production expenses of a power system, since it basically assumes that forced outages and maintenance requirements are evenly distributed along the year, an assumption that does not hold true in reality. A more sophisticated approach that accounts for forced outages and maintenance in a more realistic manner is the following.

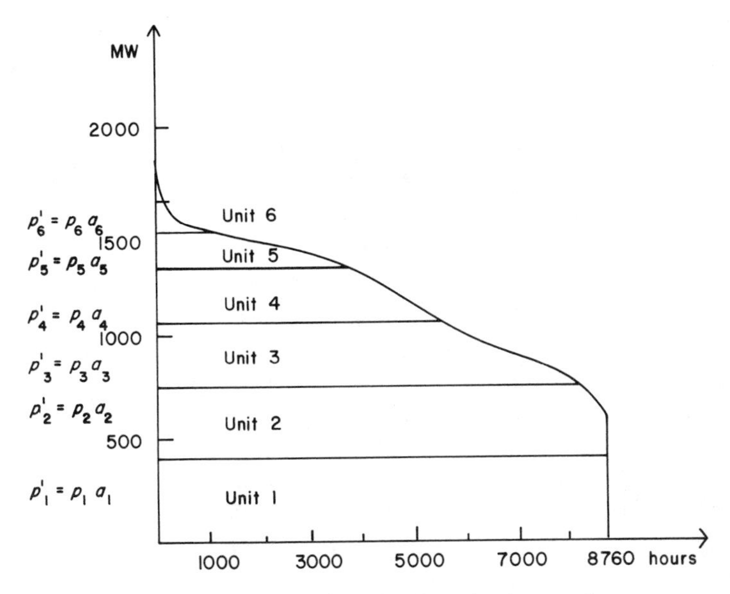

Figure 14.20. The Effective Capacity Approach.

6.4 The Probabilistic Approach to Calculate the Production Cost [5]

In this approach, forced outages are probabilistically incorporated into the cost calculation. Figure 14.21 describes the LDC of a system along with the merit order loading, with unit 1 being the most efficient unit, unit 2 the next most efficient unit, and so on. Denoting by

- $t(P_L)$ LDC of the system,
- E_i energy delivered by unit i,
- p_i capacity of unit i,
- q_i outage probability of unit i,

then, ignoring maintenance requirements, the expected energy delivered by unit 1 is given by:

$$E_1 = (1 - q_1) \int_0^{P_1} t(p_L)\, dp_L.$$

In calculating the energy delivered by the second unit in the merit order, two components of costs are to be considered. When unit 1 is available, with a probability of $(1 - q_1)$, unit 2 will be loaded to production between the loads p_1 and $p_1 + p_2$. However, when unit 1 is out of service because of forced outages, with a probability q_1, unit 2 will occupy the first position in the merit order and will then be loaded to generation between the loads 0 and p_2, as described in Figure 14.21. The expected energy delivered by unit 2 is

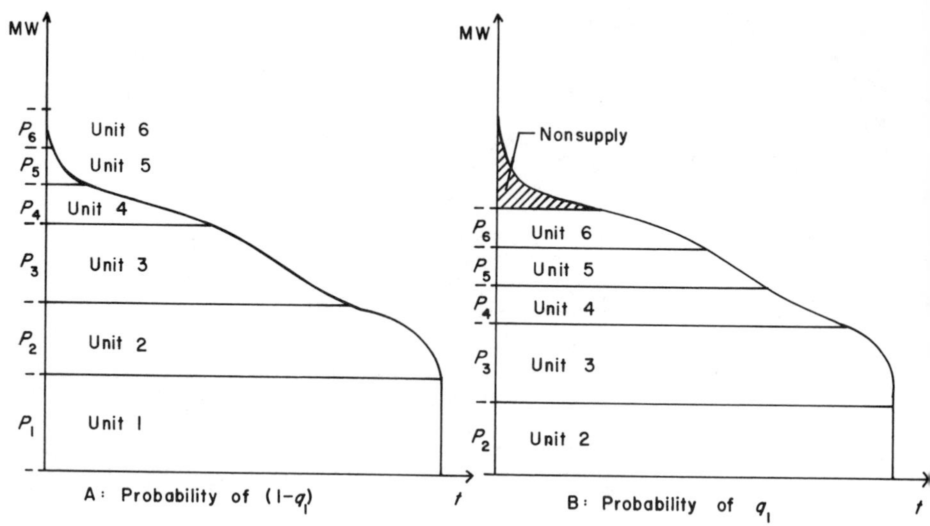

Figure 14.21. The Probabilistic Approach to Calculate Operating Expenses under the Merit Order.

therefore obtained as a weighted average, i.e.,

$$E_2 = (1 - q_2)\left[(1 - q_1)\int_{p_1}^{p_1+p_2} t(p_L)\,dp_L + q_1\int_0^{p_2} t(p_L)\,dp_L\right].$$

Substituting

$$\int_0^{p_2} t(p_L)\,dp_L = \int_{p_1}^{p_1+p_2} t(p_L - p_1)\,dp_L,$$

we have

$$E_2 = (1 - q_2)\left[(1 - q_1)\int_{p_1}^{p_1+p_2} t(p_L)\,dp_L + q_1\int_{p_1}^{p_1+p_2} t(p_L - p_1)\,dp_L\right]$$
$$= (1 - q_2)\left\{\int_{p_1}^{p_1+p_2} [(1 - q_1)t(p_L) + q_1 t(p_L - p_1)]\,dp_L\right\}.$$

Denoting the expression under the integral sign by $\bar{t}_1(p_L)$, we have

$$E_2 = (1 - q_2)\int_{p_1}^{p_1+p_2} \bar{t}_1(p_L)\,dp_L.$$

Continuing in this fashion, it can be shown that, in general,

$$E_n = (1 - q_n)\int_{\sum_{i=1}^{n-1} p_i}^{\sum_{i=1}^{n} p_i} \bar{t}_{n-1}(p_L)\,dp_L, \qquad n = 1, 2, \cdots, N,$$

where $\bar{t}_n(p_L)$ are given in terms of recursive equations by

$$\bar{t}_n(p_L) = (1 - q_n)\bar{t}_{n-1}(p_L) + q_n\bar{t}_{n-1}(p_L - p_n), \qquad n = 1, 2, \cdots, N.$$

Given the expected energy delivered by each unit, the total operating cost of the system is derived by multiplying each term by the corresponding average production cost and adding up the products for all units.

If maintenance requirements are to be taken into consideration, the period involved should be partitioned into subperiods of constant maintenance. The operating costs are then calculated separately for each subperiod and combined to yield the total operating cost for the whole period.

Random deviations of customers' load can be accounted for by starting off the calculation process (the expression for E_1) by the convolution of the deterministic and random component of customers' load, as suggested by Figure 14.13.

Appendix: Estimating Cost Curves

An essential aspect of cost calculation is to estimate the cost function for any generating unit. As mentioned above, this is done by taking measurements of fuel consumption at various output levels, and fitting a curve to these measurements. Quadratic curves are usually used to describe the cost function. In this appendix we will demonstrate the least square error method of fitting a function to a set of measurements.

Straight Line

Assuming the following measurements,

P_{a_i}	C_{i_i}
0	0.115
0.35	0.188
0.7	0.266,

we want to fit a straight line of the form

$$C_i = a_1 + b_1 P_{a_i} \tag{1}$$

to these measurements so that the means square error may be minimized.

Evaluating Eq. (1) at the three points generates the following three equations:

$$a_1 \qquad\qquad = 0.115 \tag{2a}$$

$$a_1 + 0.35b_1 = 0.188 \tag{2b}$$

$$a_1 + 0.70b_1 = 0.266. \tag{2c}$$

Multiplying each of equations (2) by its coefficient of a_i (all three are equal in this example) and adding them up yields

$$3a_1 + 1.05b_i = 0.569. \tag{3}$$

This is the so-called first normal equation of the least square error procedure. The second normal equation is obtained by multiplying each of equations (2) by its coefficient of b_1 and adding. In this case,

$$\begin{aligned} 0 \qquad\qquad\qquad &= 0 \\ 0.35a_1 + 0.1125b_1 &= 0.0658 \\ 0.70a_1 + 0.4900b_1 &= 0.1862 \\ \hline 1.05a_1 + 0.6025b_1 &= 0.252. \end{aligned} \tag{4}$$

Equation (4) is the second normal equation. Solving Eq. (3) and (4) for a_1 and b_1 yields

$$a_1 = 0.111$$

$$b_1 = 0.225,$$

and the straight line which fits the above data is, therefore,

$$C_i = 0.111 + 0.225P_{a_i}.$$

A Quadratic Function

Assume the following six data points

P_{a_i}	C_{i_i}
0	0.471
0.5	0.583
1.0	0.696
1.5	0.964
2.0	1.437
2.5	2.246

for which we want to fit a quadratic function of the form

$$C_i = \alpha_1 + \beta_1 P_{a_1} + \gamma_1 P_{a_i}^2. \tag{5}$$

Substituting these points into (5) we get:

$$\alpha = 0.471 \tag{6a}$$

$$\alpha_1 + 0.5\beta_1 + 0.25\gamma_1 = 0.583 \tag{6b}$$

$$\alpha_1 + \beta_1 + \gamma_1 = 0.696 \tag{6c}$$

$$\alpha_1 + 1.5\beta_1 + 2.25\gamma_1 = 0.964 \tag{6d}$$

$$\alpha_1 + 2.0\beta_1 + 4.00\gamma_1 = 1.437 \tag{6e}$$

$$\alpha_1 + 2.65\beta_1 + 7.022\gamma_1 = 2.246. \tag{6f}$$

The first normal equation is obtained by multiplying each of equations (6) by its coefficient of α_1 (one in this case) and adding up the results:

$$6\alpha_1 + 7.65\beta_1 + 14.522\gamma_1 = 6.397. \tag{7}$$

The second normal equation is obtained by multiplying each of equations (6) by its coefficient of β_1 and adding up the results:

$$\begin{array}{rrrcl} 0.50\alpha_1 + & 0.2500\beta_1 + & 0.1250\gamma_1 & = & 0.2915 \\ \alpha_1 + & \beta_1 + & \gamma_1 & = & 0.696 \\ 1.50\alpha_1 + & 2.2500\beta_1 + & 3.3750\gamma_1 & = & 1.446 \\ 2.00\alpha_1 + & 4.0000\beta_1 + & 8.0000\gamma_1 & = & 2.874 \\ 2.65\alpha_1 + & 7.0225\beta_1 + & 18.6096\gamma_1 & = & 5.9519 \\ \hline 7.65\alpha_1 + & 14.5222\beta_1 + & 31.1096\gamma_1 & = & 11.2594. \end{array} \tag{8}$$

The third normal equation is obtained in a similar way by multiplying each of equations (6) by its coefficient of γ_1 and adding up the results, yielding

$$14.5225\alpha_1 + 31.1096\beta_1 + 71.4405\gamma_1 = 24.5312. \tag{9}$$

Solving the three normal equations (7), (8) and (9) for $\alpha_1, \beta_1, \gamma_1$, yields

$$\alpha_1 = 0.5026$$

$$\beta_1 = -0.0917$$

$$\gamma_1 = 0.2811.$$

and the quadratic equation is

$$C_i = 0.5026 - 0.0917P_{a_i} + 0.2811P_{a_i}^2.$$

Obviously, the more data we have, the more accurate will be the resulting curve.

Note: The least square method, demonstrated above, is only one of the possible approaches to fit a curve to a set of data. Other possibilities are, for example, to minimize the sum of absolute deviations, or to minimize the largest absolute deviation between the data and the fitted curve. The fitting method to use depends pretty much on the application involved and the nature of the data.

7. An Application

7.1 Reliability Analysis

As an application of reliability analysis, we use the load data of Figure 14.7, with the following two alternatives to meet the demand for power:

(a) 20 units, 500 MW each
(b) 40 units, 250 MW each.

Table 4. $(0.9)^n$

n	$(0.9)^n$	n	$(0.9)^n$
1	0.9	21	0.1094
2	0.81	22	0.0984
3	0.729	23	0.0886
4	0.6561	24	0.0797
5	0.5905	25	0.0718
6	0.5314	26	0.0646
7	0.4783	27	0.0581
8	0.4305	28	0.0523
9	0.3874	29	0.0471
10	0.3487	30	0.0424
11	0.3138	31	0.0381
12	0.2824	32	0.0343
13	0.2542	33	0.0309
14	0.2287	34	0.0278
15	0.2059	35	0.0250
16	0.1853	36	0.0225
17	0.1667	37	0.0202
18	0.1501	38	0.0182
19	0.1351	39	0.0164
20	0.1215	40	0.0148

All units are assumed to have an outage probability of 10%. We want to find the reliability level, as measured by the LOLP, delivered by each of these alternatives.

We note that the installed capacity in both cases is 10,000 MW, which is larger than the peak demand (8000 MW). The excess of installed capacity over the peak demand is called the *reserve* of the system. The reserve capacity is actually a safety margin necessary to provide for the day-to-day variations in the operating conditions of the installed capacity due to breakdowns, maintenance, and deviations of demand from estimates. The reserve is usually measured as percentage of the installed capacity. For example, in the present problem the reserve is:

$$\frac{(10{,}000 - 8{,}000)}{10{,}000} = 25\%$$

of installed capacity.

To find the reliability of each alternative, we use the binomial expansion to develop probabilities of generation availability:

$$q = \text{outage probability} \qquad = 0.10$$

$$p = 1 - q = \text{availability probability} = 0.90$$

$$(p + q)^n = p^n + \frac{n}{1!}p^{n-1}q + \frac{n(n-1)}{2!}p^{n-2}q^2 + \cdots + \frac{n(n-1)\cdots(n-r+2)}{(r-1)!}p^{n-r+1}q^{r-1} + \cdots + p^n.$$

For example, 40 units of 250 MW = 10.000 MW capacity:

$$(p + q)^n = p^{40} + 40p^{39}q + 780p^{38}q^2 + 9880p^{37}q^3 + 91390p^{36}q^4 + 658008p^{35}q^5 + \cdots.$$

Binomial coefficients are usually available in tables. The table of $(0.9)^n$ is Table 4.

In the following we will only consider the probability that the available capacity will be less than 8000 MW, and regard this as the LOLP of the system. Probabilities of loss of load of lower load levels will be ignored, assuming they are too small to affect the answer.

The key to the LOLP calculation is to determine the probability that the available capacity (installed capacity minus capacity in outages) will be less than 8000 MW. Using the binomial expansion we have presented, this probability can be calculated as follows.

(a) 20 500 MW units:

Available Capacity	r	$p^{n-r+1}q^{r-1}$	Coefficient	Probability	Cumulative Probability
10,000	1	$(0.1215)(10^0)$	1	0.1215	0.1215
9,500	2	$(0.1351)(10^{-1})$	20	0.2702	0.3917
9,000	3	$(0.1501)(10^{-2})$	190	0.2852	0.6769
8,500	4	$(0.1667)(10^{-3})$	1140	0.1900	0.8669
8,000	5	$(0.1853)(10^{-4})$	4845	0.0898	0.9567

The probability of installed capacity < 8000 MW:

$$1.000 - 0.9567 = 0.0433.$$

8000 MW load occurs 0.05 of the time. Thus the probability of shortage at peak:

$$(0.05)(0.0433) = 0.00216$$

or, in terms of days per year,

$$\text{LOLP} \sim (0.00216)(365) = 0.7884 \text{ days/year}.$$

Given our assumption that the majority of shortages occur only at peak demand, this is a close approximation to the LOLP.

(b) 40 250 MW units:

Available Capacity	r	$p^{n-r+1}q^{r-1}$	Coefficient	Probability	Cumulative Probability
10,000	1	$(0.0148)(10^{0})$	1	0.0148	0.0148
9,750	2	$(0.0164)(10^{-1})$	40	0.0656	0.0904
9,500	3	$(0.0182)(10^{-2})$	780	0.1419	0.2223
9,250	4	$(0.0202)(10^{-3})$	9,880	0.1996	0.4219
9,000	5	$(0.0225)(10^{-4})$	91,390	0.2056	0.6275
8,750	6	$(0.0250)(10^{-5})$	658,008	0.1645	0.7920
8,500	7	$(0.0278)(10^{-6})$	3,838,380	0.1067	0.8987
8,250	8	$(0.0309)(10^{-7})$	18,643,560	0.0576	0.9563
8,000	9	$(0.0343)(10^{-8})$	76,904,685	0.0264	0.9827

The probability of peak demand exceeding the available installed capacity = 1.0 − 0.9827 = 0.0173. Probability of shortage at peak is

$$(0.05)(0.0173) = 0.000865$$

$$\therefore \text{LOLP} \sim (0.000865)(365) = 0.316 \text{ days/year.}$$

Clearly, the LOLP is greater in case (a) than (b), an obvious result, since installing a larger number of smaller units improves the reliability of the system.

If further accuracy is desired, it is necessary to calculate shortage probabilities for all levels of loads, and then sum up the results to yield the LOLP.

7.2 Operating Cost Calculations

In this section we calculate the operating expenses for the two alternatives using the effective capacity approach. Assuming an availability of 0.90 for each unit, there will be enough capacity in both alternatives to satisfy the demand for power. (Note: This is in contrast to the previous conclusion that there is some probability of loss of load! However, this is not a contradiction since we are using two entirely different methods to account for the reliability of the system. Also, this might indicate how rough the prediction of reliability in the effective capacity approach is.)

Since the units in both alternatives are of the same type and size, they should have the same average production cost. Using the merit order, it is thus immaterial which unit is loaded first and which one next. Hence the calculation of the operating cost in this case actually amounts to estimating the energy under the LDC. This energy was calculated in section 4, however, and is equal to 48,180,000 MWh.

Given the average production cost for each alternative, the operating costs can be figured by multiplying the average cost by the calculated energy output. Because of economies of scale, the average production cost for a 500 MW unit is expected to be smaller than the average production cost for a 250 MW unit, implying that the production expenses for alternative (a) (20 500-MW units) are smaller than the production expenses for alternative (b) (40 250-MW units). Note that the reverse is true for reliability, i.e., we obtained better reliability for case (b), than for case (a), because the more units we have in a system, the better is the reliability. A trade-off analysis between cost and reliability is usually advocated in order to determine which alternative, either (a) or (b), should be preferred.

References

[1] T. W. Berrie, "The economics of system planning in bulk electricity supply," in *Public Enterprise*, R. Turvey, Ed. Penguin Modern Economics, 1968.

[2] R. Billinton, *Power System Reliability Evaluation*. New York: Gordon and Breach, 1970.

[3] O. I. Elgerd, *Electric Energy Systems Theory: An Introduction*. New York: McGraw-Hill, 1971.

[4] J. Vardi, J. Zahavi, and B. Avi-Itzhak, "The combined load duration curve and its derivation," *IEEE Trans. Power Apparatus and Systems*, vol. PAS-96, no. 3, pp. 978–983, June 1977.

[5] J. Zahavi, J. Vardi, and B. Avi-Itzhak, "Operating cost calculation of an electric power generating system under incremental loading procedure," *IEEE Trans. Power Apparatus and Systems*, vol. PAS-96, no. 1, pp. 285–292, Jan. 1977.

Notes for the Instructor

Objectives. This chapter is intended to be used in more practically oriented mathematical modeling courses, or in courses on probabilistic models. The purpose of the chapter is to present students with a large and realistic problem and to describe various mathematical models used to evaluate the performance of such a system. In particular, cost and reliability performance are discussed.

Prerequisites. Basic knowledge of probability concepts and some elementary calculus. No background whatsoever is required on power system engineering. A short introduction on the basic characteristics of supply and demand for power is provided to make the module self-contained.

Time. Six to seven class meetings. The scope of the chapter is also large enough to make it suitable as a final or term project for a course if so desired.